TURBULENT DIFFUSION IN THE ENVIRONMENT

GEOPHYSICS AND ASTROPHYSICS MONOGRAPHS

AN INTERNATIONAL SERIES OF FUNDAMENTAL TEXTBOOKS

VOLUME 3

TURBULENT DIFFUSION IN THE ENVIRONMENT

by

G. T. CSANADY

University of Waterloo

D. REIDEL PUBLISHING COMPANY

DORDRECHT-HOLLAND/BOSTON-U.S.A.

Library of Congress Catalog Card Number 72–92527

ISBN 90 277 0261 6

Published by D. Reidel Publishing Company,
P.O. Box 17, Dordrecht, Holland

Sold and distributed in the U.S.A., Canada, and Mexico
by D. Reidel Publishing Company, Inc.
306 Dartmouth Street, Boston,
Mass. 02116, U.S.A.

Printed in The Netherlands by D. Reidel, Dordrecht

FOREWORD

The rather excessive public preoccupation of the immediate past with what has been labeled the 'environmental crisis' is now fortunately being replaced by a more sustained and rational concern with pollution problems by public administrators, engineers, and scientists. It is to be expected that members of the engineering profession will in the future widely be called upon to design disposal systems for gaseous and liquid wastes which meet strict pollution control regulations and to advise on possible improvements to existing systems of this kind. The engineering decisions involved will have to be based on reasonably accurate quantitative predictions of the effects of pollutants introduced into the atmosphere, ocean, lakes and rivers. A key input for such calculations comes from the theory of turbulent diffusion, which enables the prediction of the concentrations in which pollutants may be found in the neighborhood of a release duct, such as a chimney or a sewage outfall. Indeed the role of diffusion theory in pollution prediction may be likened to the role of applied mechanics ('strength of materials') in the design of structures for adequate strength. At least a certain group of engineers will have to be proficient in applying this particular branch of science to practical problems.

At present, training in the theory of turbulent diffusion is available only at the graduate level and then only in a very few places. In view of the complexity of the subject, it is likely to remain in the graduate school, but it will in all likelihood be more widely taught. One prime obstacle to offering the subject has been the lack of a suitable synthesis of this still quite rapidly evolving field. I have taught a course on turbulent diffusion in the environment at the University of Waterloo on several occasions, to engineering graduate students specializing in 'environmental fluid mechanics.' The present text is an outcome of the lectures and was originally intended to be a synthesis of the subject.

Reviewing the completed manuscript I find with some dismay that the original intention has been at best incompletely realized and that the text reads in many places like a monograph which I take to mean a summary of what the author knows about a given subject. Certainly my own contributions to the subject have been given quite undue weight for the simple and I hope excusable practical reason that they were familiar and the illustrations were easily lifted. Many interesting and important contributions had to be ignored because in this first attempt at a synthesis I could only hope to cover the very skeleton of the subject. I apologize to all those who would legitimately expect to find their work referred to in a text of this kind and do not find it. There must have been a thousand or more papers from which I learned something, but to quote all of these would have turned the writing of this book into an exercise in scientific journalism. Perhaps somebody else could use my book as a stepping stone to preparing a more balanced and comprehensive synthesis.

My thanks are due to many of my colleagues with whom I have had the pleasure of debating various turbulent diffusion problems, for having helped to clear up many issues. Of these I would especially single out some I came into contact with at the Travelers Research Center in Hartford, Drs G. R. Hilst, K. D. Hage and the late Prof. Ben Davidson.

G. T. CSANADY

University of Waterloo,
Waterloo, Ont., Canada

February, 1972

TABLE OF CONTENTS

CHAPTER III. TURBULENT DIFFUSION: ELEMENTARY STATISTICAL THEORY AND ATMOSPHERIC APPLICATIONS

CHAPTER IV. 'RELATIVE' DIFFUSION AND OCEANIC APPLICATIONS

CHAPTER V. DISPERSION IN SHEAR FLOW

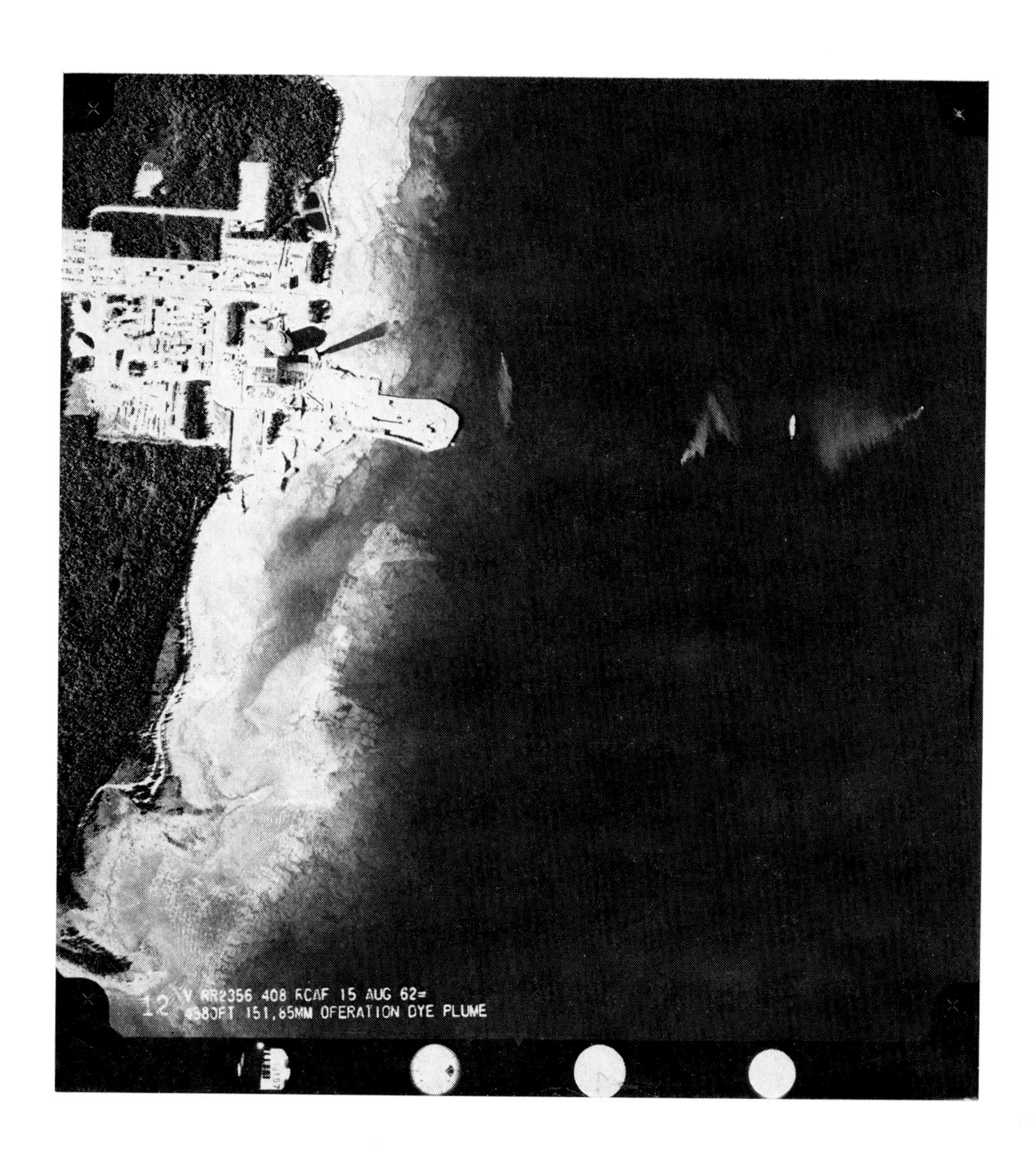

Aerial photo of 3 dye plumes generated from 3 point sources in Lake Huron off Douglas Point Generating Station. That all three move in different directions and diffuse at different rates illustrates possible complexity of diffusion phenomena in the natural environment. Outermost source was about 1 km from shore.

CHAPTER I

MOLECULAR DIFFUSION

1.1. Introduction

In pollution problems we are concerned with the fate of gaseous or liquid 'waste' materials, i.e., substances either definitely harmful or merely undesirable from the point of view of our species, which, for lack of an economically acceptable alternative, are discharged into the atmosphere or into various bodies of water. In the vast majority of cases the relative *concentration*, expressed as the fraction of volume occupied, of a pollutant is quite small, at least where such concentrations seriously concern us, so that from the point of view of theory we may regard the air-pollutant or water-pollutant mixture as a very dilute solution or suspension. In consequence, the simplest formulations of diffusion laws may be used in which the bulk-velocity due to mass transfer is ignored. (To appreciate the simplification involved read pp. 449–455 of Eckert and Drake (1959).)

In a gas or liquid at rest dissolved or suspended impurities mix with their surroundings as a consequence of molecular agitation. When the molecules of the dissolved or suspended substance are similar in size to those of the host substance the process is usually described simply as 'diffusion,' when they are much larger, as 'Brownian motion.' From our point of view there is no advantage in distinguishing between the two cases, the only perceptible difference being in mixing rates described adequately by the appropriate value of the *diffusivity*, and we shall refer to these phenomena simply as *molecular diffusion*.* The terminology is intended to distinguish this dispersal process from the much more efficient one caused by random bulk movements, known as turbulent diffusion which dominates most atmospheric and oceanic dispersal problems. In the present chapter we shall discuss molecular diffusion, partly because it plays a certain role in environmental diffusion, but mainly because it provides relatively simple mathematical models of diffusion phenomena useful in a qualitative way in understanding the more complex turbulent diffusion process.

1.2. Concentration

One of the basic hypothesis on which fluid dynamics is built is to regard the fluid a continuum, i.e., to consider only elements so large that the density of the fluid becomes a continuous function of space coordinates, in spite of the nearly 'discrete' contributions molecules make to density. A careful discussion of the conditions under which

* Following the first two chapters, in which we discuss the two phenomena in detail and separately.

liquids and gases may be treated as continua is contained, e.g., in Prandtl and Tietjens (1957, pp. 7–10). The essential point is that we define density over a sample volume large compared to the cube of the mean free path. In the theory of molecular diffusion we start exactly with the same step and define the concentration χ of the diffusing substance as the mass ΔM contained in the sample volume ΔV, centered at a point (x, y, z) where we wish to define concentration:

$$\chi = \frac{\Delta M}{\Delta V} \quad [\mathrm{g\,cm^{-3}}], \tag{1.1}$$

where ΔV is large compared to a^3, if a is an average distance between the diffusing molecules or particles, so as to include a large number of these in ΔV.

This definition is not without its difficulties, because if the solution or suspension is very dilute, a is an appreciable distance, and may be comparable to distances over which concentration *changes* interest us. However, on a finer scale, measured values of the concentration would be subject to random fluctuations as individual diffusing particles entered or left a sampling volume not large compared to a. In practice this difficulty rarely arises in connection with molecular diffusion, although the theoretical aspects of the fluctuation problem in Brownian motion have been subject to considerable study (for a review see Chandresekhar (1943)). We cannot enter here into a detailed discussion of this interesting problem and will assume that the continuum theory definition in Equation (1.1) is adequate for our purposes.

A direct consequence of Equation (1.1) is that changes in concentration over a distance of order a are 'small', i.e., to a high degree of approximation:

$$\chi(x + a) - \chi(x) = a\,\frac{\partial \chi}{\partial x} \tag{1.2}$$

whichever way the x-axis is chosen.

1.3. Flux

As is well known, mixing through diffusion tends to equalize the distribution of any impurity over the available volume. Turning this around, we may also say that diffusion takes place when concentration differences are created in a liquid or gas by whatever method. The *rate* of diffusion may be measured locally by the value of the *flux*, being the mass of the diffusing substance passing an area element ΔA centered at the point in question, (x, y, z), in unit time:

$$F = \frac{\Delta M/\Delta t}{\Delta A} \quad [\mathrm{g\,s^{-1}\,cm^{-2}}], \tag{1.3}$$

where ΔM is the total mass of particles passing the area element during time interval Δt.

Since an area element may be oriented in any direction the question arises how the value of the flux at a point depends on the orientation of the area-element. To this end, consider a tetrahedral element of the fluid (Figure 1.1) with three of the faces of the tetrahedron falling into coordinate planes, the fourth face having area δA and being oriented perpendicular to its outward unit normal **n**, the normal having projections on the coordinate axes of lengths n_x, n_y and n_z. The areas of the three sides of the tetrahedron in planes perpendicular to the x, y and z axes, are, respectively, $n_x\delta A$, $n_y\delta A$ and $n_z\delta A$. Let us call the fluxes across these three areas F_x, F_y and F_z, positive along the positive axes, the flux through the area δA, F_n, positive along positive **n**. The net mass transport into the tetrahedron during a short interval Δt is then

$$\Delta t(F_x n_x \delta A + F_y n_y \delta A + F_z n_z \delta A - F_n \delta A) = \delta M\,. \tag{1.4}$$

This mass increase will cause a corresponding change in concentration; if δV is the volume of the tetrahedron, $\delta M = \delta V\,(\partial\chi/\partial t)\,\Delta t$. If we now let the linear dimensions of the tetrahedron become arbitrarily small, δM goes to zero with the *third* power of a characteristic length (say the x-axis intercept), because $\partial\chi/\partial t$ may safely be assumed to be finite. On the other hand, δA tends to zero as the square of the characteristic length and the terms on the left-hand side of Equation (1.4) dominate. Dividing by δA and dropping δM on the right we have then

$$F_n = F_x n_x + F_y n_y + F_z n_z\,. \tag{1.5}$$

Thus the flux through the inclined face is the same as the projection to **n** of the vector having Cartesian components F_x, F_y, F_z. We conclude that flux is a vector quantity and that the rate of the diffusion in a given direction is measured locally by the relevant projection of the flux vector, $\mathbf{F}\cdot\mathbf{n}$.

The passage to the limit of zero length scale in Equation (1.4) again raises the question of the validity of the continuum-theory approach. It is sufficiently accurate to reduce the characteristic length to order a, the average distance between diffusing particles, and postulate that ΔM is large compared to the mass of an individual dif-

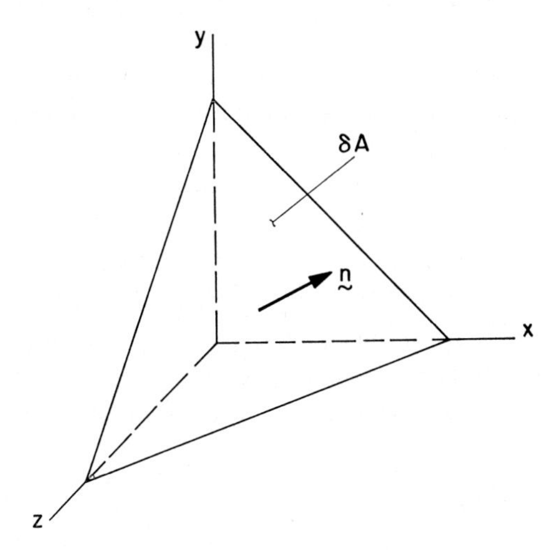

Fig. 1.1. Tetrahedral fluid element.

fusing particle. This means that Δt has to be large compared to the average 'flight-time' τ of a particle, defined as the time in which it crosses the distance a. A consequence of this definition of flux is then that the time-scale of any concentration changes must also be large compared to the flight-time or 'relaxation time' τ, if we are to use the above continuum-theory approach in the description of such changes.

It may be pointed out once more that the continuum theory approach to molecular diffusion 'works' in most practical cases, so that the fundamental difficulties referred to above are of little consequence. They are pointed out here because in turbulent diffusion these difficulties can no longer be swept under the rug and they force us to adopt the much more complicated approach of regarding both concentration and flux as random variables. Strictly speaking, of course, they are also random variables in molecular diffusion (as is clear from the above, if we do not apply appropriate smoothing), but here we are content to deal with the *mean* values of these variables, approximated to a satisfactory degree by the above described spatial smoothing procedure.

1.4. Fick's Law

Consider now how the flux of the diffusing substance arises through molecular or Brownian motion. Diffusing molecules or particles cross the area ΔA oriented perpendicular to **n** in both directions. However, if the concentration changes in the direction **n**, there are more particles coming from the 'richer' side of ΔA and it is the effective 'excess' transport that produces the flux. To approach this problem in the simplest possible way assume that the particles coming from side (1) of Figure 1.1, have an average velocity u_1 (in the positive **n** direction), those crossing the opposite way, velocity u_2, along negative **n**. In time Δt the first particles reach a distance $u_1 \Delta t$ beyond ΔA and fill a volume $u_1 \Delta A \Delta t$. Let their average concentration in this volume be χ_1; then the one-way transport along positive **n**, per unit area and unit time is $u_1\chi_1$.* Conversely, the one-way transport along negative **n** is $u_2\chi_2$ and the net flux becomes $(u_1\chi_1 - u_2\chi_2)$. If there are no temperature differences we may assume $u_1 = u_2 = u_m$ (= a characteristic velocity magnitude of molecular or Brownian motion). We may also assume that χ_1 and χ_2 are equal to the local concentrations at the 'origin' of the particles flying one way or the other, i.e., to χ at some average distance l_m from ΔA, from where the particles 'come.' This distance is equal in order of magnitude to that length over which the velocity of a given particle persists on the average in a given direction. When diffusing particles are molecules of like size to the host substance, the 'mixing length' l_m is equal in order of magnitude to the mean free path as is known from the kinetic theory of gases. In Brownian motion the mixing length may be greater than the mean free path of the host substance molecules, but it can still reasonably be assumed to be a small enough distance for Equation (1.2) to hold, therefore:

$$\chi_2 - \chi_1 = l_m \frac{\partial \chi}{\partial n}.$$

* The reader should note that, quite generally, $\mathbf{u}\chi$ gives flux due to a local bulk fluid velocity $\mathbf{u}$.

Hence the flux becomes:

$$F_n = - u_m l_m \frac{\partial \chi}{\partial n} \quad [\mathrm{g\,s^{-1}\,cm^{-2}}]. \tag{1.6}$$

The product $u_m l_m$ is conveniently combined into the material constant 'diffusivity';

$$D = u_m l_m . \tag{1.7}$$

Taking into account the vector character of the flux we may summarize our results in the following relationship known as 'Fick's law':

$$\mathbf{F} = - D \nabla \chi , \tag{1.8}$$

where ∇ is the gradient operator.

The above argument can be made a good deal more rigorous (for diffusing *molecules*) with the aid of the kinetic theory of gases (Chapman and Cowling, 1951), which also supplies a quantitative estimate of diffusivity D in function of temperature, molecular weight, etc. For our purposes it is sufficient to regard Equation (1.8) a phenomenological 'law', exactly analogous to Fourier's law of heat conduction, on which a theory of macroscopic phenomena may be built. The diffusivity D is, in this light, a material constant forming an empirical input into the theory. While direct evidence for the validity of Equation (1.8) in molecular diffusion is scarce, many conclusions based on Fick's law have been verified and there is no reason to doubt its validity within the limitations of the concepts involved (discussed in the previous section). Some diffusivities of interest are given in Table I.1.

TABLE I.1

Some molecular diffusivities of interest [$\mathrm{cm^2\,s^{-1}}$] from Rohsenow and Choi (1961)

CO_2 in air	0.137
Water vapor in air	0.220
Hydrocarbons in air	0.05–0.08
CO_2 in water	1.8×10^{-5}
N_2 in water	2.0×10^{-5}
NaCl in water	1.24×10^{-5}

1.5. Conservation of Mass

Fick's law provides a relationship between the spatial distribution of concentration and flux. A second relationship between the same two variables (resulting in a mathematical 'closure' of the problem) may be obtained by an application of the principle of continuity or conservation of mass.

Consider a volume of fluid V, bounded by surface S, in and out of which a substance is being transported by diffusion and (for generality) bulk fluid motion (Figure 1.2).

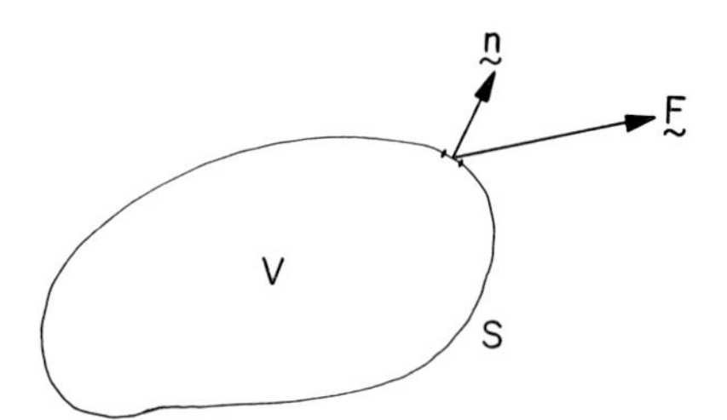

Fig. 1.2. Control volume of fluid.

Let the concentration of the transported substance be $\chi(x, y, z, t)$ so that the total mass within V is

$$M(t) = \int_V \chi \, \mathrm{d}V \quad [\mathrm{g}]. \tag{1.9}$$

Let the flux vector at any point on the boundary surface be $\mathbf{F}(x, y, z, t)$: note that whether this arises out of diffusion or bulk fluid motion or both, the argument of Section (1.2) establishing the vector character of flux remains valid. Thus the total flux out of the volume V is

$$J(t) = \int_S \mathbf{F} \cdot \mathbf{n} \, \mathrm{d}S \quad [\mathrm{g\,s^{-1}}], \tag{1.10}$$

where $\mathbf{n}$ is the outward normal to the surface element $\mathrm{d}S$.

Since mass must be conserved, in the absence of any sinks or sources we may write

$$J = -\frac{\partial M}{\partial t}. \tag{1.11}$$

By an application of the divergence theorem Equation (1.10) may be converted into a volume integral:

$$\int_S \mathbf{F} \cdot \mathbf{n} \, \mathrm{d}S = \int_V \nabla \cdot \mathbf{F} \, \mathrm{d}V. \tag{1.12}$$

Thus Equation (1.11) may be expressed as

$$\int_V \left(\frac{\partial \chi}{\partial t} + \nabla \cdot \mathbf{F} \right) \mathrm{d}V = 0, \tag{1.13}$$

where $\nabla \cdot$ is the divergence operator.

This equation must be valid for any portion of the volume V, as well as for the total, therefore the integrand is zero point by point:

$$\frac{\partial \chi}{\partial t} = -\nabla \cdot \mathbf{F}. \tag{1.14}$$

The result is valid whether flux is due to fluid motion or diffusion or both. If the flux is caused by molecular diffusion alone we apply Fick's law (Equation (1.8)) to yield

$$\frac{\partial \chi}{\partial t} = \nabla \cdot (D \nabla \chi) \tag{1.15}$$

which is a partial differential equation for the unknown χ. Where D is constant this 'classical diffusion equation' becomes

$$\frac{\partial \chi}{\partial t} = D \nabla^2 \chi = D \left[\frac{\partial^2 \chi}{\partial x^2} + \frac{\partial^2 \chi}{\partial y^2} + \frac{\partial^2 \chi}{\partial z^2} \right], \tag{1.16}$$

where ∇^2 is the Laplacian operator.

In the case when also laminar flow contributes to flux the total is

$$\mathbf{F} = \mathbf{u}\chi - D\nabla\chi . \tag{1.17}$$

For our purposes it is sufficient to regard air and water as incompressible fluids. Then the divergence of the flow-contribution to flux is

$$\nabla \cdot (\mathbf{u}\chi) = \mathbf{u} \cdot \nabla \chi \tag{1.18}$$

so that the diffusion equation becomes

$$\frac{\partial \chi}{\partial t} + \mathbf{u} \cdot \nabla \chi = \frac{\mathrm{d}\chi}{\mathrm{d}t} = D \nabla^2 \chi , \tag{1.19}$$

where the operator $\mathrm{d}/\mathrm{d}t$ denotes the 'total derivative' of χ, i.e., temporal change following the bulk motion of the fluid.

1.6. Instantaneous Plane Source

The classical diffusion equation or the mathematically identical heat conduction equation has a considerable mathematical literature, two prominent texts being Carslaw and Jaeger (1959) and Crank (1956). For purposes of illustration we shall discuss here some solutions of particular relevance to atmospheric and oceanic diffusion problems.

Consider first the idealized case of a uniform medium at rest, ($\mathbf{u}=0$, $D=$const.). It is easily seen that the following is a solution of Equation (1.16):

$$\chi = \frac{A}{\sqrt{t}} e^{-x^2/4Dt} \tag{1.20}$$

with $A=$const.

The physical interpretation of this solution is easily built up from certain of its mathematical properties. Equation (1.20) describes a one-dimensional phenomenon, not being dependent on y or z – say diffusion along the axis of a duct or pipe with conditions uniform across the section, or in the atmosphere along the vertical, assuming homogeneous conditions in the horizontal. The total amount of material in a column of unit cross sectional area is given by the integral:

$$Q = \int_{-\infty}^{\infty} \chi \, \mathrm{d}x = 2A\sqrt{D} \int_{-\infty}^{\infty} \exp\left(-\frac{x^2}{4Dt}\right) \mathrm{d}\left(\frac{x}{2\sqrt{Dt}}\right)$$

$$= 2A\sqrt{\pi D} \tag{1.21}$$

which is independent of time, expressing conservation of mass. Thus Equation (1.20) describes the diffusion of an amount of material, Q[g] per unit area; replacing the constant A in Equation (1.20) by the physically more meaningful Q from Equation (1.21) we have:

$$\chi = \frac{Q}{2\sqrt{\pi D t}} e^{-x^2/4Dt} . \tag{1.22}$$

The asymptotic behavior of χ as $t \to 0$ is:

$$\begin{aligned} \chi &\to 0 \qquad |x| > 0 \\ \chi &\to \infty \qquad x = 0 . \end{aligned} \tag{1.23}$$

These properties, plus Equation (1.21) show that as $t \to 0$, χ becomes proportional to the 'delta function' (see Lighthill (1962)):

$$\chi = Q\delta(x) \quad (t = 0) . \tag{1.24}$$

The delta function is the mathematical description of such idealizations as the 'concentrated force' or, in our case, the 'concentrated source.' Thus Equation (1.22) describes the diffusion of an amount of material (Q g cm^{-2}) which was initially concentrated in a thin sheet at $x=0$, in an 'instantaneous plane source.' At subsequent instances the material spreads out as illustrated in Figure 1.3.

The functional form of the distribution $\chi(x)$ at $t>0$ is known as a 'Gaussian' or 'normal' curve, with the maximum concentration at the center of gravity which remains at the origin. The moment of inertia or second moment of this distribution may be used to characterize its 'spread':

$$\int_{-\infty}^{\infty} \chi x^2 \, \mathrm{d}x = 2QDt . \tag{1.25}$$

If we divide this by the total diffusing material, we obtain a kind of mean square distance to which the particles have diffused:

$$\sigma^2 = 2Dt . \tag{1.26}$$

The square root of this, i.e., the length σ serves as a convenient scale of the width of the distribution, and is known as the 'standard deviation.' Expressed in terms of this scale (which may be seen to increase with the square root of time) the distribution

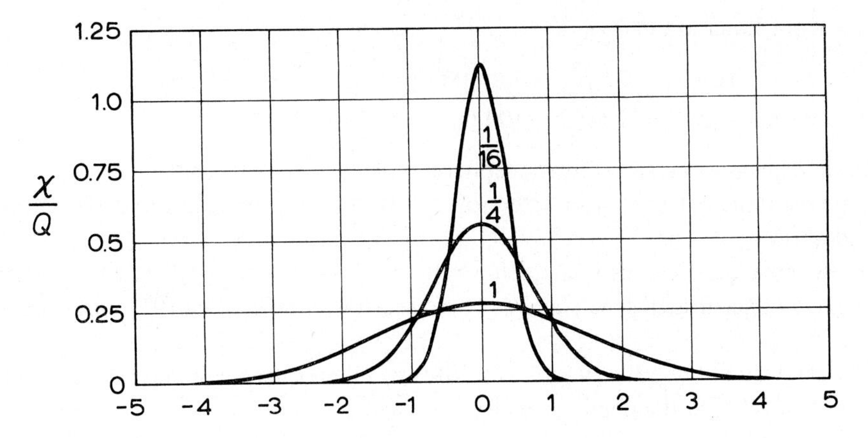

Fig. 1.3. Concentration – distance curves for an instantaneous plane source. Numbers are values of Dt (Crank, 1956).

becomes:

$$\chi = \frac{Q}{\sqrt{2\pi}\sigma} e^{-x^2/2\sigma^2}. \tag{1.27}$$

This is perhaps the easiest form in which to remember a Gaussian distribution, with the additional advantage that it also applies in turbulent diffusion (where σ is, however, a different function of time). A 'normalized' form, in terms of nondimensional variable, is

$$\frac{\chi\sigma}{Q} = \frac{1}{\sqrt{2\pi}} \exp\left\{-\frac{1}{2}\left(\frac{x}{\sigma}\right)^2\right\} \tag{1.28}$$

which is independent of time.

1.7. Some Simple Examples

(a) Diffusion of CO_2 in N_2 at 15 °C, $D = 0.158$ cm^2 s^{-1}

Source strength $Q = 3$ g cm^{-2} (about equal to atmosphere's total content of CO_2). How fast will this spread out by molecular diffusion? (if released as a concentrated sheet).

From Equation (1.26) we calculate:

t	3.16 s	8.8 h	10 yr
σ	1 cm	1 m	100 m

By contrast, suspension of dust in air, 1 μ diameter,*

$$D = 2.2 \times 10^{-6} \text{ cm}^2 \text{ s}^{-1} \text{ (Brownian motion)}$$

* 1 μ = 1 micrometer = 10^{-4} cm.

Again, from Equation (1.26)

t	63 h	72 yr	720000 yr
σ	1 cm	1 m	100 m

Hence a fine dust layer in the atmosphere in the absence of turbulent mixing would be quite permanent, CO_2 not so. Exercise: calculate maximum concentrations at times named above.

(b) Reverse question: estimate diffusivity from observed spread. Say at 1 h after release as a concentrated sheet, layer is 10 m 'deep', as measured by 10% concentration levels.

This definition of depth or width of a diffusing cloud is frequent and is illustrated in Figure (1.4); if W is the depth or width, we have

$$\chi(W/2) = 0.1\chi(0)$$

or

$$\frac{Q}{\sqrt{2\pi}\sigma} e^{-W^2/8\sigma^2} = 0.1 \frac{Q}{\sqrt{2\pi}\sigma}$$

from which

$$W^2/8\sigma^2 = \ln 10 = 2.3$$
$$W = 4.3\sigma.$$

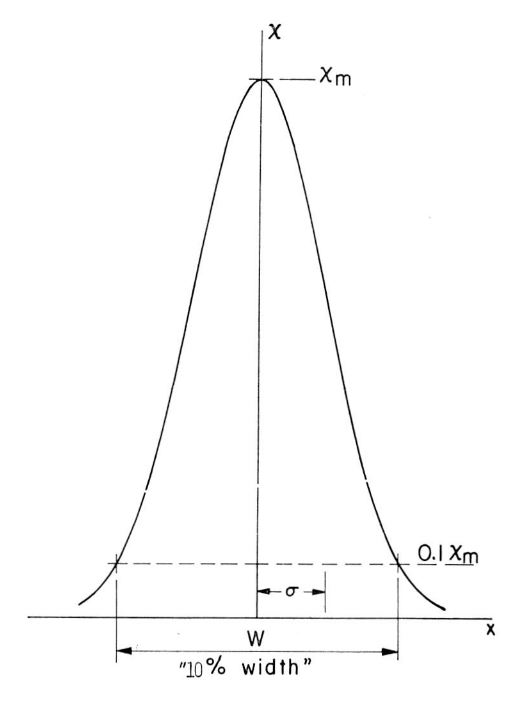

Fig. 1.4. Definition of cloud 'width'. The concentration distribution shown is Gaussian of standard deviation σ.

Often in experimental work the crude rule is adopted that $W \simeq 4\sigma$, if the accuracy of determining W is low.

With the above data, $W = 10$ m,

$$\sigma = 10/4.3 = 2.32 \text{ m}.$$

Now since $2\,Dt = \sigma^2$ at $t = 3600$ s,

$$D = 7.43 \text{ cm}^2 \text{ s}^{-1}$$

which is rather higher than normal molecular diffusivities.

1.8. Diffusion of Finite Size Cloud

The 'concentrated source' is a useful mathematical model but is in some ways over-idealized. The initial concentration of any discharged admixture, in particular, is a practically important variable, which cannot be incorporated into a model which predicts $\chi \to \infty$ as $t \to 0$. We therefore consider the following problem, still in one dimension: a 'plug' or a 'cloud' of 'marked fluid' of length b is released at $t = 0$, meaning that

$$\left.\begin{array}{ll} \chi = \chi_0 & \text{at} \quad -b/2 < x < b/2 \\ \chi = 0 & \text{outside this region} \end{array}\right\} \text{at} \quad t = 0.$$

Conditions along y and z are supposed uniform again. The cloud spreads out at both ends in virtue of molecular diffusion (diffusivity D). What is the subsequent history of concentration?

This problem may be treated by integrating the 'simple' source solution, assuming that each layer x' to $x' + dx'$ $(-b/2 < x' < b/2)$ acts as a concentrated source of strength $Q = \chi_0 dx'$ g cm^{-2}. The diffusion equation being linear, we may simply superimpose the solutions:

$$\chi = \frac{1}{\sqrt{2\pi}\sigma} \int_{-b/2}^{b/2} \chi_0 \exp\left\{-\frac{(x - x')^2}{2\sigma^2}\right\} dx'. \tag{1.29}$$

If $\chi_0 = $ const. the integration may be carried out and results in

$$\chi = \frac{\chi_0}{2}\left\{\text{erf}\left[\frac{(b/2 + x)}{\sqrt{2}\sigma}\right] + \text{erf}\left[\frac{(b/2 - x)}{\sqrt{2}\sigma}\right]\right\}, \tag{1.30}$$

where

$$\text{erf } p = \frac{2}{\sqrt{\pi}} \int_0^p e^{-q^2} dq \quad (\text{'error function'}).$$

Extensive tables of the error function are widely available (see e.g., Abramowitz and Stegun, 1964).

The time-dependence of the solution in Equation (1.30) is contained in σ (see Equation (1.26)). Figure 1.5 illustrates the behavior of the solution at increasing values of the parameter $2\sqrt{2\sigma/b}$, which is proportional to $\sqrt{t}$.

In practice the most important problem is the decay of the maximum (center) concentration χ_m. From Equation (1.30) with $x=0$ we find:

$$\chi_m = \chi_0 \operatorname{erf}(b/2\sqrt{2\sigma})\,. \tag{1.31}$$

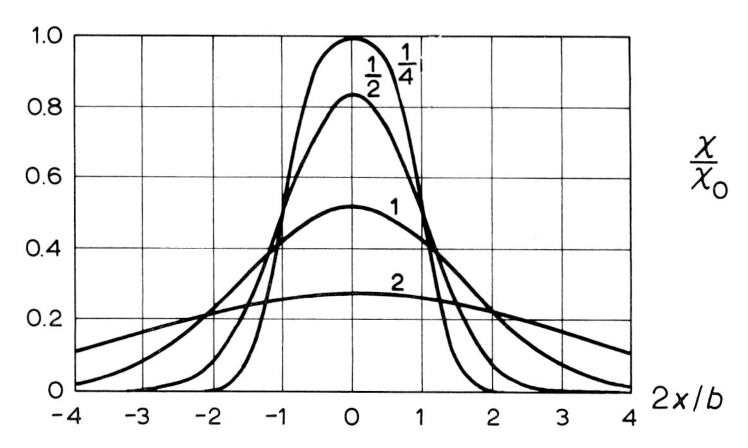

Fig. 1.5. Concentration-distance curves for an instantaneous source of finite extent. Numbers on curves are values of $2\sqrt{2\sigma/b}$ (Crank, 1956).

At short times after release the argument of the error function in Equation (1.31) is very large and the asymptotic value of erf p may be used:

$$\operatorname{erf}(b/2\sqrt{2\sigma}) \simeq 1 \quad t \to 0$$

$$\text{or} \quad \chi_m \simeq \chi_0\,. \tag{1.32}$$

Indeed this last result is nearly true if

$$\frac{b}{2\sqrt{2\sigma}} \geqslant 2$$

or

$$\sigma \leqslant \frac{b}{4\sqrt{2}} = 0.177b\,. \tag{1.33}$$

Substituting the definition of σ we find from Equation (1.33) that Equation (1.32) holds for a period:

$$t_i = 0.0156\, b^2/D\,. \tag{1.33a}$$

For example, if $b=1$ m, $D=0.1$ cm^2 s^{-1}, $t_i=1560$ s or approximately $\frac{1}{2}$ h. For this length of time after release the maximum concentration does not differ appreciably from the initial concentration χ_0. Later the 'dilution factor' χ_0/χ_m begins to increase and when the cloud becomes large compared to its initial size the error function may be approximated by its argument:

$$\frac{\chi_m}{\chi_0} = \operatorname{erf}\left(\frac{b}{2\sqrt{2}\sigma}\right) = \frac{2}{\sqrt{\pi}}\frac{b}{2\sqrt{2}\sigma} = \frac{b}{\sqrt{2\pi}\sigma} \quad (b/\sigma \ll 1). \tag{1.34}$$

This is exactly equal to the reciprocal dilution factor of a concentrated source, if we write $Q=b\chi_0$ for the total amount of material released. Thus in the later stages of diffusion the concentrated source model is adequate, but the initial circumstances could only be elucidated by the slightly more complex distributed source model.

1.9. 'Reflection' at Boundary

Consider now the case when the diffusing cloud encounters a plane boundary at $x=h$ (Figure 1.6). Individual diffusing particles either (a) stick to the boundary, or (b) bounce back from it. For simplicity we shall assume here that *all* particles bounce back, so that the cloud is 'reflected' by the boundary. Parenthetically we may remark here that the physics of this problem (whether particles adhere to a solid surface or not) is quite complicated, actual surfaces being rarely perfect absorbers or reflectors. The exact behavior depends not only on the quality of the surface, but also on the properties of the diffusing particles. 'Natural' surfaces (e.g., grass covered ones) are

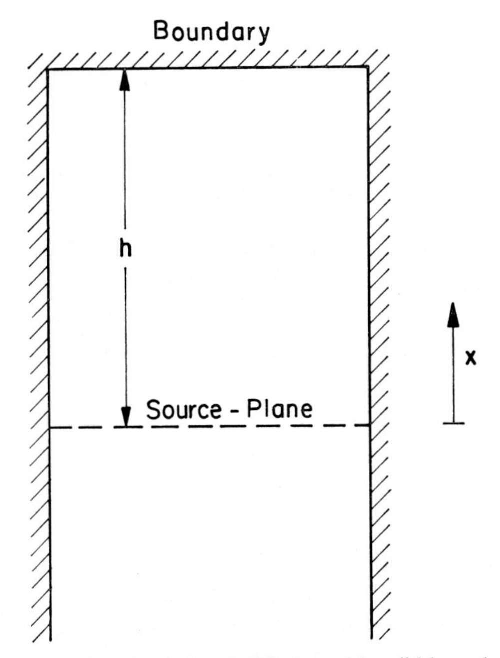

Fig. 1.6. One-dimensional diffusion with solid boundary.

often good absorbers for particles and good reflectors for gases, but there are exceptions, for instance iodine gas which is readily absorbed by vegetation.

If the boundary is a perfect reflector, exactly as many particles arrive at $x=h$ as leave the boundary, the net flux there being zero:

$$F_x(h) = 0 = -D \left.\frac{\partial \chi}{\partial x}\right|_h . \tag{1.35}$$

The mathematical device by which the boundary condition in Equation (1.35) is usually satisfied, is the introduction of a 'mirror-image' source at $x=2\,h$, of equal strength to the source at $x=0$. The two clouds diffuse identically, and midway in between at $x=h$ the flux from the one exactly cancels the flux from the other. Of course at $x>h$ the concentration field has no physical reality.

The combined concentration field of the 'real' and the 'image' sources is clearly

$$\chi = \frac{Q}{\sqrt{2\pi}\sigma}\left\{\exp\left[-\frac{x^2}{2\sigma^2}\right] + \exp\left[-\frac{(x-2h)^2}{2\sigma^2}\right]\right\}, \tag{1.36}$$

where $\sigma(t)$ is still given by Equation (1.26).

The behavior of the cloud as it grows is illustrated in Figure 1.7. The peak concentration shifts to the boundary approximately when $\sigma=0.9\,h$, i.e. at a time $t=$ $=0.4\,h^2/D$. Physically this occurs because the particles are 'crowded' at the boundary by the reflection process. The concentration at the boundary itself is (Figure 1.7b):

$$\chi_h = \frac{2Q}{\sqrt{2\pi}\sigma} \exp\left(-\frac{h^2}{2\sigma^2}\right) \tag{1.37}$$

which is, at large t, equal to *twice* the maximum concentration in a cloud diffusing in the absence of the boundary.

1.10. Two- and Three-Dimensional Problems

In the last few sections some important one-dimensional diffusion problems were discussed. As we increase the number of dimensions to two and three little that is new is added, the diffusion taking place more or less independently in different directions. To approach the two-dimensional problem consider first the following solution of the classical diffusion equation:

$$\chi = \frac{Q}{2\pi\sigma^2} \exp\left\{-\frac{x^2}{2\sigma^2} - \frac{y^2}{2\sigma^2}\right\} \tag{1.38}$$

with $\sigma^2 = 2Dt$, as before.

The reader will readily verify that this expression satisfies Equation (1.16). The following properties of this solution may be noted:

(a) At a given time t the $\chi=$const. contours are circles centered at the origin. A cross section along any diameter yields a concentration profile exactly as in one-dimen-

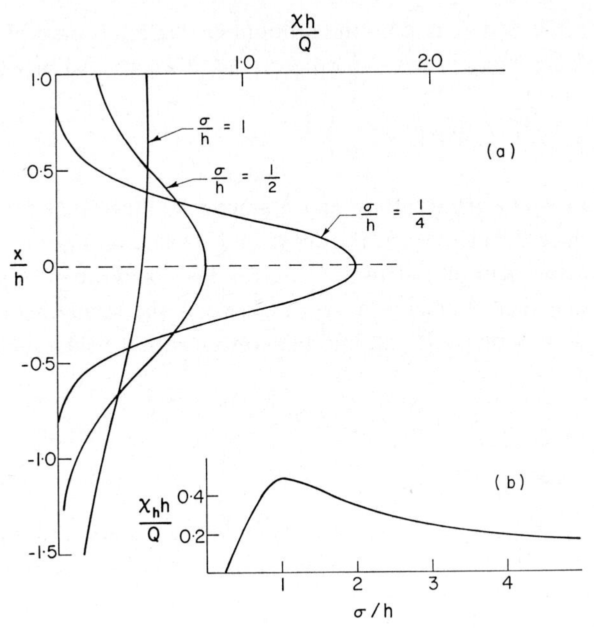

Fig. 1.7. (a) Concentration distribution in cloud released as concentrated sheet at $x = 0$, in presence of a reflecting boundary at $x = h$. Three successive stages of cloud growth are shown, characterized by increasing σ/h. – (b) Variation of concentration at barrier as function of cloud size.

sional diffusion, so that the spreading out of the cloud still proceeds according to Figure 1.3.

(b) The total amount of material in suspension is per unit depth of the two-dimensional field:

$$\int_{-\infty}^{\infty}\int \chi \, \mathrm{d}x \, \mathrm{d}y = Q = \text{const.} \quad [\mathrm{g\ cm}^{-1}]. \tag{1.39}$$

Thus this solution represents diffusion from an 'instantaneous line source' of strength Q g cm^{-1} (note the somewhat different physical dimension than that used before), which is initially concentrated in the z-axis, along which conditions are uniform. The spreading out along x and y are seen to be 'independent' in the sense that, except for the constant Q, Equation (1.38) is a product of two plane source (one dimensional) solutions:

$$\frac{1}{\sqrt{2\pi}\sigma} \exp\left(-\frac{x^2}{2\sigma^2}\right) \cdot \frac{1}{\sqrt{2\pi}\sigma} \exp\left(-\frac{y^2}{2\sigma^2}\right).$$

Similarly, a product of three one-dimensional solutions provides a description of diffusion from an 'instantaneous point source':

$$\chi = \frac{Q}{(\sqrt{2\pi}\sigma)^3} \exp\left\{-\frac{x^2}{2\sigma^2} - \frac{y^2}{2\sigma^2} - \frac{z^2}{2\sigma^2}\right\}, \tag{1.40}$$

where still $\sigma = \sqrt{2Dt}$ and Q is now the amount of material released at the origin at $t=0$ in [g]. Writing $r^2 = x^2 + y^2 + z^2$ Equation (1.40) may also be expressed as

$$\chi = \frac{Q}{(2\pi)^{3/2}\,\sigma^3} \exp\left(-\frac{r^2}{2\sigma^2}\right) \tag{1.40a}$$

so that the cloud again grows along any diameter as illustrated for the one-dimensional case in Figure 1.3. (However, the decay of χ is faster in time, $\chi_m = \text{const.}\, t^{-3/2}$.)

The concentration field of distributed sources may be written by simply summing contributions from their elements. If $q(x', y', z')$ is the strength of the distributed (instantaneous) source per unit volume, the concentration field is given by

$$\chi = \frac{1}{(\sqrt{2\pi}\sigma)^3} \int q\,(x', y', z') \exp\left\{-\frac{(x-x')^2 + (y-y')^2 + (z-z')^2}{2\sigma^2}\right\} \times \\ \times\, \mathrm{d}x'\, \mathrm{d}y'\, \mathrm{d}z'\,. \tag{1.41}$$

For $q = \text{const.}$ over a finite volume very similar results are obtained as in Section 1.8, but in two or three dimensions. The reader should have no difficulty in writing down the appropriate expressions.

1.11. Continuous Sources

A number of important problems may be modeled by the assumption that a source emits material continuously. For a point source, let the rate of emission be $q(t')$ g s^{-1}, such that, in the short interval t' to $t' + \mathrm{d}t'$ an amount $q\, \mathrm{d}t'$[g] is emitted. Each of these 'puffs' generates its own cloud and the total concentration field is obtained by a summation of contributions from the individual puffs. At fixed time t the standard deviation σ differs from puff to puff, being given by $\sqrt{2D(t-t')}$. The combined field, if, the emission takes place from $t'=0$ to t, by an integration of Equation (1.40a), is:

$$\begin{aligned} \chi(x, y, z, t) &= \frac{Q}{8\,(\pi D)^{3/2}} \int_0^t \exp\left\{-\frac{r^2}{4D\,(t-t')}\right\} \frac{\mathrm{d}t'}{(t-t')^{3/2}} \\ &= \frac{Q}{4\pi D r} \operatorname{erfc}\left(\frac{r}{\sqrt{4Dt}}\right), \end{aligned} \tag{1.42}$$

where the 'complementary error function' is erfc $\lambda = 1 - \text{erf}\, \lambda$. As $t \to \infty$ the last result reduces to

$$\chi = \frac{Q}{4\pi D r} \tag{1.43}$$

which describes the concentration field of a point source maintained indefinitely. It may be noted in passing that no such asymptotic distribution is obtained for continuous one or two-dimensional sources, the concentration then increasing continuously with time at any point in the field.

1.12. Source in Uniform Wind

All the previous results pertained to diffusion of marked fluid in a uniform medium at rest. In connection with environmental diffusion a somewhat more realistic model is one in which the medium moves at a constant and uniform velocity, U (to be called 'wind' velocity, although it could also be 'current'), in a direction which will be taken to coincide with the x-axis.

Mathematically, this case differs from the uniform-medium case in only a trivial manner: in a coordinate system moving with the wind the medium is still at rest. Hence applying the coordinate transformation $x' = x - Ut$ we have at once for the instantaneous point source in a wind:

$$\chi = \frac{Q}{(\sqrt{2\pi}\sigma)^3} \exp\left\{-\frac{(x-Ut)^2 + y^2 + z^2}{2\sigma^2}\right\} \tag{1.44}$$

still with $\sigma = \sqrt{2Dt}$.

It may also be verified that this expression satisfies the differential Equation (1.19) with one convective term:

$$\frac{\partial \chi}{\partial t} + U\frac{\partial \chi}{\partial x} = D\nabla^2 \chi . \tag{1.45}$$

Similar results hold for instantaneous line or plane sources.

The extension to a *continuous* point source in a wind supplies a particularly important model (a very crude model of a chimney plume). The concentration field now consists of a series of 'puffs' (point source clouds) with their centers stretched out along the x-axis as the growing puffs are convected downwind (Figure 1.8). For a source maintained indefinitely, if the emission rate is a constant q g s^{-1} we obtain the combined concentration field of the many puffs by integration, as in the previous section:

$$\chi = \int_0^\infty \frac{q\, \mathrm{d}t'}{8[\pi D(t-t')]^{3/2}} \exp\left\{-\frac{[x - U(t-t')]^2 + y^2 + z^2}{4D(t-t')}\right\}$$

$$= \frac{q}{4\pi D r} \exp\left[-\frac{U}{2D}(r-x)\right], \tag{1.46}$$

where $r^2 = x^2 + y^2 + z^2$. This concentration distribution (which is clearly independent of time) is illustrated in Figure 1.9 which shows iso-concentration lines in a meridional section of the plume (x–z plane, say), in a nondimensional representation valid for all D, U and q. The plume is seen to be 'slender', i.e., excepting the region near the source, points with significant concentration levels are characterized by $z \ll x$.

For the portion of the plume sufficiently far downstream we may thus simplify the expression in Equation (1.46) by writing

$$r = \sqrt{x^2 + y^2 + z^2} \simeq x\left(1 + \frac{y^2 + z^2}{2x^2}\right). \tag{1.47}$$

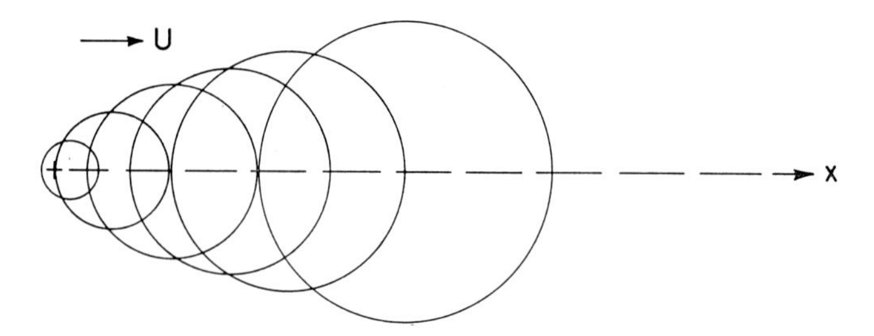

Fig. 1.8. Superposition of puffs in uniform wind.

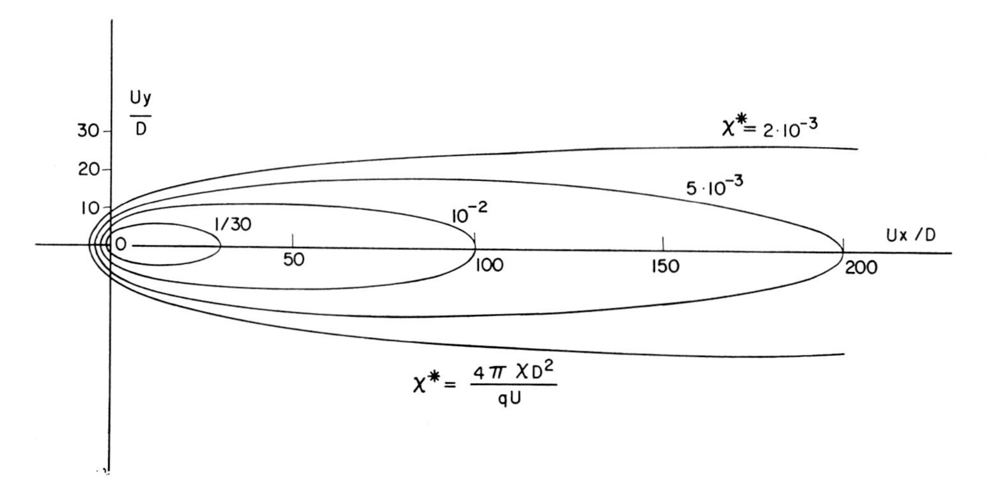

Fig. 1.9a. Lines of constant concentration in longitudinal section of continuous plume.

Therefore to the lowest order in small quantities:

$$r \simeq x \qquad r - x \simeq \frac{y^2 + z^2}{2x}. \tag{1.48}$$

This gives with Equation (1.46):

$$\chi = \frac{q}{2\pi U \sigma^2} \exp\left(-\frac{y^2 + z^2}{2\sigma^2}\right), \tag{1.49}$$

where we have now written

$$\sigma = \sqrt{2D \frac{x}{U}} \tag{1.50}$$

this being the standard deviation of the puff centered at x, which was released at a time $t = x/U$ earlier. The concentration distribution in a cross section of the plume (y–z plane) is thus seen to be identical with that in a puff at $t = x/U$, to the accuracy of the approximations. The quantity q/U is the amount of material carried in a slice

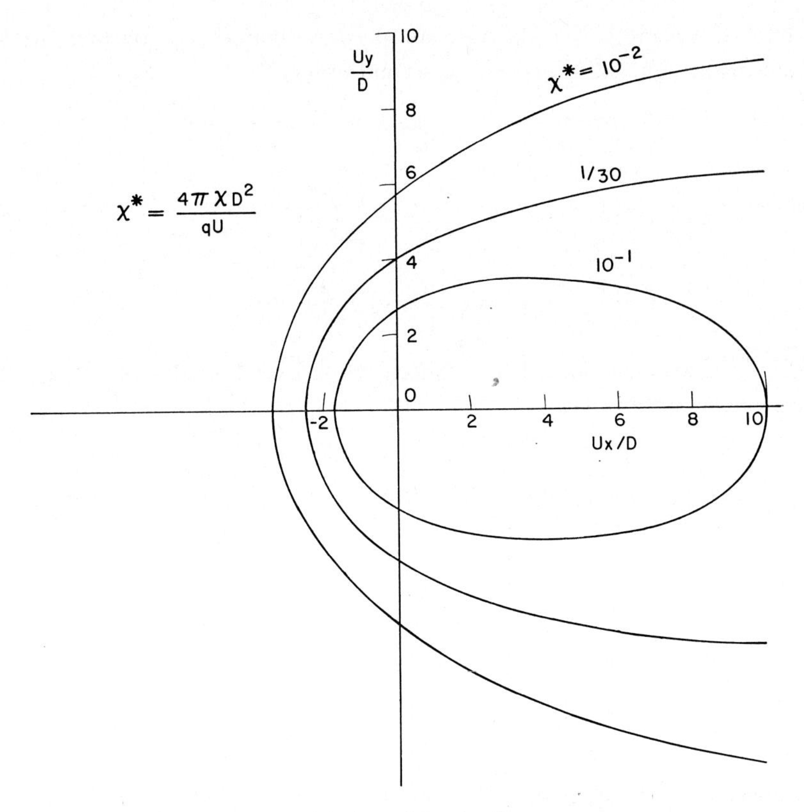

Fig. 1.9b. As Figure 1.9a, detail near source.

of the plume per unit length, or in other words $q\ dx/U$ may be regarded as the puff strength. The meaning of the result in Equation (1.49) is therefore that (sufficiently far downwind) individual puffs do not interfere with each other, diffusion along the x-direction is negligible. Indeed, Equation (1.49) is the solution of the modified diffusion equation:

$$U \frac{\partial \chi}{\partial x} = D \left(\frac{\partial^2 \chi}{\partial y^2} + \frac{\partial^2 \chi}{\partial z^2} \right) \tag{1.51}$$

which is Equation (1.19) without the time-dependent term *and* the flux-term along x, $D\partial^2\chi/\partial^2 x$.

Appendix to Chapter I

The reader may find the following additional remarks on the material of this chapter useful or enlightening. They would have taken us too far from the main line of argument in the text.

A. PHYSICAL MEANING OF THE LAPLACE OPERATOR

The most important (most highly differentiated) term in the diffusion Equation (1.16) is the Laplace operator. To illustrate its physical significance consider the Taylor expansion of the concentration field to second order:

$$\chi = \chi_0 + x \left.\frac{\partial \chi}{\partial x}\right|_0 + y \left.\frac{\partial \chi}{\partial y}\right|_0 + z \left.\frac{\partial \chi}{\partial z}\right|_0 + \tfrac{1}{2}\left[x^2 \left.\frac{\partial^2 \chi}{\partial x^2}\right|_0 + y^2 \left.\frac{\partial^2 \chi}{\partial y^2}\right|_0 + z^2 \left.\frac{\partial^2 \chi}{\partial z^2}\right|_0\right] + xy \left.\frac{\partial^2 \chi}{\partial x \partial y}\right|_0 + yz \left.\frac{\partial^2 \chi}{\partial y \partial z}\right|_0 + xz \left.\frac{\partial^2 \chi}{\partial x \partial z}\right|_0 + \cdots.$$

Calculate the average value of the concentration over a small cube of edge-length a centered at the origin:

$$\langle \chi \rangle = \frac{1}{a^3} \iiint_{-a/2}^{a/2} \chi \, \mathrm{d}x \, \mathrm{d}y \, \mathrm{d}z = \chi_0 + \frac{a^2}{24}\left[\frac{\partial^2 \chi}{\partial x^2} + \frac{\partial^2 \chi}{\partial y^2} + \frac{\partial^2 \chi}{\partial z^2}\right]_0 + \cdots$$

Therefore to second order:

$$\langle \chi \rangle - \chi_0 = \frac{a^2}{24} \nabla^2 \chi \Big|_0 .$$

The Laplacian is seen to be proportional to the difference between χ at a given point and the spatial mean value of χ in the neighborhood of that point. The diffusion equation states that the rate of change of concentration (following a fluid element, should there be motion) is directly proportional to this difference, and opposite in sign, i.e., it is a 'leveling-off' process. Also it is similar to simple damping, in that the rate of change and the magnitude of the departure from the local mean are linearly related, as velocity and damping force. By contrast, in elastic phenomena (e.g., propagation of acoustic waves) the Laplacian is proportional to a second time derivative, so that the departure from equilibrium (measured by the Laplacian) acts as a restoring force and causes an acceleration.

B. TRANSFORMATION OF COORDINATES

The Laplace operator may of course be expressed in other than Cartesian coordinates (see e.g., Morse and Feshbach (1953) for transformations to general curvilinear coordinates). Two important special cases are:

(1) Cylindrical Polars

$$x = r \cos \theta \qquad y = r \sin \theta$$

$$z = z$$

$$\nabla^2 = \frac{\partial^2}{\partial r^2} + \frac{1}{r}\frac{\partial}{\partial r} + \frac{1}{r^2}\frac{\partial^2}{\partial \theta^2} + \frac{\partial^2}{\partial z^2}.$$

(2) Spherical Polars

$$\theta = \text{'co-latitude'}$$
$$\phi = \text{longitude}$$
$$z = r\cos\theta$$
$$x = r\sin\theta\cos\phi$$
$$y = r\sin\theta\sin\phi$$
$$\nabla^2 = \frac{\partial^2}{\partial r^2} + \frac{2}{r}\frac{\partial}{\partial r} + \frac{1}{r^2}\frac{\partial^2}{\partial\theta^2} + \frac{1}{r^2}\cos\theta\frac{\partial}{\partial\theta} + \frac{1}{r^2\sin^2\theta}\frac{\partial^2}{\partial\phi^2}.$$

C. SIMILARITY IN DIFFUSION

From the diffusion Equation (1.19) it is clear that the χ-field in any diffusion problem involving fluid flow depends on the velocity-field $\mathbf{u}(x, y, z, t)$. To achieve similarity of two flow fields the well known criteria of fluid mechanics must be satisfied, i.e., the Reynolds number and any other important nondimensional parameters must be constant. More elementary conditions of flow similarity (sometimes overlooked) are similarity of geometry and of boundary conditions. These obviously apply also in case of diffusion problems. Diffusion from an infinite plane source or sink can under no circumstances be similar to diffusion from a point source.

If we suppose that these elementary conditions for the similarity of two diffusion problems are satisfied, we can deduce one more necessary condition from the diffusion equation. The terms of this equation may be made nondimensional if we divide through by $\chi^0 U/L$ where χ^0, U and L are scales of concentration, velocity, and length, respectively. The resulting nondimensional equation contains the single constant

$$\mathrm{Le} = UL/D$$

which is known as the Lewis number. Two diffusion phenomena will be similar if the Lewis number is constant, in addition to Reynolds number, governing flow similarity.

In practice the ratio of the Lewis number to Reynolds number is more frequently used as a similarity criterion. This is known as the Schmidt number:

$$\mathrm{Sc} = \mathrm{Le}/\mathrm{Re} = \nu/D.$$

Similar flow and diffusion fields require Re=const., Sc=const. In the absence of any flow, Re=0, Le=0, diffusion phenomena will be similar in any case (given similarity of boundary conditions): the diffusion equation may be made nondimensional if we use $\sigma=\sqrt{2Dt}$ as a length scale, as we have seen in Equation (1.28).

EXERCISES

1. 10 g cm^{-2} of ethanol is released in water (D=1 cm^2 s^{-1}) in a thin layer at t=0. Plot the resulting concentration profiles at t=1 s, 5 s and 20 s, and also plot χ_m vs. t. Then assume that the layer in which the ethanol is released is not 'thin' but 3 cm thick, and plot χ and χ_m on the same graphs as before.

2. A cloud is observed to be 1200 m 'wide' (10% width) $\frac{1}{2}$ h after its release. Assuming that the classical diffusion equation describes its diffusion, calculate the diffusivity if (a) the initial dimensions of the cloud were negligible; and (b) the cloud was already 800 m wide when released.

References

Abramowitz, M. and Stegun, I. A.: 1964, *Handbook of Mathematical Functions*, National Bureau of Standards, U.S. Government Printing Office, Washington, D.C.

Carslaw, H. S. and Jaeger, J. C.: 1959, *Conduction of Heat in Solids*, Oxford Univ. Press, London.

Chandrasekhar, S.: 1943, *Rev. Mod. Phys.* **15**, 1.

Chapman, S. and Cowling, T. C.: 1951, *The Mathematical Theory of Non-Uniform Gases*, Cambridge Univ. Press, London.

Crank, J.: 1956, *The Mathematics of Diffusion*, Oxford Univ. Press, London.

Eckert, E. R. G. and Drake, R. M.: 1959, *Heat and Mass Transfer*, McGraw-Hill Book Co. Inc. New York.

Lighthill, M. J.: 1962, *Fourier Analysis and Generalized Functions*, Cambridge Univ. Press, London.

Morse, P. M. and Feshbach, H.: 1953, *Methods of Theoretical Physics*, McGraw-Hill Book Co. Inc. New York.

Prandtl, L. and Tietjens, O. G.: 1957, *Fundamentals of Hydro- and Aeromechanics*, Dover Publications, New York.

Rohsenow, W. M. and Choi, H. Y.: 1961, *Heat, Mass and Momentum Transfer*, Prentice Hall Inc., New York.

CHAPTER II

STATISTICAL THEORY OF DIFFUSION AND BROWNIAN MOTION

2.1. Introduction

In the first chapter we discussed molecular diffusion from an 'Eulerian' point of view, establishing the diffusion equation from a consideration of concentration and flux at fixed points in space. In the course of these discussions we pointed out in Section 1.4 that flux arises through the random molecular or Brownian movements of diffusing particles. It is also possible to approach the diffusion problem from a Lagrangian viewpoint, focusing on the history of such random movements executed by the diffusing particles. To arrive at useful results, *statistical* properties of the random motions have to be considered, so that this second approach may also be labeled a 'statistical' one, in contrast to the 'phenomenological' theory of Chapter I. In the present chapter we explore this alternative approach.

When applied to molecular or Brownian diffusion, the Lagrangian or statistical method yields few new results, even if it results in considerably increased physical insight. In the end we are led back to the classical diffusion equation, with some added information on diffusivities for Brownian particles and on boundary conditions at solid surfaces. However, the theoretical framework of the statistical theory remains of first order importance for the treatment of the more difficult problem of turbulent diffusion. Historically, the first successful attack on turbulent diffusion by Taylor (1922) was indeed directly based on the statistical theory of Brownian motion, the latter having been developed a little earlier by Einstein (1905).

2.2. Dispersion Through Random Movements

It is more or less intuitively evident that random movements of a suitably defined ensemble of particles lead to their dispersion in space. Consider a 'marked' particle, such as a dyed element of fluid or a suspended foreign particle, and suppose that such a particle executes random movements, for whatever dynamical reasons, collisions with fluid molecules, turbulent stresses, etc. Let the particle be at position $\mathbf{x}'$ at time $t=0$, and at $\mathbf{x}$ at a later time t. Its displacement $\mathbf{x}-\mathbf{x}'$ is then a random function of time which may be described in terms of a spatial probability-density function $P(\mathbf{x}-\mathbf{x}', t)$, such that $P(\mathbf{x}-\mathbf{x}', t)\, d\mathbf{x}$ is the (small) probability that the displacement vector will end in the volume element $d\mathbf{x}$ surrounding the point $\mathbf{x}$, at time t. We suppose at first that the field within which the diffusion takes place is homogeneous so that $P(\ )$ does not depend explicitly on $\mathbf{x}'$, any release point being equivalent to any other. An alternative interpretation of $P(\mathbf{x}-\mathbf{x}', t)$, due to Batchelor (1949), and appropriate for diffusing *continuous elements*, is to regard it the probability that the point $\mathbf{x}$ will be immersed in the diffusing element in question at time t. In the case of diffusing discrete particles, the volume element is supposed large compared to particle diameter.

If one had the task of approximating $P(\mathbf{x}-\mathbf{x}', t)$ experimentally, one could approach it by releasing a large ensemble of particles at $\mathbf{x}'$, and determining their density distribution in surrounding space at subsequent times. The fraction of particles found in a given volume element would be regarded as an experimental approximation to $P(\mathbf{x}-\mathbf{x}', t)$. (The probability that a particle *is* at $\mathbf{x}$ is at once the probability that it got there.) If we define as an ensemble mean 'concentration' χ the mass of particles within volume element $\mathrm{d}\mathbf{x}$ in such an experiment, we have therefore

$$\chi(\mathbf{x}, t \mid \mathbf{x}') = QP(\mathbf{x} - \mathbf{x}', t), \tag{2.1}$$

where Q is the total mass released and it is indicated that the particles were released at $\mathbf{x}'$.

In carrying out such an experiment, we would take pains to ensure that a stable distribution $P(\mathbf{x}-\mathbf{x}', t)$ is in fact observed, i.e., that on repeating the experiment the second observed distribution $\chi(\mathbf{x},t \mid \mathbf{x}')$ differs only in negligible ways from the first determination. In order to accomplish this, all possible displacements would have to be represented in their stable proportion in the observed ensemble, which is only possible if the latter contains a sufficiently large number of *independently* diffusing particles. Such may not be the case, for example, if particles are discharged serially in quick succession so that most of them have similar velocity and displacement histories. A serial release must be continued long enough so that there is no resemblance between the displacement histories of a given particle and most of the others, or that any batch of 'correlated' histories form only a negligible fraction of the total. Similarly, if the particles are all released simultaneously, but distributed over a small neighborhood of the point $\mathbf{x}'$, their subsequent motions may be very nearly identical (they are, if such motions are mainly due to turbulence in the fluid). On repeating such an experiment, a completely different displacement history is obtained for the new batch. To arrive at a stable distribution $P(\mathbf{x}-\mathbf{x}', t)$ a large number of independently diffusing particles must be observed, which requires many such puff-releases in turbulent flow. In Brownian motion the random movements of particles are due to collisions with molecules of the host substance and are more or less independent even in an instantaneously released cluster of many particles, unless the initial concentration is exceptionally high. Thus single cluster experiments yield at once a stable ensemble-average concentration field, which is related to particle displacement probabilities as indicated by Equation (2.1).

We may therefore regard the theory developed in Chapter I as relating to the 'ensemble average' concentration field of diffusing molecules or Brownian particles in a fluid at rest or in laminar flow. In turbulent diffusion we shall have to distinguish between such an ensemble average concentration field and an instantaneously observable one. For either molecular or turbulent diffusion Equation (2.1) shows a relationship of the ensemble mean field to particle displacement probabilities. In virtue of this relationship, certain properties of the diffusion process may be elucidated by focusing on the kinematics of diffusing particle movements. Such results complement those obtained directly on the concentration field, by the methods of Chapter I. Particularly in turbulent diffusion theory, these considerations on particle motions turn out to be of key importance in arriving at a satisfactory understanding of complex phenomena. In view of its central importance, we shall therefore refer to Equation (2.1) as our 'fundamental theorem', which links the 'P-field' to the 'χ-field'.

2.3. Diffusion with Stationary Velocities

In the above discussion we implicitly used the concept of a 'random process' or 'stochastic process,' as these are usually called in the statistical literature (see e.g., Bartlett, 1956). In such processes a variable (say the velocity or displacement of a diffusing particle) is a function of a parameter (time), but the realized value for any given time is random and may only be specified in terms of a probability distribution. When this probability distribution is *not* a function of time, the process is known as a 'stationary stochastic process' and has some relatively simple properties. Other stochastic processes are 'evolutionary'. In Brownian motion, when the temperature and the composition of the ambient fluid are constant, the *velocity* of a diffusing particle is a stationary process. In turbulent diffusion the same applies provided that the turbulent field is *homogeneous* (in temperature, mean velocity and turbulent intensity).

In a stationary process the ensemble mean value of the variable, of its square, any chosen products, etc. is independent of time. Without loss of generality we may assume the mean value of the velocity to be zero (by measuring velocities relative to a frame of reference moving with any mean motion of the ensemble). Thus if we consider for concreteness the x-component of the velocity vector $\mathbf{u}(u, v, w)$, we may write:

$$\begin{aligned} \bar{u}(t) &= 0 \\ \overline{u^2}(t) &= \text{const.} = \overline{u^2}, \end{aligned} \tag{2.2}$$

where overbars denote ensemble averages. Similar relationships hold, of course, for the other two components v, w.

An important function characterizing stationary stochastic processes is their autocorrelation function $R(\tau)$. This may be regarded as measuring the 'persistence' of a given value (once realized) of the random variable concerned. In the case of the velocity-history of a diffusing particle, once the particle possesses a certain velocity, a short time τ later it is still likely to have a velocity of a similar magnitude and sign. For a given time-delay τ a mean product may be formed, $\overline{u(t)u(t+\tau)}$ (velocity 'covariance'), which evidently tends to $\overline{u^2}$ as $\tau \to 0$. When there are no organized flow-structures (such as standing waves) affecting the motion of the particle, after a long enough time τ it 'forgets' its velocity at time t, in the sense that the realized values $u(t+\tau)$ will be quite independent of $u(t)$. The covariance $\overline{u(t)u(t+\tau)}$ then simply becomes the product of the mean values $\bar{u}(t)$ and $\bar{u}(t+\tau)$, which are zero by hypothesis.

A further property of the velocity covariance follows from the fact that the process is stationary. This means that the ensemble-average product $\overline{u(t)u(t+\tau)}$ is independent of what instant t is chosen as the 'base' for computing this covariance. Thus:

$$\overline{u(t)\,u(t+\tau)} = \overline{u(0)\,u(\tau)} = \overline{u(-\tau)\,u(0)} \text{ etc.} \tag{2.3}$$

which also shows that the velocity-covariance is an even function of τ. A convenient nondimensional representation of the covariance is the 'autocorrelation function' $R(\tau)$:

$$R(\tau) = \frac{\overline{u(t)\,u(t+\tau)}}{\overline{u^2}}. \tag{2.4}$$

In a stationary process this is independent of time t and is an even function. At $\tau=0$ its value is $R(0)=1.0$, while as $\tau\to\infty$ we have $R(\tau)\to 0$. It may be shown that $R(\tau)$ always lies between -1.0 and $+1.0$:

$$-1.0 \leqslant R(\tau) \leqslant 1.0 \quad (\text{all } \tau). \tag{2.5}$$

The *displacement* of an individual diffusing particle is related to its velocity of course by

$$x(t) = \int_0^t u(t')\,\mathrm{d}t', \tag{2.6}$$

where we have again taken the x-components for concreteness. Taking ensemble-averages on both sides of this equation we find that if $\bar{u}(t)=0$ (as we supposed) then also $\bar{x}(t)=0$. Also because realized random processes $u(t')$ may not even resemble each other, the total displacements $x(t)$ may differ in a random manner, i.e., $x(t)$ is again a stochastic process. However, observation shows that large displacements become more probable as time passes. In other words, the process $x(t)$ is *not* stationary, but 'evolutionary'. In virtue of Equation (2.6) the properties of this (displacement) process may be related to the simpler properties of the velocity-history. From Equation (2.6) the rate of change of $x^2(t)$ may be written, in a given realized process:

$$\frac{d\left[x^2(t)\right]}{\mathrm{d}t} = 2x\frac{\mathrm{d}x}{\mathrm{d}t} = 2\int_0^t u(t)\,u(t')\,\mathrm{d}t'. \tag{2.7}$$

If we now take ensemble averages on both sides (which may be taken *inside* the differentiation or integration signs, being a simple arithmetic averaging process) we find the velocity-covariance in the integrand. We may substitute the velocity-autocorrelation from Equation (2.4) and write

$$\frac{\mathrm{d}\overline{x^2}}{\mathrm{d}t} = 2\overline{u^2}\int_0^t R(\tau)\,\mathrm{d}\tau \tag{2.8}$$

having made use of the fact that $R(\tau)$ is an even function, in transforming the integration variable from t' to $\tau = t' - t$. Equation (2.8) is a most important basic result in the theory of diffusion by random movements (Brownian motion *or* turbulence), being known in turbulent diffusion theory as 'Taylor's theorem' (Taylor, 1922).

We return now to an ensemble of independently diffusing particles which we discussed in the previous section. For simplicity, consider an ensemble released at the *origin*, $\mathbf{x} = 0$ at $t = 0$. We may define as a measure of 'spread' along the x-axis the radius of inertia of the mean concentration distribution:

$$\sigma_x^2 = \frac{1}{Q} \int\int\int x^2 \chi(x, y, z, t)\, \mathrm{d}x\, \mathrm{d}y\, \mathrm{d}z\,, \tag{2.9}$$

where the integration is to extend over all available space, and we have explicitly written out the dependence on the three axes. The length σ_x defined by this equation is an appropriate measure of 'cloud size,' as we have repeatedly seen in Chapter I. If now $\chi(x, y, z, t)$ results from the release at the origin of independently diffusing particles, the fundamental theorem (Equation (2.1)) shows that

$$\sigma_x^2 = \int\int\int x^2 P(x, y, z, t)\, \mathrm{d}x\, \mathrm{d}y\, \mathrm{d}z = \overline{x^2} \tag{2.10}$$

so that the radius of inertia σ_x of the concentration field $\chi(\mathbf{x}, t)$ equals the rms particle displacement $(\overline{x^2})^{1/2}$. The latter may be calculated from Taylor's theorem if certain statistical information is available on the velocity-history of diffusing particles. For the radius of inertia σ_x of an *initially concentrated cloud* we have therefore by Taylor's theorem (Equation (2.8)):

$$\frac{d\sigma_x^2}{\mathrm{d}t} = 2\overline{u^2} \int_0^t R(\tau)\, \mathrm{d}\tau\,. \tag{2.11}$$

On a further time-integration we find

$$\begin{aligned} \sigma_x^2(t) &= 2\overline{u^2} \int_0^t \int_0^{t'} R(\tau)\, \mathrm{d}\tau\, \mathrm{d}t' \\ &= 2\overline{u^2} \int_0^t (t - \tau)\, R(\tau)\, \mathrm{d}\tau \end{aligned} \tag{2.12}$$

where the second form follows on partial integration.

2.4. Brownian Motion

The sustained irregular motion performed by small grains or particles of colloidal size immersed in a fluid and observable with the aid of a microscope was described in

1826 by the botanist Robert Brown. Its nature remained a puzzle and a subject of controversy for a long time. Finally, a paper by Einstein (1905) showed convincingly that this 'Brownian motion' is maintained by collisions with the molecules of the surrounding fluid. Under normal conditions, in a liquid, a Brownian particle will suffer on the order of 10^{21} collisions per second. In each collision a randomly oriented impulse is exerted on the particle, the average magnitude of which depends on the absolute temperature. The impulses each produce a fine 'kink' in the path of the drifting particle, the end result being effectively curvilinear irregular motion. In most cases of interest the concentration of Brownian particles is low enough so that they encounter quite different molecules and perform their own motion independently of each other. The random wanderings of many Brownian particles result in their dispersal. As we already remarked, the statistical theory of Brownian motion provides a useful background for an attack on the more difficult problem of turbulent diffusion. For this reason we give in the remainder of this chapter a brief treatment of Brownian diffusion, based largely on Chandrasekhar (1943), Sutton (1953), Fuchs (1964), and Sommerfeld (1964.)

If we assume that the Brownian particle is much denser than the surrounding fluid (e.g., dust in air), a theory of Brownian motion may be built up on the following simple dynamical hypotheses:

(1) The Brownian particle is much larger than the molecular structure of the surrounding fluid, thus it experiences a viscous resistance which is directly proportional to the relative velocity of fluid and particle (by a well known result of Stokes, for most Brownian particles). For the viscous force per unit mass we shall write $-\beta\mathbf{u}$, where $\mathbf{u}$ is the particle velocity vector (relative to the fluid) and β is a constant of dimension s^{-1}.

(2) At the same time, the particle is small enough to be affected by molecular collisions. The collisions are supposed to exert small random impulses (small compared to a typical value of the particle's velocity) independent of the velocity. For the random acceleration we shall write $\mathbf{A}(t)$.

Through the above two hypotheses we divided the forces exerted by the fluid on the particle into two categories: (a) a continuum-type viscous force, and (b) a random component arising from the discontinuous nature of matter. While this is to some extent artificial, it is a successful idealization, because the conclusions based on the theory agree very well with observation.

Ignoring the force of gravity (which, as is easily shown, merely causes a slow downward drift at the constant 'free fall velocity', $f=g/\beta$) the equation governing Brownian motions becomes thus ('Langevin's equation'):

$$\frac{\mathrm{d}\mathbf{u}}{\mathrm{d}t} = -\beta\mathbf{u} + \mathbf{A}(t). \tag{2.13}$$

In this equation the coefficient of the viscous term β may for a spherical particle be calculated from Stokes' law

$$\beta = 6\pi a\mu/m \quad [s^{-1}], \tag{2.14}$$

where a is the radius of the particle, μ viscosity of the fluid, and m the mass of the particle. The reciprocal of β is a kind of 'relaxation time' of the viscous effects. To gain an idea of its order of magnitude, consider a spherical particle of density $\varrho \simeq 1$ g cm^{-3}, floating in air, $\mu = 16 \times 10^{-2}$ g s^{-1} cm^{-1}. If the radius of the particle is $a = 10^{-4}$ cm (limit of visibility) we have

$$\beta^{-1} = \frac{4\pi}{3} \varrho a^3/(6\pi a\mu) = 1.4 \times 10^{-8} \text{ s}$$

so that the viscous relaxation time, even for such a relatively large particle (as far as Brownian motion is concerned) is only of the order of 0.01 μs. The time-scale of diffusion phenomena normally concerning us is very much larger than this.

Integration of Langevin's equation yields, if $\mathbf{u}_0$ is the velocity of the particle at time $t=0$:

$$\mathbf{u} = \mathbf{u}_0 \, \mathrm{e}^{-\beta t} + \mathrm{e}^{-\beta t} \int_0^t \mathrm{e}^{\beta t'} \, \mathbf{A}(t') \, \mathrm{d}t' . \tag{2.15}$$

This result shows the velocity to consist of two additive components, one dependent on the initial velocity $\mathbf{u}_0$, while the other is not, because the random acceleration $\mathbf{A}(t')$ is independent of the velocity. Thus at time $t \gg \beta^{-1}$, when $\mathrm{e}^{-\beta t} \simeq 0$, the particle has 'forgotten' its initial velocity $\mathbf{u}_0$, in the sense that the contribution that depends on $\mathbf{u}_0$ has decayed to zero. The remaining component is the end-result of a large number of independent random impulses, a fact which may be profitably exploited in probabilistic arguments.

2.5. Dispersion of Brownian Particles

It is now clear that, as a result of continuous molecular bombardment, the velocity of a diffusing Brownian particle at any instant is random in magnitude and direction, its velocity-history being thus a stochastic process. The time-integral of the velocity-history is the displacement-history, which is necessarily again a stochastic process.

The kinetic energy of diffusing Brownian particles is derived from collisions with molecules. According to the 'law of the equipartition of energy' of statistical mechanics, the energy associated with one component of the velocity (say the x-component, u), if the temperature and composition of the fluid in which the dispersion takes place are constant, is

$$\tfrac{1}{2} m \overline{u^2} = \tfrac{1}{2} kT , \tag{2.16}$$

where m is the mass of the Brownian particle, the overbar denotes an ensemble average, k is the Boltzmann constant, T absolute temperature. By Equation (2.16) $\overline{u^2}$ is constant in time, so that the random process $u(t)$ is 'stationary' at least to the second order. The same holds of course for the other two components of the velocity.

The velocity autocorrelation function for Brownian particles may be calculated without difficulty from the integrated form of Langevin's equation (Equation (2.15)).

Only the part dependent on the initial velocity $\mathbf{u}_0$ makes a contribution:

$$R(\tau) = \frac{\overline{u(0)\, u(\tau)}}{\overline{u^2}} = e^{-\beta\tau}. \tag{2.17}$$

The typical 'persistence time' for the velocities of Brownian particles is thus equal to the viscous relaxation time β^{-1}, for which we have given some estimates above.

If we substitute the correlation function arrived at in Equation (2.17) into Taylor's theorem (Equation (2.12)) we find after carrying out the integrations:

$$\sigma_x^2 = 2\overline{u^2}\left[\frac{t}{\beta} - \frac{1}{\beta^2}(1 - e^{-\beta t})\right]. \tag{2.18}$$

When $t \gg \beta^{-1}$ this becomes indistinguishable from

$$\sigma_x^2 = \frac{2\overline{u^2}}{\beta}\, t. \tag{2.19}$$

This result is of the same form as we found from the classical diffusion equation so that we may identify as our 'diffusivity':

$$D = \frac{\overline{u^2}}{\beta}. \tag{2.20}$$

For spherical particles we have β given by Equation (2.12) while $\overline{u^2}$ follows from the equipartition of energy (Equation (2.16)); thus:

$$D = \frac{kT}{6\pi a \mu}. \tag{2.21}$$

This is Einstein's result which has been confirmed experimentally in numerous ways. The restriction that the result holds for diffusion times $t \gg \beta^{-1}$ is not serious in practice (for Brownian diffusion) because β is of the order of a small fraction of a second, as has been pointed out above.

For spherical particles of unit density floating in air some values of the Brownian diffusivity D and the relaxation time β^{-1} are shown in Table (II.1) (after Fuchs (1964)).

2.6. Simple Random Walk Model

So far we have dealt with two important but restricted aspects of the displacement probability distribution $P(x, y, z, t)$, (for Brownian particles) namely its first and second moments and found that while $\bar{x}(t)$ remained zero (in a fluid at rest) $\overline{x^2}(t)$ grew in direct proportion to time. In order to arrive at the full distribution $P(x, y, z, t)$ some more complicated calculations are necessary, in which we once more make use of the smallness of the relaxation time β^{-1}.

Given that a particle 'forgets' its initial velocity in a time of order β^{-1}, it is possi-

TABLE II.1

Diffusivities in Brownian motion (spherical particles of unit density in air).

a, cm	β^{-1}, s	D cm^2 s^{-1}
10^{-7}	1.33×10^{-9}	1.28×10^{-2}
2×10^{-7}	2.67×10^{-9}	3.23×10^{-3}
5×10^{-7}	6.76×10^{-9}	5.24×10^{-4}
10^{-6}	1.40×10^{-8}	1.35×10^{-4}
2×10^{-6}	2.97×10^{-8}	3.59×10^{-5}
5×10^{-6}	8.81×10^{-8}	6.82×10^{-6}
10^{-5}	2.28×10^{-7}	2.21×10^{-6}
2×10^{-5}	6.87×10^{-7}	8.32×10^{-7}
5×10^{-5}	3.54×10^{-6}	2.74×10^{-7}
10^{-4} (1 μ)	1.31×10^{-5}	1.27×10^{-7}
2×10^{-4}	5.03×10^{-5}	6.10×10^{-8}
5×10^{-4}	3.08×10^{-4}	2.38×10^{-8}
10^{-3}	1.23×10^{-3}	1.38×10^{-8}

ble to divide the total displacement of a Brownian particle into a number of (nearly) independent 'steps', taking place over a time period Δt, where $\Delta t \gg \beta^{-1}$. The x-component (say) of the j-th step is then

$$x_j = \int_{j\Delta t}^{(j+1)\Delta t} u(t')\,dt'. \tag{2.22}$$

The persistence of particle velocities only affects each step for a period of β^{-1} seconds at the beginning and at the end, which is a negligible effect if Δt is long enough. Thus each step is effectively taken at random, independently of any previous step, and the particle may be said to execute a 'random walk.'

The probability distribution of a particle executing random steps is a standard problem of probability theory with an extensive literature (see e.g., Feller, 1957; Cramer, 1946; Bartlett, 1956). On account of the complexity of the general theory, only an elementary discussion of it is given in the following few sections (Chandrasekhar, 1943).

The main features of the problem are brought out most simply in an analysis of random walk along a straight line, of unit step length, the probability of either a forward or a backward step being exactly 1/2. Thus after taking N steps the particle *could* be at any of the points (it is released at the origin):

$$-N, -N+1, -N+2, \ldots -1, 0, +1, \ldots N-1, N.$$

The question is, with what probability does a particle reach a given point $-N < m < +N$. Let that probability be denoted by $P(m, N)$: in this simple case it can be calculated by enumerating all the possible outcomes of a random walk consisting of N steps and determining which ones of these will result in the particle finishing up at point m.

Denoting forward steps by $+$, backward steps by $-$, each possible outcome is a combination of steps of the form:

$$++---+-++++\cdots-+-$$

or

$$---+--+++\cdots=++-$$

etc.

Total $=N$ steps.

Since the realized direction at each step has probability 1/2, any individual sequence has probability $(1/2)^N$ the steps being all independent. The required probability $P(m, N)$ is therefore $(1/2)^N$ times the number of distinct sequences which will lead to the point m after N steps. Let the number of forward steps in such a sequence be f, the number of backward steps b. For the particle to arrive at m after N steps we have to have

$$f+b=N$$
$$f-b=m$$

therefore $f=N+m/2$, $b=N-m/2$. The number of different sequences consisting of exactly f forward and b backward steps is $N!/(f!\cdot b!)$.

Thus

$$P(m, N)=\frac{N!}{[\frac{1}{2}(N+m)]!\,[\frac{1}{2}(N-m)]!}\left(\frac{1}{2}\right)^N=$$
$$=\binom{N}{(N+m/2)}\cdot\left(\frac{1}{2}\right)^N, \tag{2.23}$$

where, in the last expression, the first bracket denotes a binominal coefficient. This probability distribution is known as a 'Bernoulli distribution'; its mean and mean-square may be shown to be

$$\bar{m}=\sum_{m=-N}^{N} mP(m, N)=0 \tag{2.24}$$
$$\overline{m^2}=\sum_{m=-N}^{N} m^2P(m, N)=N\,.$$

The increasing spread of this distribution is illustrated in Table II.2. Note that m has to be even or odd as N is even or odd.

In Brownian motion the case of greatest interest is when N is very large. In this case the probability near the fringes ($|m|\rightarrow N$) becomes very small and the practically significant part is where $|m|\ll N$. For this case we can simplify our result for $P(m, N)$ by making use of Stirling's formula:

$$\ln(n!)=(n+\tfrac{1}{2})\ln n-n+\tfrac{1}{2}\ln 2\pi+0(1/n)\;(n\rightarrow\infty)\,. \tag{2.25}$$

After a little algebra we find then that

$$P(m, N)=\sqrt{\frac{2}{\pi N}}\exp\left(-\frac{m^2}{2N}\right) \tag{2.26}$$

TABLE II.2

Bernoulli probability distribution for the first few steps

N	m: −6	−5	−4	−3	−2	−1	0	1	2	3	4	5	6
0							1.0						
1						0.5		0.5					
2					0.25		0.5		0.25				
3				0.125		0.375		0.375		0.125			
4			0.0625		0.25		0.375		0.25		0.0625		
5		0.03125		0.15625		0.3125		0.3125		0.15625		0.03125	
6	0.015625		0.09375		0.234375		0.3125		2.34375		0.09375		0.015625

or in other words, a Gaussian distribution with standard deviation $\sqrt{N}$. The convergence of the Bernoulli distribution to the Gaussian is in fact reasonably rapid; Table II.3 illustrates the comparison for the relatively small $N=10$. The percentage error arising from using the asymptotic formula would in this case be quite small except at the very fringes.

TABLE II.3
Comparison of Bernoulli and Gaussian probability distributions for $N=10$

$\pm$m	0	2	4	6	8	10
Bernoulli	0.246	0.205	0.117	0.044	0.010	0.001
Gaussian	0.252	0.207	0.113	0.042	0.010	0.002

The 'discrete' distribution we have obtained here may be converted into a continuous one if we assume that the individual steps are small compared to the length Δx over which we wish to define particle concentrations. Let the step-length be l and let us write

$$m = \frac{x}{l}, \tag{2.27}$$

where x is the displacement from the origin. The total probability of finding a particle over a range Δx of the x-axis (centered at x) is then approximately $P(\mathrm{m}, N)\cdot(\Delta x/2l)$ (because neighboring discrete probability points are separated by *two* step-lengths, see Table II.2). In a diffusing cluster (of independently moving particles) of total mass Q the material contained within this range Δx would then be

$$\Delta M = QP(m, N)\cdot\frac{\Delta x}{2l} \tag{2.28}$$

from which the concentration distribution is

$$\chi = \frac{\Delta M}{\Delta x} = \frac{Q}{2l}\cdot\sqrt{\frac{2}{\pi N}}\exp\left(-\frac{x^2}{2Nl^2}\right). \tag{2.29}$$

The total number of steps N may be related to diffusion time t if we suppose that the particles suffer n displacements per unit time; then $t=N/n$ and $u=\frac{1}{2}nl$ becomes a kind of 'diffusion velocity'. Writing

$$D = \tfrac{1}{2}nl^2 = ul \tag{2.30}$$

we can transform our last result into:

$$\chi(x, t) = \frac{Q}{2\sqrt{\pi Dt}}\exp\left(-\frac{x^2}{4Dt}\right) \tag{2.31}$$

which is exactly as we found for one-dimensional diffusion from the classical diffusion equation. Equation (2.30) therefore defines diffusivity in the case of this simple random walk model.

2.7. Reflecting Barrier

Suppose now that at $m=m_1(>0)$ there is a barrier of such a kind that whenever the particle arrives at m_1 its next step takes it invariably back to (m_1-1). Such a barrier may be said to 'reflect' the arriving particles.

The path a particle may take under such conditions is conveniently illustrated in an (m, N) plane (Figure 2.1). Starting at $m=0$ $N=0$ the particle moves along any succession of diagonals, except that it turns 90° every time it reaches $m=m_1$.

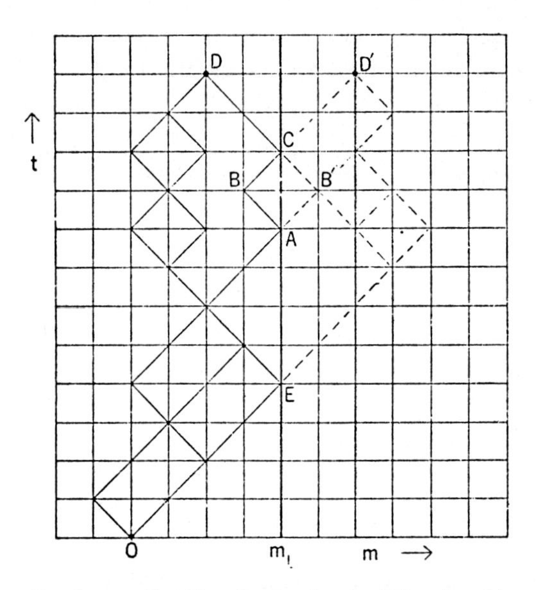

Fig. 2.1. Random walk with reflecting barrier (Chandrasekhar, 1943).

Compared to our previous problem of the unrestricted random walk the change is that at $m=m_1$ the probability of the next step being backward or forward is no longer 1/2-1/2, but 1.0 that it is backward. The sequences leading to $m<m_1$ and involving a backward step at $m=m_1$ have been counted in writing down Equation (2.23). Starting off from this equation as a base, we have to multiply the probability of such a sequence by 2, of those involving n reflections by 2^n, because the factor $(1/2)^N$ in Equation (2.23) for any individual sequence is no longer quite appropriate. An alternative view is to add to the probability $P(n, m)$ calculated for the unrestricted walk the probability $P(2m_1-m, N)$ of arriving at the 'image' point $(2m_1-m)$ after N steps (also in the absence of the reflecting wall), i.e.:

$$P(m, N; m_1) = P(m, N) + P(2m_1 - m, N). \tag{2.32}$$

We can verify the truth of this assertion in the following manner: consider first a path like OED which has suffered just one reflection at m_1. By reflecting this path about the vertical line through m_1 we obtain a trajectory leading to the image point $(2m_1 - m)$ and conversely, for every trajectory leading to the image point, having crossed the line through m_1 once, there is exactly one which leads to m after a single reflection. Thus, instead of counting twice each trajectory reflected once, we can add a uniquely defined trajectory leading to $(2m_1 - m)$. Consider next a trajectory like OABCD which leads to m after two reflections. A trajectory like this should be counted four times. But there are two 'reflected' trajectories, OAB′CD′ and OABCD′, leading to the image point, and a third, OAB′CD, leading to m but which we must exclude because it involves crossing the barrier. These three additional trajectories together with the actual OABCD give exactly four possible sequences of steps leading either to m or its image $2m_1 - m$ in the absence of the reflecting barrier. In this manner the arguments can be extended to prove the general validity of Equation (2.32).

If we pass to the limit of large N, Equation (2.32) becomes, using Equation (2.26):

$$P(m_1N; m_1) = \sqrt{\frac{2}{\pi N}}\left[\exp\left(-\frac{m^2}{2N}\right) + \exp\left\{-\frac{(2m_1 - m)^2}{2N}\right\}\right]. \tag{2.33}$$

Converting this to a continuous distribution by means of the same assumptions as in the last section we find, if a barrier is placed at $x = h = m_1 l$:

$$\chi = \frac{Q}{2\sqrt{\pi D t}}\left[\exp\left(-\frac{x^2}{4Dt}\right) + \exp\left\{-\frac{(2h - x)^2}{4Dt}\right\}\right]. \tag{2.34}$$

It may be noted here that at $x = h$

$$\left.\frac{\partial \chi}{\partial x}\right|_h = 0 \tag{2.35}$$

on account of the symmetry of the two parts of the distribution in Equation (2.34). Physically, this corresponds to the condition of zero flux, which in our model was ensured by the 'reflection' condition.

2.8. Absorbing Barrier

Let the barrier at $m = m_1$ be now so constituted that whenever a particle arrives at it, it becomes incapable of executing further steps, it is 'removed from the game'. Such a barrier may be termed a 'perfect absorber'. Two practically important questions arise: one, what is the probability distribution of particles for $m < m_1$, $P(m, N; m_1)$ analogous to our last result, and two, at what rate are particles arriving at m_1 and being removed from the game.

In counting the number of distinct sequences which lead to $m < m_1$ we now have to exclude all those including even a single arrival at m_1. If we again first count all possible sequences leading to m in the absence of the barrier, we then have to exclude a certain number of 'forbidden' sequences. From the argument of the previous section it is clear (Figure 2.1) that every such forbidden sequence leading to m uniquely defines another sequence leading to the image point $(2m_1 - m)$, because by reflecting about the line $m = m_1$ the part of the forbidden trajectory above its last point of contact with the barrier we are led to a trajectory ending at the image point. Therefore the number of forbidden sequences is equal to the number of sequences leading to the image point in the unrestricted case, i.e.:

$$P(m, N; m_1) = P(m, N) - P_n(2m_1 - m, N). \tag{2.36}$$

For large N this becomes

$$P(m, N; m_1) = \sqrt{\frac{2}{\pi N}} \left[\exp\left(-\frac{m^2}{2N}\right) - \exp\left(-\frac{[2m_1 - n]^2}{2N}\right)\right]. \tag{2.37}$$

Passing to the continuous description as in the last sections we obtain

$$\chi = \frac{0}{2\sqrt{\pi D t}} \left[\exp\left(-\frac{x^2}{4Dt}\right) - \exp\left(-\frac{[2h - x]^2}{4Dt}\right)\right]. \tag{2.38}$$

This result gives $\chi(x=h)=0$, an expression of our basic hypothesis that particles arriving at the barrier are instantly removed from suspension.

To answer our question regarding the arrival rate of particles at the barrier we have to count the number of possible sequences of N steps that take the particle to m_1 without having been in contact with the barrier before. This is again equal to the number of all possible sequences leading to m_1 in an unrestricted walk, less a number of forbidden sequences. To calculate the latter we observe that the particle arriving at m_1 after N steps must have been (in the unrestricted case) either at $(m_1 - 1)$ or at $(m_1 + 1)$ after $(N-1)$ steps. Sequences of the latter kind are all forbidden, but so are some of the former kind, because they may have involved contact with the barrier at an earlier step. Such sequences can again be reflected about their last point of contact with the barrier and they then end up at $(m_1 + 1)$ after $(N-1)$ steps, which of course is the image point of $(m_1 - 1)$. Thus the number of forbidden sequences is simply *twice* the number leading to $(m_1 + 1)$ in $(N-1)$ steps, the total number of permitted sequences leading to m_1 after N steps being then:

$$\frac{N!}{\left(\frac{N-m_1}{2}\right)!\left(\frac{N+m_1}{2}\right)!} - 2\frac{(N-1)!}{\left(\frac{N-m_1}{2}\right)!\left(\frac{N-m_1-2}{2}\right)!} = \frac{m_1}{N}\frac{N!}{\left(\frac{N-m_1}{2}\right)!\left(\frac{N+m_1}{2}\right)!} \tag{2.39}$$

This is (m_1/N) times the total number of sequences leading to m_1 in the unrestricted walk, so the 'arrival probability' is given by

$$a(m_1, N) = \frac{m_1}{N} P(m_1, N). \tag{2.40}$$

At large N this is equal to

$$a(m_1, N) = \frac{m_1}{N}\sqrt{\frac{2}{\pi N}}\exp\left(-\frac{m_1^2}{2N}\right). \tag{2.41}$$

In formulating a continuous description of this process we have to phrase our question: What quantity of diffusing particles arrives at the barrier between time $t - \Delta t/2$ and $t + \Delta t/2$ where Δt is long compared to the time taken to traverse an individual step? If we again assume that n steps are taken per unit time, this means that we are interested in the total arrivals between step numbers $nt - n\Delta t/2$ and $nt + n\Delta t/2$, which is approximately, from Equation (2.41), writing $N = nt$:

$$\frac{1}{2}\frac{\Delta t}{t} m_1 \sqrt{\frac{2}{\pi nt}}\exp\left(-\frac{m_1^2}{2nt}\right),$$

where the factor $\frac{1}{2}$ comes from the recognition that particles may arrive at m_1 only at every *second* step, even or odd according as m_1 is even or odd. The *intensity* of deposition of particles at the barrier then becomes in g s^{-1}, dividing by Δt and substituting again as before $h = m_1 l$, $D = \frac{1}{2} nl^2$:

$$I = \frac{h}{t}\cdot\frac{Q}{2\sqrt{\pi Dt}}\exp\left(-\frac{h^2}{4Dt}\right). \tag{2.42}$$

We readily verify that $I(t)$, satisfies the relationship

$$I(t) = -D\left.\frac{\partial \chi}{\partial x}\right|_{x=h} \tag{2.43}$$

which shows that the rate of deposition is calculated as flux at the barrier in classical diffusion theory, an entirely consistent result.

2.9. Connection of Random Walk to Diffusion Equation

The method of enumerating possible sequences of displacements becomes quite complex when an attempt is made to deal with two- or three-dimensional problems. Yet, the asymptotic solutions (for $N \to \infty$) are relatively simple and identical to those obtained from the diffusion equation. The question arises then: Is it not possible to establish this connection to the diffusion equation *before* solving the individual random walk problem? It turns out that such a connection can indeed be established, the assumptions one is forced to make in the process being particularly illuminating.

Consider a cluster of particles undergoing Brownian motion in a uniform homogeneous fluid. Let Δt denote a time interval long enough for a particle to suffer a large number of displacements, but still short enough for the total displacement to be small compared to the spatial scale of concentration variations. During Δt each particle is then subjected to a random total displacement, the probability of a given displacement Δx being distributed according to a Gaussian law:

$$\psi(\Delta x, \Delta t) = \frac{1}{2\sqrt{\pi D \Delta t}} \exp\left(-\frac{(\Delta x)^2}{4D\Delta t}\right). \tag{2.44}$$

This distribution may be termed the 'transition probability' that a particle at point x will move in Δt to $x + \Delta x$. If the cluster of particles was released at the origin at $t = 0$, the concentration at (x, t) is proportional to the probability $P(x, t)\Delta x$ of finding an individual particle between $x - \Delta x/2$ and $x + \frac{1}{2}\Delta x$, as we have seen in earlier sections. From the probability distribution $P(x, t)$ and the transition probability $\psi(\Delta x, \Delta t)$ we wish to calculate the probability distribution of particles Δt later, $P(x, t + \Delta t)$. This is easily accomplished, because by our earlier assumptions concerning Brownian motion the transition probability ψ is independent of the past history of the particle and, in the case of a homogeneous field, it is independent of particle position as well. Thus we have simply:

$$P(x, t + \Delta t) = \int_{-\infty}^{\infty} P(x - \Delta x, t)\, \psi(\Delta x, \Delta t)\, \mathrm{d}\Delta x. \tag{2.45}$$

For finite Δx, Δt this is an integral equation, but since we assumed Δx to be small on the spatial scale of variations in $P(x, t)$, we may expand $P(x - \Delta x, t)$ in a Taylor series:

$$P(x - \Delta x, t) = P(x, t) - \Delta x \frac{\partial P}{\partial x} + \frac{(\Delta x)^2}{2} \frac{\partial^2 P}{\partial x^2} - + \cdots. \tag{2.46}$$

Substituting Equations (2.44) and (2.46) into (2.45) we find after integration:

$$P(x, t + \Delta t) = P(x, t) + D\Delta t \frac{\partial^2 P}{\partial x^2} + 0(\Delta t^2). \tag{2.47}$$

Hence, to first order in Δt:

$$\frac{\partial P}{\partial t} = \frac{P(x, t + \Delta t) - P(x, t)}{\Delta t} = D \frac{\partial^2 P}{\partial x^2}. \tag{2.48}$$

Noting that P is proportional to concentration χ this is seen to be identical with the one-dimensional diffusion equation. Exactly the same argument, carried out in three dimensions, leads to the three-dimensional diffusion equation (Chandrasekhar, 1943). Note also that the precise (Gaussian) form of the transition probability distribution is irrelevant, as long as its second moment is $2D\Delta t$, a result that already follows from Section 2.4, if $\Delta t \gg \beta^{-1}$.

Physically, the main result is that when a large number of particles execute random flights without persistence of velocity and without mutual interference in a homogeneous field, the result is a diffusion process with constant diffusion coefficient. From our previous sections we note once more the boundary conditions:

$$\begin{aligned} &\chi = 0 \quad \text{on a 'perfectly absorbing surface'} \\ &\frac{\partial \chi}{\partial n} = 0 \ \text{normal to an element of a perfectly reflecting surface}. \end{aligned} \tag{2.49}$$

A further important practical piont is that, just as the solution of the diffusion equation provides an asymptotic solution to random walk problems, conversely, an approximate solution to the diffusion equation with complex boundary conditions may be obtained from actual 'coin tossing' trials (forward or backward step according to heads or tails). This is referred to as the 'Monte Carlo method' of solving the diffusion equation. Of course, actual coin tossing may be replaced by a table of random numbers stored in the memory of a computer. This approach of using random numbers for the empirical study of theoretical problems has long been familiar to statisticians and is illustrated, e.g., by Bartlett (1956).

2.10. Deposition on Vertical Surfaces

The model of the absorbing barrier treated above connects directly to a problem of some practical importance, the deposition of Brownian particles on surfaces which come into contact with the aerosol. The surface in question may be for example the membrane of the human nose or the inside walls of an instrument used to measure particle concentrations. The fact that particles may stick to surfaces has some practical consequences; for this reason we shall discuss relevant solutions of the diffusion equation.

Generally, unless a surface is jarred or unless a strong air current is present, Brownian particles will stick to a wall, so that the absorbing wall boundary condition is appropriate. We prescribe therefore $\chi=0$ at the wall and seek a solution of the diffusion equation satisfying this boundary condition. As a first example, consider a vertical wall perpendicular to the x-axis (Figure 2.2), which is suddenly brought into contact with aerosol of uniform concentration χ_0. Our initial and boundary conditions are therefore:

$$\begin{aligned} &\text{at} \quad t=0 \quad \chi=\chi_0 \quad \text{for} \quad x>0 \\ &\qquad t>0 \quad \chi=0 \quad \text{for} \quad x=0. \end{aligned} \tag{2.50}$$

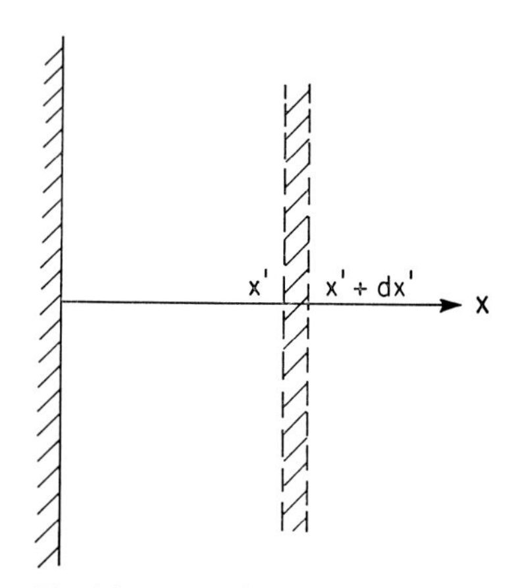

Fig. 2.2. Deposition on vertical wall.

To find a solution, each slice of the initial aerosol lying between x' and $x'+\mathrm{d}x'$ may be considered a concentrated area source of strength $\chi_0 \mathrm{d}x'$. Each such source gives rise to a concentration field:

$$\chi(x, x', t) = \frac{\chi_0 \,\mathrm{d}x'}{\sqrt{2\pi}\sigma}\left[\exp\left\{-\frac{(x-x')^2}{2\sigma^2}\right\} - \exp\left\{-\frac{(x+x')^2}{2\sigma^2}\right\}\right] \tag{2.51}$$

The second exponential herein is the mirror image term, which appears with the negative sign in accordance with the absorbing boundary condition (Equation (2.38)). To arrive at the combined field of all the individual slices we integrate:

$$\chi(x, t) = \frac{\chi_0}{\sqrt{2\pi}}\int_0^\infty \left[\exp\left\{-\frac{(x-x')^2}{2\sigma^2}\right\} - \exp\left\{-\frac{(x+x')^2}{2\sigma^2}\right\}\right]\frac{\mathrm{d}x'}{\sigma} = \chi_0 \operatorname{erf}\left(\frac{x}{\sqrt{2}\sigma}\right). \tag{2.52}$$

In both Equations (2.51) and (2.52) (as in Chapter I) σ stands for:

$$\sigma^2 = 2Dt\,. \tag{2.53}$$

The resulting concentration distribution is sketched in Figure 2.3. The width of the depleted region near the wall grows with the square root of time (as σ). The flux at the wall, giving the intensity of deposition is:

$$I = -D\left.\frac{\partial\chi}{\partial x}\right|_0 = -\chi_0\frac{D}{\sigma}\sqrt{\frac{2}{\pi}}\;[\mathrm{g\ s\ cm^{-2}}]. \tag{2.54}$$

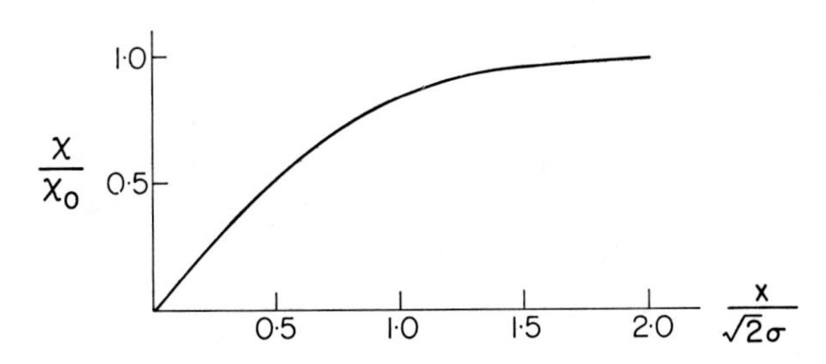

Fig. 2.3. Concentration distribution near absorbing wall.

The negative sign shows a flux to the left, i.e., toward the wall. The total accumulated deposit up to a certain time t is

$$A = \int_0^t I(t')\,\mathrm{d}t' = 2\chi_0\sqrt{\frac{Dt}{\pi}} = \sqrt{\frac{2}{\pi}}\chi_0\sigma\;[\mathrm{g\ cm^{-2}}]. \tag{2.55}$$

This is equivalent to completely removing Brownian particles to a distance of $\sqrt{2/\pi}\,\sigma$ from the wall. The order of magnitude of this distance may be judged using some diffusivities in Table II.1. We find the results listed in Table II.4, which show that this absorption 'boundary layer' is usually quite thin.

As is often done in mass-transfer problems, we may interpret the rate of deposition I as the product of the undisturbed concentration χ_0 and a 'deposition velocity' v_d. According to Equation (2.54) v_d is in this problem equal to $(2/\pi)^{1/2}D/\sigma$, i.e. it decreases as $t^{-1/2}$.

TABLE II.4
Absorption distance, $\sqrt{2/\pi}\,\sigma$, in centimeters, for spherical particles of unit density in air

Time / Particle Radius	0.1 μ	1 μ	10 μ
0.1 s	5.3×10^{-4}	1.3×10^{-4}	0.14×10^{-4}
1 s	16.8×10^{-4}	4.0×10^{-4}	0.44×10^{-4}
10 s	53.2×10^{-4}	12.7×10^{-4}	1.38×10^{-4}

2.11. Deposition on Horizontal Surfaces

When the surface in contact with the aerosol is horizontal, the vertical drift of the particles due to gravity may have to be taken into account. Under conditions of practical interest this is a slow, orderly vertical motion on which Brownian movements are simply superimposed. The drag force arising from the vertical drift velocity ('free fall velocity') of the particles then exactly balances their weight.

As we have seen before, the drag-force on a small spherical particle is given by the Stokes relationship:

$$F_d = -\,6\pi\mu a V \tag{2.56}$$

wherein V is the velocity of the particle relative to the fluid, a is particle radius, $\mu = \nu\varrho$ viscosity of the fluid. To calculate the free fall velocity ($V = f$) of such a particle, we have to substitute its weight for the drag force,

$$F_d = -\,mg = -\,\tfrac{4}{3}a^3\pi\varrho_p g\,, \tag{2.57}$$

where ϱ_p = particle density. In this manner we find

$$f = \tfrac{2}{9}\,\frac{ga^2\varrho_p}{\mu}\,. \tag{2.58}$$

At Reynolds numbers fa/ν greater than about 1 this relationship no longer holds and empirical data on the drag coefficient must be used. These result in the free fall velocities shown in Figure 2.4 (Fuchs, 1964).

Consider now a cluster of Brownian particles released instantaneously from a plane

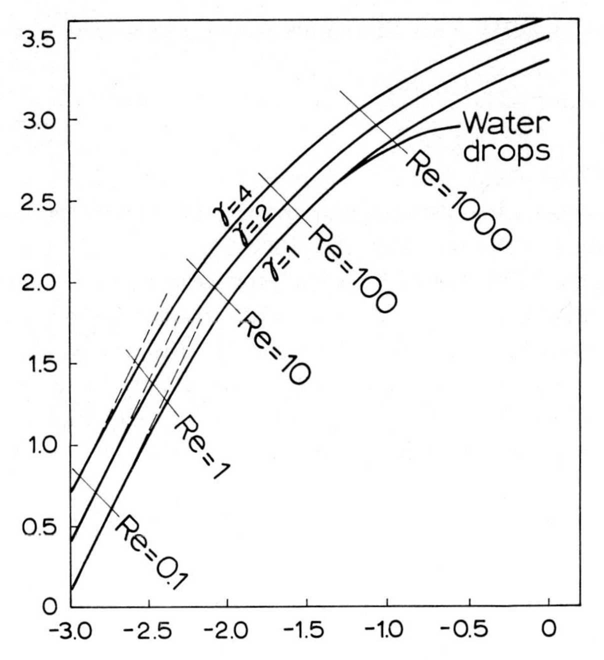

Fig. 2.4. Free-fall velocity of spherical particles in air. Abscissa is 10-base logarithm of radius in cm, ordinate 10-base logarithm of free-fall velocity in cm s^{-1}. γ stands for density of particles in g cm^{-3} (Fuchs, 1964).

source a distance h above a horizontal boundary. The cluster will drift vertically downward, while it will also diffuse vertically. This is a one dimensional diffusion problem again (given a plane source and a parallel plane boundary), but the diffusion equation now contains one convective term. Laying the x-axis vertically upward (so that the drift velocity is $-f$) we have:

$$\frac{\partial \chi}{\partial t} - f\frac{\partial \chi}{\partial x} = D\frac{\partial^2 \chi}{\partial x^2}. \tag{2.59}$$

Before the cloud encounters the boundary, it diffuses about its center exactly as it would without free fall. At successive instants, the position of its center will be at

$$x = h - ft$$

so that the concentration field should be given by:

$$\chi = \frac{Q}{\sqrt{2\pi}\sigma} \exp\left\{-\frac{(x-h+ft)^2}{2\sigma^2}\right\}, \tag{2.60}$$

where Q is source strength in g cm^{-2}, and $\sigma^2 = 2Dt$. It may be verified directly that this is a solution of Equation (2.59), but of course it does not satisfy the boundary condition at the boundary $x=0$. The result may also be written as a product:

$$\chi = \exp\left\{-\frac{2f(x-h)t + f^2t^2}{2\sigma^2}\right\} \times \left\{\frac{Q}{\sqrt{2\pi}\sigma} \exp\left[-\frac{(x-h)^2}{2\sigma^2}\right]\right\} \tag{2.60a}$$

the second bracketed factor being the solution without free fall $f=0$.

The boundary condition, assuming an absorbing boundary, is $\chi=0$ at $x=0$. It is unfortunately not possible to satisfy this using a mirror image term in the ordinary sense: reflecting the cloud about $x=0$ would produce an upward 'falling' image, which is unrealistic (it would only satisfy Equation (2.59) if we changed the sign of f). However, it may be verified that Equation (2.59) remains satisfied if we change $+h$ to $-h$ in the *second* factor only in Equation (2.60a), i.e., reflect only the factor independent of f. It is then easy to subtract this modified solution from Equation (2.60a) to yield an expression satisfying the boundary condition:

$$\begin{aligned}\chi(x,t) &= \frac{Q}{\sqrt{2\pi}\sigma} \exp\left\{-\frac{2f(x-h)t + f^2t^2}{2\sigma^2}\right\} \\ &\times \left[\exp\left\{-\frac{(x-h)^2}{2\sigma^2}\right\} - \exp\left\{-\frac{(x+h)^2}{2\sigma^2}\right\}\right].\end{aligned} \tag{2.61}$$

The flux at the boundary, i.e., the intensity of deposition is now

$$\begin{aligned}I &= -f\chi|_0 - D\left.\frac{\partial\chi}{\partial x}\right|_0 = -D\left.\frac{\partial\chi}{\partial x}\right|_0 = \\ &= \frac{h}{t}\left[\frac{Q}{\sqrt{2\pi}\sigma} \exp\left\{-\frac{(h-ft)^2}{2\sigma^2}\right\}\right].\end{aligned} \tag{2.62}$$

The bracketed factor is the concentration field in the absence of a boundary and h/t is a 'deposition velocity' applied to the undisturbed concentration field. Equation (2.62) may be put into the following nondimensional form:

$$\frac{Ih^2}{DQ} = \sqrt{\frac{2}{\pi}\frac{h^3}{\sigma^3}} \exp\left\{-\frac{h^2}{2\sigma^2}\left(1 - \frac{hf}{2D}\cdot\frac{\sigma^2}{h^2}\right)^2\right\}. \tag{2.62a}$$

On the right there are two independent variables: σ/h, which is a nondimensional time-variable (proportional to $\sqrt{t}$) and $hf/2D$ which is the only place where the free fall velocity occurs, a kind of nondimensional fall velocity. According to whether this variable is small, of order unity, or large, quite different deposition rate histories may prevail, as illustrated in Figure 2.5.

EXERCISES

1. A sphere of radius R is suddenly brought into contact with an aerosol of uniform concentration χ_0, diffusivity D, and negligible free fall velocity. Calculate the concentration field in the neighborhood of the sphere.

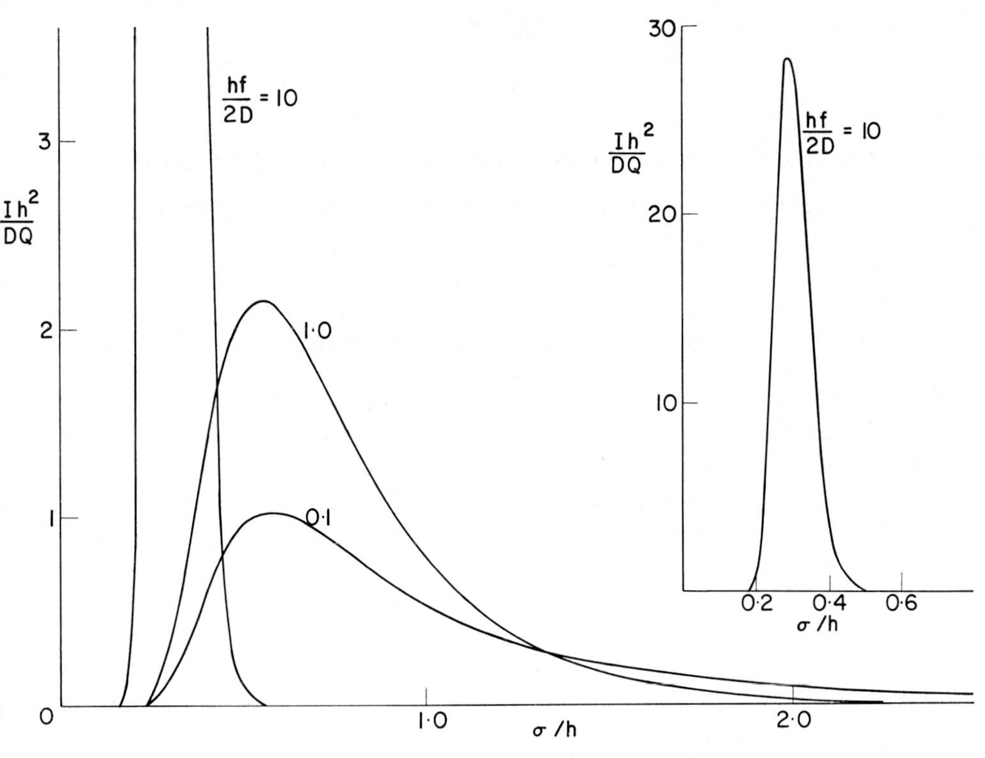

Fig. 2.5. Rate of deposition of heavy particles on horizontal surface.

2. An aerosol of initial concentration χ_0, diffusivity D and free fall velocity f is suddenly brought into contact with a horizontal wall which forms the (a) top, and (b) bottom of the half-space occupied by the aerosol. Calculate the concentration field, paying particular attention to asymptotic conditions ($t \to \infty$) in case (b).

References

Bartlett, M. S.: 1956, *An Introduction to Stochastic Processes*, Cambridge Univ. Press, London, 312 pp.
Batchelor, G. K.: 1949, *Australian J. Sci. Res.* **2**, 437.
Chandrasekhar, S.: 1943, *Rev. Mod. Phys.* **15**, 1.
Cramer, H.: 1946, *Mathematical Methods of Statistics*, Princeton University Press, 573 pp.
Einstein, A.: 1905, *Ann. Physik* **17**, 549.
Feller, W.: 1957, *An Introduction to Probability Theory and its Applications*, John Wiley and Sons, New York, 461 pp.
Fuchs, N. A.: 1964, *The Mechanics of Aerosols*, Pergamon Press.
Sommerfeld, A.: 1964, *Thermodynamics and Statistical Mechanics*, Academic Press, 401 pp.
Sutton, O. G.: 1953, *Micrometeorology*, McGraw Hill Book Co. New York, 333 pp.
Taylor, G. I.: 1922, *Proc. London Math. Soc.* **A 20**, 196.

CHAPTER III

TURBULENT DIFFUSION: ELEMENTARY STATISTICAL THEORY AND ATMOSPHERIC APPLICATIONS

3.1. Fundamental Concepts of Turbulence

The main difficulty of our subject stems from the fact that the spreading of any pollutants in the atmosphere or in various bodies of water takes place in a flow field that is almost invariably turbulent. Turbulent motion has proved to be one of the most untractable problems of the physical sciences, a full understanding of which is not in sight yet. Important progress has been made in recent decades, however, and some of the basic concepts of turbulence theory are now well established. Without attempting here a detailed discussion of turbulent flow, we shall give a brief glossary of the fundamental physical concepts involved because they will frequently enter our later discussions. For a more detailed introduction to the subject the reader is referred to Batchelor (1953), Townsend (1956), Schubauer and Tchen (1959), Lin (1959), Hinze (1959) and Lumley and Panofsky (1964).

In a general way, one may say that matter in the gaseous or liquid state is often subject to two kinds of *random movements*: one on the molecular scale, the thermal agitation of molecules, and one on a macroscopic scale, turbulence. In everyday life one may observe turbulence on a number of occasions, for example in a billowing smoke plume.

The main differences between molecular agitation and turbulent movements are: (1) a difference of scale, the typical 'sweep' of a single turbulent movement being ordinarily very large compared to one molecular free path, and (2) the constraint imposed by continuity. The fairly large parcels of fluid partaking of turbulent movements can only move by displacing other fluid which eventually has to fill in the space vacated by the moving parcel. Consequently one may picture turbulent flow as consisting of a number of closed flow structures of diverse shapes and sizes, called 'eddies,' although it would be a mistake to think of these as resembling regular vortices or indeed regular flow structures of any kind.

The question arises: What measurable variables can be used to characterize the turbulent 'state?' On observing the fine details of the motion by means of a rapid response instrument one is immediately led to the concept of turbulent 'intensity', or mean-square velocity in a given coordinate direction, $\overline{u'^2}$, $\overline{v'^2}$ and $\overline{w'^2}$, where u', v', w' are the 'turbulent' components of the velocity, i.e., those fluctuating quantities superimposed on a (supposedly steady) local mean velocity. If the magnitude of the mean velocity is U, one may also speak of 'relative intensity' or 'gustiness,' the components of which are defined by

$$
\begin{aligned}
i_x &= \frac{(\overline{u'^2})^{1/2}}{U} \\
i_y &= \frac{(\overline{v'^2})^{1/2}}{U} \\
i_z &= \frac{(\overline{w'^2})^{1/2}}{U}.
\end{aligned}
\tag{3.0}
$$

The square root of one mean square velocity, say $u_m = \overline{(u'^2)}^{1/2}$, may also be regarded as a 'characteristic velocity' of the eddying movements.

It transpires that a 'characteristic velocity' alone is not adequate to describe even in the crudest form the more important effects of turbulence. In addition to it, we must also define a 'characteristic size' or 'scale' L of the eddies. In view of the irregularity of turbulent movements, it is difficult to define precisely what one may mean by the 'typical eddy size' particularly so since one cannot even say with any kind of definiteness what an eddy is. Experimentally the method of defining this 'scale' L is the following: consider the record of velocity fluctuations u' taken at point A and another record taken simultaneously at the neighboring point B. If the two points are very close together, the records are nearly identical, if they are very far apart they are completely unrelated. In between it is possible to define a distance at which the relationship (the covariance of the fluctuations) drops to a negligible value. This distance (or an agreed fraction of it) is defined as the 'scale of turbulence.' The statistical concepts on which this definition is based are discussed in the texts referred to earlier; here it should suffice to accept the fact that such an average eddy size can be defined.

The inadequacy of a single length to describe the turbulent movements in, say, a billowing smoke plume should be obvious to the most casual observer, there clearly being eddies of vastly different sizes in this type of flow. What this 'scale' reflects is a kind of average size for those relatively large and fast flow structures which mainly determine the mean square turbulent velocity, or, in other words, which contain most of the turbulent energy. One customarily describes this state of affairs by the statement that the 'energy containing eddies' are characterized by the length scale L and the velocity scale u_m.

Elaborating on the discussion above, on careful and prolonged observation of turbulent flow one concludes that, speaking a little more accurately than before, turbulence apparently consists of a variety of 'eddies,' characterized by different length and velocity scales. In fact it is possible to isolate eddies of a certain size and determine their average characteristic velocity (or the average kinetic energy per unit mass carried by them). Doing this for a number of different sizes one arrives at a 'spectrum' of turbulent energy over the eddy sizes. Without going into details, it is sufficient to accept here that eddies of different sizes are characterized by different velocities and hence carry varying contributions to the turbulent energy.

The generation of turbulence may be more or less satisfactorily explained by considering the hydrodynamic stability of sheared flow at high Reynolds numbers. One finds that certain fairly large eddies have a tendency to grow and extract energy from the main flow. In their turn these large eddies are also unstable flow structures and smaller eddies are capable of 'feeding' on them and thus of extracting their energy. Still smaller eddies may then grow on the last group, with the process continuing until viscosity sets a limit to it, since a small enough eddy will be characterized by a low enough Reynolds number to ensure stability. In such a small but stable eddy direct energy conversion into heat takes place. This theoretical picture of an 'energy cascade' down decreasing eddy sizes has been put forward several decades ago and there is enough evidence for its correctness nowadays to accept it as substantially true, even if one should hesitate to attach too concrete a physical picture to an 'eddy' of a given 'size.'

With the aid of statistical theory and appropriate experiments some of the details of this cascade process may be elucidated. One finds that the energy content (or velocity amplitude) of eddies diminishes as one goes down the scale, most of the turbulent energy being concentrated in the larger eddies, obtaining their energy supply directly from the main flow. These have been called before the 'energy containing eddies,' their characteristic size being the *scale* of turbulence. At the other end of the cascade process are the 'dissipative' eddies, characterized by their own typical size which will not concern us very greatly in diffusion problems. In between there are eddies receiving energy from the larger ones and handing it down to the smaller ones. Practically all the energy ultimately converted into heat thus goes through a number of degrading steps. Clearly, the rate of total energy dissipation per unit mass ε is an important variable in determining the internal structure of turbulence. While the energy content varies with eddy size, the rate of handing down the energy is practically constant for a fairly broad range of eddies at the top and the middle of the 'cascade.' Several properties of turbulent flow are determined by the eddies in this intermediate size range (known as the 'inertial subrange') and important conclusions may be drawn from the fact that their properties depend only on the dissipation rate ε.

Without going into details we remark here that in geophysical situations turbulence is often of an even more complex structure, because it may receive its energy supply from a number of diverse sources, not only one well defined main shear flow. For example, as was discussed by Ozmidov (1965), oceanic turbulence is generated by winds and currents of vastly different spatial scales and by thermal influences, such as the cooling of the surface or by solar heat input. The resulting spectral distribution of energy is complex and leads to various complications in the interpretation of diffusion experiments carried out in such an environment.

3.2. Field Measurements of Concentration and Dosage

In attempting to determine the concentration of some unwanted 'pollutant' or of a deliberately introduced 'tracer' in the natural environment a small sample volume is usually drawn into an appropriate measuring instrument, or a small volume *in situ* is made to interact with the instrument (e.g., illuminated by UV light to cause fluorescence). The mass of the pollutant or tracer within the sample volume may then be determined for example by chemical analysis, spectroscopic analysis, fluorometry, radioactivity count, or by visual counting of discrete particles to mention only the more common techniques. The results of such measurements are often expressed, at least in the first instance, in a variety of units (% by volume, particles per liter, etc.). Conversion to the standard unit of g cm^{-3} is of course possible but unfortunately not always carried out.

Dosage measurements are very similar, except that the ambient fluid (air or water) is sampled at a *continuous* rate for a period, and the concentration determined in the final sample. This is equivalent to observing the time-integral of concentration, known as dosage D:

$$D = \int_{-\infty}^{t} \chi \, \mathrm{d}t' = \frac{M}{q_s} \quad [\text{g s cm}^{-3}], \tag{3.1}$$

where M is the mass of the pollutant or tracer collected before time t and q_s is the rate at which ambient fluid was sampled ($\text{cm}^3 \, \text{s}^{-1}$).

The quantity defined by Equation (3.1) may be termed 'partial dosage.' As $t \to \infty$, D becomes 'total dosage,' observable during the entire passage of some diffusing cloud at a sampling instrument. In virtue of the definition in Equation (3.1) the concentration is sometimes also referred to as 'dosage rate,' $\mathrm{d}D/\mathrm{d}t$, but we will avoid this terminology. When speaking of 'dosage' most experimentalists mean 'total dosage' and this is the sense in which we shall use the word. To determine the concentration of many common pollutants in a given volume is a relatively slow and difficult process, and for this reason dosage measurements are in practice often preferred.

Dosage measurements are particularly common in *atmospheric* experiments, because (nearly) instantaneous concentration measurements are very difficult to carry out. One instrument that has found a great deal of use is the 'rotorod,' the principal part of which is a small metal arm rotated at high speed. The arm sweeps a considerable volume of air (order 20 l min^{-1} for example) and collects by impaction the fine tracer particles in suspension in this volume. A good description of this technique may be found, e.g., in Leighton *et al.* (1965). The (fluorescent) particles on the arm are subsequently counted under the microscope and the dosage expressed as: particles counted, divided by air sampling rate, particle-min l^{-1} which can be converted into g s cm^{-3}, the c.g.s. unit of dosage, if desired.

The instruments determining concentration or dosage may be fixed to the ground or to an instrument tower, balloon tether line, etc., or else they may be mounted in a boat, truck or aircraft for the purpose of taking concentration profiles while crossing (in theory instantaneously, in practice fairly rapidly) a diffusing cloud. The advantage of this latter method is that one can follow the diffusing cloud regardless of wind or current shifts, while in experiments with fixed instruments the cloud frequently 'misses the grid.' There is, however, a difference in the statistical properties of observed concentration as between 'fixed' and 'moving' samplers and in the theoretical treatment we shall have to define carefully what precise physical measurement we have in mind when we talk about observed 'mean' concentration or dosage.

3.3. The Statistical Approach to Environmental Diffusion

In environmental diffusion the biological effects of pollutants, insecticides, etc., on humans, plants and animals are ultimately of concern and therefore the sampling volume is usually chosen appropriately: a volume of the order of 1 liter (linear dimension 10 cm) yields a useful measure of local *concentration*, while a 1 in a few seconds is an appropriate sampling rate for the determination of *dosage*. In this way all necessary detail of the concentration or dosage field is documented, even though a finer structure of the concentration variations could presumably be observed by a finer subdivision of the field. The finest concentration structure is determined by the small-scale eddies of turbulence and their interaction with molecular diffusion (Batchelor, 1959). In the atmosphere, near ground level, appreciable changes in concentration probably occur over distances of the order of 1 cm and less. Such details are, however, practically irrelevant and become smoothed out by the usual sample volume.

On the other hand it is also clear that the degree of spatial smoothing so introduced is insufficient to remove the randomness from individual concentration readings. As is easily observed in the behavior of smoke plumes, for example, a puff of 'marked fluid' will spread into its immediate neighborhood after release in an irregular fashion as it is moved and distorted by many turbulent eddies in its region of the environment. At any fixed point in the wake of the source more and less concentrated parcels pass in irregular succession and any instantaneous crossing of the puff also provides a highly irregular profile, all on a scale usually much larger than 10 cm. Indeed from what we said about the basic properties of turbulence it is clear that concentration variations will occur on the scale of the 'typical' eddy size, L, which in atmospheric and oceanic applications is usually 10 m or larger. In order to remove the randomness by spatial smoothing, as we did in dealing with molecular diffusion, we would have to specify a sampling volume large compared to L. This leads to such gross indices of pollution as the total pollutants in the atmosphere of a city, or in an entire lake. While such indices are of some interest, they hardly offer adequate information from which pollution hazards, etc., could be determined. Nor is it possible to draw such a large sample volume directly through an instrument to determine the space-average concentration.

In analyzing dispersal through Brownian motion we have arrived at the classical diffusion equation because neighboring particles had independent displacement-histories and a cluster of them at once yielded an ensemble of independent realizations of a one-particle release. By contrast, an initially small cluster of particles released into a *turbulent* field is acted upon by many eddies of large spatial scale, meaning that all particles in the cluster are moved through approximately the same distance by each such large random flow structure. Even quite small eddies, as a rule, move many diffusing particles together. Thus the motion of neighboring particles is far from independent and individual cluster-diffusion experiments do not result in anything like the smooth and regular distributions which ensemble averages provide, but exhibit quite unpredictable irregularities on a variety of scales.

We recognize at the outset therefore that practically observable concentrations or dosages of a pollutant at some location in the environment are in general random variables, about which we will only be able to make probabilistic predictions. Fully satisfactory predictions of this sort would specify with what frequency given concentrations are exceeded and what the probability is that a given high value of concentration lasts for more than 10 s, 1 min, etc. Such information could be given in the form of suitably defined probability distributions (single and joint probabilities). Experimentally, the relevant information would have to be collected by carrying out large numbers of identical trials, a practically very difficult task on account of the natural variability of winds and currents. At present very little experimental information exists on such probability distributions, nor is the theory of turbulent diffusion far enough developed to predict them with any confidence. Effectively the only quantity about which we have adequate evidence, both theoretical and experimental, is the first moment of the concentration probability distribution, i.e., its 'expectation.' Brief reflection shows that this expectation is also equal to the mean concentration of an ensemble of independently diffusing particles, $\chi(\mathbf{x}, t)$, which we considered in deriving our 'fundamental theorem,' (Equation (2.1)). We have already seen that this mean value may be connected to relatively simple kinematic properties of particle displacements through our 'fundamental theorem.' This theorem therefore will be the basis of most of our subsequent approach, meaning that most of our results will relate to the ensemble-mean concentration field. In Chapter VII we shall briefly take up the question of 'fluctuations' or departures from this mean, if only to show how little this aspect of the subject has developed so far.

The 'ensemble mean' concentration is a relatively subtle concept the theoretical advantages of which fail to impress experimentalists unless there is a tolerable resemblance between theory and experiment. One way to ensure such resemblance is to tailor the precise definition of this mean (and of the particle displacement probabilities to which it is proportional) to the particular experimental situation. Thus one may consider the average of many concentration readings taken at a *fixed point* $\mathbf{x}$ in space, or else that obtained at fixed distances $\mathbf{y}=\mathbf{x}-\mathbf{c}$ from the *center of gravity* $\mathbf{c}$ of a diffusing cloud. The two approaches lead respectively to the theory of 'absolute' and 'relative' diffusion, and provide more or less satisfactory models for different experimental findings.

Apparently the simplest physical situation in which an ensemble mean concentration may be satisfactorily approximated experimentally is the *continuous* release of tracer at a constant rate into a statistically steady and (nearly) homogeneous field of turbulence of constant mean velocity, such as one may find in the core region of fully developed pipe flow. At any *fixed point* $\mathbf{x}$ in the wake of such a maintained source the observable ('instantaneous') concentration varies in a random manner but under the conditions mentioned the concentration history at fixed **x** is a *stationary* process. By the *ergodic property* of such processes the ensemble-mean concentration at given time t, is equal to the *time-average* concentration obtained by averaging the concentrations, observed in a single experiment over a sufficiently long period (long compared to the lifetime of the typical turbulent eddy). Thus by the fundamental theorem this *time-average* concentration field at a fixed point (accessible to direct physical measurement) may be related to particle displacement probabilities, after taking suitable account of the continuous nature of the source. This leads us to what might be called the 'elementary statistical theory' of turbulent diffusion, which is the subject of the next few sections. The importance of this theory rests partly on the fact that a continuous point source in a uniform wind models fairly satisfactorily such prime pollution sources as factory chimneys, although in applying the theory to similar practical situations many approximations have to be made.

3.4. 'Lagrangian' Properties of Turbulence

In order to exploit our fundamental theorem, much as we have done in studying Brownian motion, we focus on the kinematics of particle movements in an idealized field of homogeneous and stationary turbulence. The random movements of a diffusing particle (such as a dye molecule or a dust particle) in such a field consist of the movements of continuum elements (irregular 'bulk' motion) and a superimposed Brownian or molecular motion*. Both components of course contribute to dispersion; however, there is under ordinary circumstances a large difference in scale between even the smallest turbulent eddy and the mean free path, so that these two motions proceed independently, mean square displacements or velocities being simply the sum of squares arising from either source (the correlation between the two being negligible). Moreover, the molecular contribution to displacements is often negligible compared to the turbulent continuum (bulk) motions (for a careful discussion of this point see Saffman (1960)). For the mean square displacements along, say the x-axis we may thus write

$$\sigma_x^2 = \sigma_{xt}^2 + 2Dt\,, \tag{3.2}$$

where σ_{xt}^2 is the mean square displacement due to turbulent bulk motions alone. In most environmental applications the molecular contribution 2 Dt is negligible and $\sigma_x \simeq \sigma_{xt}$; in the following we shall assume this and ignore the distinction between σ_x and σ_{xt}. There may be cases, however, when one must revert to the more accurate Equation (3.2).

Once we ignore molecular motions, the random movements of individual *fluid* elements (in which particles are suspended) alone become responsible for particle displacements (a process we might label 'pure' turbulent diffusion). In continuum fluid mechanics, a description of fluid motions in terms of the velocities or displacements of

* We assume in this chapter that there is no organized motion (e.g. gravitational drift) of the particle relative to the fluid element in which it is suspended.

individual tagged elements of the fluid is known as a 'Lagrangian' description, in contrast to an 'Eulerian' one in which attention is focused on volume elements fixed in space. Our concern is with the Lagrangian properties of turbulence, but these are unfortunately less well known than Eulerian ones.

In a stationary and homogeneous field of turbulence the 'Eulerian' mean velocity and the mean square turbulent velocities (as well as other, more complex products) are constant in space and time. Such fixed-point velocities are a result of many eddies randomly moving a given volume element fixed in space. When we focus on a tagged fluid element, its velocities are also determined by random eddies, but in a somewhat different way, because the element 'travels with' the eddies. On intuitive grounds it seems reasonable to expect that the velocity history of a wandering continuum-element will also be a *stationary* process, with constant mean and mean square velocities, which equal the Eulerian quantities characterizing the turbulent field. A theoretical investigation, however, uncovers some subtle difficulties. Lumley (1962) has proven that Lagrangian and Eulerian velocity distributions are identical, provided that the homogeneous and stationary field of turbulence is either unbounded or bounded by solid walls. This does not encompass all possibilities: in the atmosphere, a layer of nearly homogeneous turbulence may be regarded to be bounded above by a nonturbulent medium. At the free upper boundary fluid particles may escape, for example, with relatively high velocities, and be replaced by other particles having relatively low velocities. The physical reason for the higher escape velocities may be buoyant forces on slightly warmer (and lighter) elements. The equivalence of Eulerian and Lagrangian velocity distributions does not apply to such a case. It certainly applies, however, to turbulent flow within the core region of a pipe, for example, where turbulence is stationary and nearly homogeneous.

As we have seen in Section 2.3, the mean square dispersion of independently moving particles may be related to their velocity autocorrelation $R(\tau)$. Neglecting molecular velocities and in the absence of any organized drift this may be approximated by the Lagrangian velocity autocorrelation of the moving *fluid element* in which the particle is suspended. Such an autocorrelation is a 'two-point' Lagrangian quantity (involving velocities of a tagged fluid element at time t and $t+\tau$) and it *cannot* strictly be related to any Eulerian quantity (Lumley, 1962). Empirically, a certain resemblance was noted between the shape of this autocorrelation coefficient and the corresponding one for the fixed point (Eulerian) record of a turbulent velocity component. As we shall see later, a practical method (due to Hay and Pasquill, 1959) of atmospheric diffusion prediction is based on this resemblance, with an empirically determined rescaling-factor for the time axis. It appears, however, that the success of the method is based partly on the insensitivity of the prediction to the exact shape of the correlation function (as well as on the possibility of allowing for the nonstationary character of atmospheric turbulence, which is the method's main advantage).

The actual shape of the Lagrangian correlation function $R(\tau)$ for homogeneous and stationary turbulence has not so far been predicted theoretically in a satisfactory way. The only theoretical argument which has had modest success is that the velocity history of a fluid element is approximately a 'Markov process.' Such processes have been studied extensively (Bartlett, 1956); they are characterized by the fact that values of the random variable in question at times greater than t depend on the realized value at t, but not on any previous values. Owing to the inertia of fluid elements, this is not quite true of an element's velocity history, but the discrepancy is only important for small lag-times τ. At larger lag-times, the random impulses exerted by the turbulent field dominate the correlation function and a Markov process becomes a satisfactory theoretical model. This at least appears to be true under the assumptions we have made above, that the field of turbulence is stationary and homogeneous and provided that it does not contain organized periodic (wave-like) motions. In atmospheric applications more complex circumstances often prevail which may require a modification of the theoretical model. In laboratory experiments, however, satisfactory agreement with theory is obtained if we assume a Markov process, except in regard to phenomena influenced by the shape of the autocorrelation function at small τ.

The autocorrelation function of a Markov process is of the exponential-decay type:

$$R(\tau) = e^{-\tau/t_L}, \tag{3.3}$$

where t_L is an appropriate time scale. In applying this formula to the velocity history of a fluid element t_L becomes the 'Lagrangian time scale.' We observe that a similar correlation function was derived in Section 2.4 for Brownian motion, which is therefore also a Markov process. The time scale in Brownian motion was $t_L = \beta^{-1}$, which we have estimated before to have a magnitude in the microseconds or less. By contrast, the Lagrangian time scale in turbulent flow is a characteristic of the energy containing eddies and is thus determined by the length scale L and the velocity scale u_m of those eddies, i.e., $t_L = \text{const.}\ L/u_m$, where the constant is of order unity. In typical laboratory experiments t_L is of the order of 10^{-1} s, in the atmosphere 100 s.

The correlation function $R(\tau)$ is an even function, so that the argument τ in Equation (3.3) should more correctly be the absolute value $|\tau|$. This shows up at once the shortcomings of the Markov process model: the discontinuous tangent at the origin corresponds to infinite acceleration or zero inertia. From the properties of $R(\tau)$ summarized before in Section 2.3 (valid for all stationary stochastic processes) it follows that its expansion near $\tau = 0$ should be

$$R(\tau) = 1 - \left(\frac{\tau}{t_m}\right)^2 + \cdots \quad (\tau \to 0), \tag{3.4}$$

where t_m is a Lagrangian 'microscale,' which is also the intercept of the osculating parabola of $R(\tau)$ at $\tau = 0$ with the τ-axis. In Figure 3.1 these properties of $R(\tau)$ are illustrated for the case where a Markov process model is nearly correct.

3.5. Consequences of Taylor's Theorem

If the velocity history of a diffusing particle is a stationary process, the arguments of Section 2.3 are valid and the mean square dispersion along a given coordinate axis

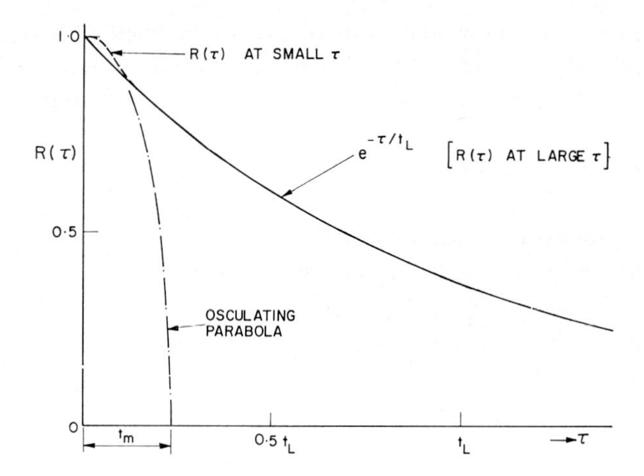

Fig. 3.1. Properties of Lagrangian correlation function.

of an ensemble of independently diffusing particles is given by Taylor's (1922) theorem. Along the y-axis, for example, from Equation (2.12):

$$\sigma_y^2 = 2\overline{v^2} \int_0^t (t - \tau)\, R(\tau)\, \mathrm{d}\tau . \tag{3.5}$$

Some important asymptotic properties of σ_y follow from the behavior of $R(\tau)$ at small and large values of the lag-time τ. As $\tau \to 0$, $R(\tau) \to 1$ (to order τ^2, see Equation (3.4)) and Equation (3.5) yields

$$\sigma_y^2 = \overline{v^2} t^2 - 0(t^4) \quad (t \to 0) \tag{3.5a}$$

so that σ_y increases initially *linearly* with time (and not with the square root of time as in molecular or Brownian diffusion). As $\tau \to \infty$, on the other hand, $R(\tau) \to 0$; if we assume that the correlation tends to zero fast enough for $\int_0^\infty \tau R(\tau)\, \mathrm{d}\tau$ to exist we find

$$\sigma_y^2 = 2\overline{v^2} t_0 (t - t_1) \quad (t \to \infty), \tag{3.5b}$$

where

$$t_0 = \int_0^\infty R(\tau)\, \mathrm{d}\tau$$

$$t_1 = \frac{1}{t_0} \int_0^\infty \tau\, R(\tau)\, \mathrm{d}\tau \quad (\text{if } t_0 \neq 0)$$

so that t_0 is the area under the $R(\tau)$ curve, t_1 its center of gravity, provided that t_0 is not zero.

Under certain special circumstances involving the presence of wave-like motions the area under the $R(\tau)$ curve, t_0, may become zero (area of negative loops exactly equals the area of positive loops). A correlation function of this type is, for example

$$R(\tau) = e^{-m\tau}\left(\cos n\tau - \frac{m}{n}\sin n\tau\right) \tag{3.6}$$

(note that this again does not satisfy Equation (3.4)). The first moment $\int_0^\infty \tau R(\tau)\,d\tau$ of such a function is usually negative and we may write

$$\sigma_y^2 = 2\overline{v^2}t_2^2 = \text{const.} \quad \begin{matrix}(t\to\infty)\\(t_0 = 0)\end{matrix} \tag{3.7}$$

with

$$t_2^2 = -\int_0^\infty \tau R(\tau)\,d\tau\,.$$

For the special case of an exponential-decay type correlation function (Equation (3.3)) we find $t_1 = t_0 = t_L$. Substituting into Taylor's theorem we find the mean square dispersion already given in Section 2.5 (Equation (2.18), substituting t_L for β^{-1}). The behavior of σ_y^2 calculated in this manner is illustrated in Figure 3.2.

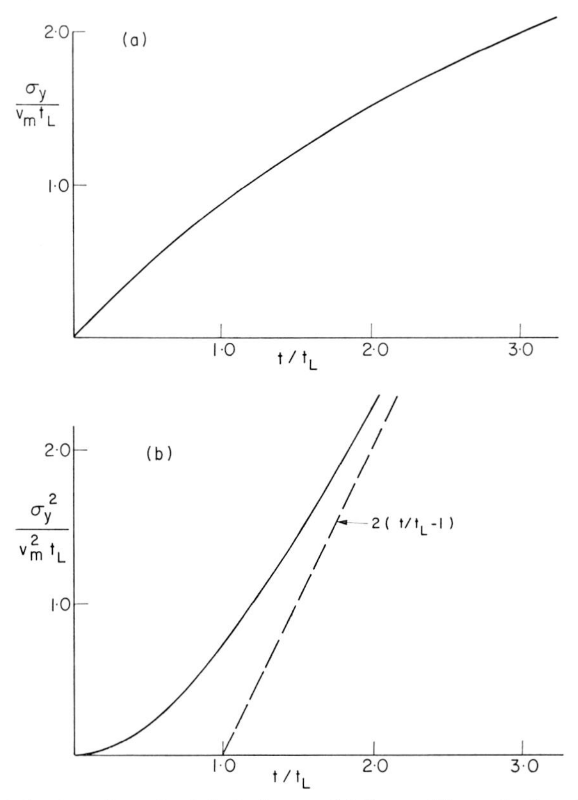

Fig. 3.2. Behavior of standard deviation σ_y, (a) linear, (b) quadratic ordinate scale.

A possible (if quite arbitrary) alternative Lagrangian correlation function which has the same area under the $R(\tau)$ curve as the function given in Equation (3.3) is

$$R(\tau) = \exp\left\{-\frac{\pi}{4}\frac{\tau^2}{t_L^2}\right\}. \tag{3.8}$$

This is an even function and satisfies the physical constraint at the origin that $R(\tau)$ should have a continuous tangent (implying finite accelerations). The resultant dispersion is

$$\sigma_y^2 = 2\overline{v^2} t_L \left[t \operatorname{erf}\left(\frac{\sqrt{\pi}}{2}\frac{t}{t_L}\right) + \frac{2}{\pi} t_L \exp\left(-\frac{\pi}{4}\frac{t^2}{t_L^2}\right) - \frac{2}{\pi} t_L \right]. \tag{3.9}$$

The asymptotic behavior is again as per Equations (3.5a) and (3.5b), with $t_0 = t_L$, $t_1 = (2/\pi) t_L$. The function in Equation (3.9) yields a behavior very much as illustrated in Figure 3.2. It may be shown also with other examples that the behavior of σ_y^2 is relatively insensitive to the exact shape of the correlation curve $R(\tau)$. This remains the case as long as $R(\tau)$ is positive everywhere: when wave-like motions are present, $R(\tau)$ acquires strong negative loops leading to a significantly different behavior of diffusing clouds, possibly involving oscillations of cloud size. Some hypothetical correlation functions of this type were investigated by Frenkiel (1953). In many geophysical situations turbulence and waves coexist and theoretical models similar to Frenkiel's may be of some use in describing diffusion in such fields. So far this possibility does not seem to have been exploited, except as will be described in Chapter VI, in connection with diffusion in stably stratified flow.

3.6. The Form of the Particle-Displacement Probability Distribution

The mean square dispersion discussed in some detail above is the second spatial moment of the particle-displacement probability distribution. To describe fully the concentration field by our 'fundamental theorem' we need the entire distribution $P(x, y, z, t)$, not only its second moments. To the extent that the wanderings of a marked fluid element in a turbulent field may be regarded as a 'random walk,' we would expect the distribution $P(\;)$ to be a Gaussian one (in the three spatial dimensions). Experimental evidence on diffusion in a homogeneous field indeed shows this to be the case to a high degree of approximation. Because its second moments completely specify a Gaussian distribution, the problem of describing the mean concentration field is thus for practical purposes solved and we could at once proceed to discuss the applications of this result. However, the question why a Gaussian distribution is observed in experiments is not yet satisfactorily answered and a brief discussion of the problem may be of interest.

The correspondence between the wanderings of an individual fluid element in a turbulent field and random walk models of the type discussed in the previous chapter is rather imperfect. Fluid elements possess a continuous motion and their velocity is correlated for periods of the order t_L. Thus an individual 'step' in random walk for such elements would have to be defined as the total displacement over a period rather longer than t_L, so that successive steps may be regarded as independent. A succession of a large number of independent steps leads to a Gaussian distribution by the Central Limit Theorem, but this would only occur at diffusion times large compared to t_L, which are certainly not of the greatest interest in, say, atmospheric diffusion. At the same time, as Batchelor (1949) has pointed out, the probability distribution of displacements at very short times $t \ll t_L$ is proportional to the distribution of turbulent velocity components at a point, which is again known to be Gaussian. This still does not explain why the distribution is Gaussian in the most interesting intermediate range (t of order t_L), but at least suggests that major departures are unlikely.

An alternative basis for 'proving' that the probability distribution $P(x, y, z, t)$ is Gaussian is to assume that the *Fourier components* into which the velocity history $u(t)$ of an individual particle may be resolved are independent random variables ('principle of random phase'). That the resulting distribution is Gaussian is a fairly general result in the theory of stochastic processes (Bartlett, 1956) and was discussed in its specific application to turbulent diffusion by Ogura (1954). However, it has been pointed out by Batchelor (1953) and others that individual Fourier components in turbulent flow are definitely not independent, so that the above argument does not strictly apply. On the other hand, many successful theoretical arguments in turbulent flow assume that only Fourier components of comparable wavelengths are interrelated, more distant ones (in wave number space) being independent. Thus there are presumably enough independent components present in the typical turbulent flow to produce much the same result as if all such components were independent. No rigorous theoretical demonstration of this point seems to exist, and the justification for the Gaussian nature of the particle probability distribution remains vague.

3.7. Mean Concentration Field of Continuous Sources

Accepting that the probability distribution $P(x, y, z, t)$ is Gaussian we may now write down the ensemble-average concentration field of an 'instantaneous point source' without difficulty. Let us assume that the axes x, y, z are 'principal axes' of the dispersion in the sense that $\overline{xy} = \overline{yz} = \overline{xz} = 0$. Then the three mean square dispersions $\sigma_x^2 = \overline{x^2}$, $\sigma_y{}^2 = \overline{y^2}$, $\sigma_z{}^2 = \overline{z^2}$ fully determine the distribution, the center of gravity of which remains at rest in a coordinate system moving with any constant velocity the fluid has. Let such a mean velocity be U, directed along the x-axis; then relative to a *fixed* coordinate system the ensemble-average concentration field of an instantaneous point source is, in virtue of the fundamental theorem:

$$\chi(x, y, z, t) = \frac{Q}{(2\pi)^{3/2}\,\sigma_x\sigma_y\sigma_z} \cdot \exp\left\{ -\frac{(x - Ut)^2}{2\sigma_x^2} - \frac{y^2}{2\sigma_y^2} - \frac{z^2}{2\sigma_z^2} \right\}, \qquad (3.10)$$

where the standard deviations σ_x, σ_y, σ_z are functions of time (Equation (3.5)). These are usually unequal in ordinary turbulent flow: even if the Lagrangian correlations are similar, the mean square turbulent velocities usually differ. Dispersion along the 'longitudinal' (x) axis is often markedly faster than along y and z.

In order to verify this equation experimentally a large number of puff-trials would have to be carried out, from which ensemble-average concentrations could be determined at fixed points. Such experiments are nearly impossible to carry out in the atmosphere or ocean and in any case the verification of this particular formula (Equation (3.10)) would have relatively little practical value for reasons discussed in greater detail in the next chapter. As we have already remarked, the main practical use of the 'elementary' statistical theory we are discussing is in connection with *continuous* sources, because the theoretical ensemble mean concentration field provides a model for the realized *time-average* concentrations. To arrive at the appropriate theoretical formula, the composite field of a continuous source may be built up, exactly as in molecular diffusion, as a linear superposition of many instantaneous sources, released at successive instants t'. That such a simple superposition is possible also in turbulent diffusion follows from the observation made by Batchelor (1949) that fluid elements wander about in a way that is unaffected by their neighbors being 'marked' or 'unmarked.' Generally speaking, diffusion by turbulent movements remains a linear phenomenon and composite fields of multiple sources may be described by simply adding the fields of simple sources, as in molecular diffusion. In this instance the superposition yields the composite field:

$$\chi(x, y, z) = \frac{q}{(2\pi)^{3/2}} \int_0^\infty \exp\left\{-\frac{(x - Ut')^2}{2\sigma_x^2} - \frac{y^2}{2\sigma_y^2} - \frac{z^2}{2\sigma_z^2}\right\} \frac{dt'}{\sigma_x \sigma_y \sigma_z}, \tag{3.11}$$

where again q is release rate in g s^{-1} or equivalent.

We cannot now carry out the integration until the functional dependence of the standard deviations σ_x etc. on diffusion-time t' is specified. At *short* distances from the source the instantaneous-source ensemble average cloud grows according to Equation (3.5a), i.e., linearly with time:

$$\begin{aligned} \sigma_x &= u_m t \\ \sigma_y &= v_m t, \\ \sigma_z &= w_m t \end{aligned} \tag{3.12}$$

where u_m, etc., are rms turbulent velocities. For this simple dependence on time Equation (3.11) may be integrated and yields (Frenkiel, 1953):

$$\begin{aligned} \chi(x, y, z) = {} & \frac{q u_m}{(2\pi)^{3/2} v_m w_m r^2} \exp\left(-\frac{U^2}{2u_m^2}\right) \times \\ & \times \left[1 + \sqrt{\frac{\pi}{2}} \frac{Ux}{u_m r} \exp\left(\frac{U^2 x^2}{2u_m^2 r^2}\right) \operatorname{erfc}\left(-\frac{1}{\sqrt{2}} \frac{Ux}{u_m r}\right)\right], \end{aligned} \tag{3.13}$$

where we have written

$$r^2 = x^2 + \frac{u_m^2}{v_m^2} y^2 + \frac{u_m^2}{w_m^2} z^2 .$$

Figure 3.3 after Frenkiel shows the mean concentration distribution for $u_m = v_m = w_m$ and $u_m/U = 0.5$ in a section across the cloud which is in this case axially symmetric. For the more usual, rather smaller values of the relative turbulent intensity u_m/U the iso-concentration lines on the upstream side of the source crowd together even more. Normally, this relative turbulent intensity is of the order of 0.1, in which case appreci-

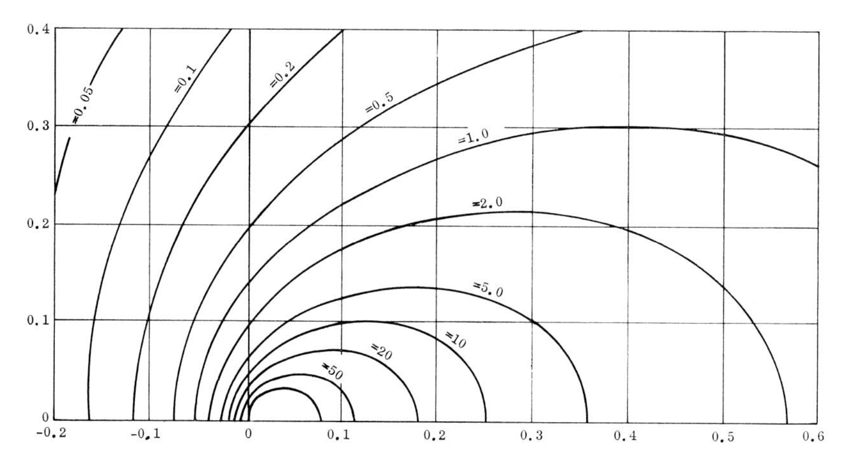

Fig. 3.3. Contours of constant mean concentration in meridional section across continuous point source. Abscissa and ordinate are x/x_0 and y/x_0 (with $x_0 = Ut_L$, a suitable length scale). Figures or contours indicate values of nondimensional concentration, $\chi U x_0^2/q$ (Frenkiel, 1953).

able concentrations are confined to a narrow cone downstream of the source. In such 'slender' plumes we have approximately $r \simeq x$ (r having been defined just above), and since $U/u_m \gg 1$,

$$\operatorname{erfc}\left(-\frac{1}{\sqrt{2}}\frac{Ux}{u_m r}\right) \simeq 2.$$

The second term in the brackets in Equation (3.13) becomes then quite large compared to unity and one finds after a few simple approximations

$$\chi(x, y, z) = \frac{q}{2\pi\sigma_y\sigma_z U} \exp\left\{-\frac{y^2}{2\sigma_y^2} - \frac{z^2}{2\sigma_z^2}\right\}, \tag{3.14}$$

where we have substituted back from Equation (3.12) the values of the standard deviations, making the identification for diffusion time $t = x/U$. This distribution is identical with cross sections of *instantaneous* point source clouds which have diffused for a period $t = x/U$, the material diffusing within each slice of the plume, dx long, being $q\ dt = q\ \mathrm{d}x/U$. The same amount of material is retained in such a slice as it travels downwind, the diffusion taking place along y and z only, and not in the x-direction. Physically, the absence of turbulent transport along the x-direction is a consequence of the slenderness of the plume, mean concentration gradients along x being negligible compared to those along y and z. This is clearly an approximation, but in practice a very good one, out to $y/x=3v_m/U$ for turbulent intensities up to $v_m/U=0.2$ (for a diagram of the accuracy of this approximation see Frenkiel (1953)).

Further downstream, where the diffusion time $t=x/U$ becomes comparable to the Lagrangian time-scale, the plume grows even less rapidly than given by Equation (3.12) and the slender-plume approximation remains valid *a fortiori*. For the asymptotic case, when σ_x^2, etc., grow linearly in time the integral in Equation (3.11) may again be evaluated in terms of closed functions, but the approximation of Equation (3.14) is sufficient for practical purposes. As was already remarked in Chapter I, this distribution is exactly as found in molecular diffusion, except that the standard deviations σ_y and σ_z are now more complex functions of diffusion time $t=x/U$, the details having been discussed in Section 3.5.

Just as we have obtained the mean field of a continuous point source by the superposition of many instantaneous point clouds, it is possible to write down the ensemble-average concentration fields of instantaneous or continuous (or those operating for a finite period) point, line, area or volume sources by integrating the fields of instantaneous point sources. Indeed the procedure is very much as we have employed in Chapter I and raises the question: How can these solutions be interpreted in relation to the equation of continuity, from which the linear differential equation describing molecular diffusion was derived?

3.8. Apparent Eddy Diffusivity

Our results so far were deduced entirely from the properties of the particle-displacement probabilities, or the 'P-field,' which by our fundamental theorem is related to the 'χ-field' the latter being the field of practical interest. A different constraint, which applies to the χ-field directly, is the principle of the conservation of mass, as expressed by the equation of continuity.

The (molecular) diffusion equation of course also applies to the fine details of the turbulent diffusion process. In the considerations that follow it will be important to distinguish between the concentration distribution *realized* in a given trial and the ensemble-mean concentration field, with which most of our theories are concerned. Because we have already used the symbol χ to designate the latter, we need a different symbol for the former: let $N(\mathbf{x}, t)$ be the instantaneous, realized concentration, so that

$$\chi(\mathbf{x}, t) = \bar{N}(\mathbf{x}, t). \tag{3.15}$$

For a given trial the conservation of mass yields, according to the argument of Chapter I (Equation (1.14)):

$$\partial N/\partial t = - \nabla \cdot \mathbf{F}, \tag{3.16}$$

where the flux $\mathbf{F}$ arises from bulk fluid motion as well as molecular diffusion:

$$\mathbf{F} = \mathbf{u}N - D\nabla N. \tag{3.17}$$

In turbulent flow both fluid velocity and tracer concentration are random variables and we may separate these into (ensemble) mean and 'fluctuating' components:

$$\mathbf{u} = \bar{\mathbf{u}} + \mathbf{u}'$$
$$N = \bar{N} + N' . \tag{3.18}$$

We shall take ensemble averages on both sides of Equation (3.16). The average flux becomes, from Equation (3.17)

$$\overline{\mathbf{F}} = \bar{\mathbf{u}}\chi + \overline{\mathbf{u}'N'} - D\nabla\chi . \tag{3.19}$$

The first and the third terms on the right of this equation are simple counterparts of those appearing in molecular diffusion, but the velocity-concentration correlation is new and represents flux by turbulent movements:

$$\overline{\mathbf{F}}_t = \overline{\mathbf{u}'N'} . \tag{3.20}$$

Thus the averaged Equation (3.16) becomes:

$$\frac{\partial\chi}{\partial t} + \bar{\mathbf{u}}\cdot\nabla\chi = -\nabla\cdot\overline{\mathbf{F}}_t + D\nabla^2\chi . \tag{3.21}$$

As we have pointed out before the molecular contribution to the spreading of clouds in turbulent flow may usually be neglected. This leaves only the divergence of the turbulent flux on the right of Equation (3.21), which so far is an unspecified quantity.

Consider now the point-source solution of Equation (3.10), from which, as we have already mentioned, many other solutions may be built up by linear superposition. The left-hand side of Equation (3.21) may be calculated for this χ-field without difficulty:

$$\begin{aligned}\frac{\partial\chi}{\partial t} + \bar{\mathbf{u}}\cdot\nabla\chi &= \frac{\mathrm{d}\sigma_x}{\mathrm{d}t}\left[\frac{(x-Ut)^2}{\sigma_x^3} - \frac{1}{\sigma_x}\right]\chi + \\ &\quad + \frac{\mathrm{d}\sigma_y}{\mathrm{d}t}\left[\frac{y^2}{\sigma_y^3} - \frac{1}{\sigma_y}\right]\chi + \frac{\mathrm{d}\sigma_z}{\mathrm{d}t}\left[\frac{z^2}{\sigma_z^3} - \frac{1}{\sigma_z}\right]\chi = \\ &= \sigma_x \frac{\mathrm{d}\sigma_x}{\mathrm{d}t}\frac{\partial^2\chi}{\partial x^2} + \sigma_y \frac{\mathrm{d}\sigma_y}{\mathrm{d}t}\frac{\partial^2\chi}{\partial y^2} + \\ &\quad + \sigma_z \frac{\mathrm{d}\sigma_z}{\mathrm{d}t}\frac{\partial^2\chi}{\partial z^2} .\end{aligned} \tag{3.22}$$

This is consistent with Equation (3.21) (after neglecting the $D\nabla^2\chi$ term) if we identify as the *turbulent flux* components the following

$$\begin{aligned}F_{tx} &= -\frac{1}{2}\frac{\mathrm{d}\sigma_x^2}{\mathrm{d}t}\frac{\partial\chi}{\partial x} \\ F_{ty} &= -\frac{1}{2}\frac{\mathrm{d}\sigma_y^2}{\mathrm{d}t}\frac{\partial\chi}{\partial y} \\ F_{tz} &= -\frac{1}{2}\frac{\mathrm{d}\sigma_z^2}{\mathrm{d}t}\frac{\partial\chi}{\partial z} .\end{aligned} \tag{3.23}$$

Thus the point source solution expressed in Equation (3.10) is also a solution of the differential equation:

$$\frac{\partial \chi}{\partial t} + U\frac{\partial \chi}{\partial x} = K_x \frac{\partial^2 \chi}{\partial x^2} + K_y \frac{\partial^2 \chi}{\partial y^2} + K_z \frac{\partial^2 \chi}{\partial z^2} \tag{3.24}$$

where the 'apparent eddy diffusivities' are defined as:

$$\begin{aligned} K_x &= \frac{1}{2}\frac{d\sigma_x^2}{dt} \\ K_y &= \frac{1}{2}\frac{d\sigma_y^2}{dt} \\ K_z &= \frac{1}{2}\frac{d\sigma_z^2}{dt}. \end{aligned} \tag{3.25}$$

This result was derived by Batchelor (1949) in a slightly more general form (without assuming the x, y, z axes to be principal axes). All linear superpositions of the point-source cloud are, of course, also solutions of the linear Equation (3.24), provided only that the same K_x, etc., are appropriate.

From Taylor's theorem it follows now at once that

$$K_x = \overline{u'^2} \int_0^t R(\tau)\,d\tau \tag{3.26}$$

(and similarly for K_y and K_z), so that K_x, etc., and functions of time, except at large diffusion times, when they tend to $\overline{u'^2}\, t_L$, etc. (Figure 3.4). Because this apparent eddy diffusivity can thus be determined from turbulent properties, we may be tempted to regard it as simply a more complicated counterpart of molecular diffusivities. A difficulty is, however, caused by the fact that K_x, etc., also depend on diffusion time: if two

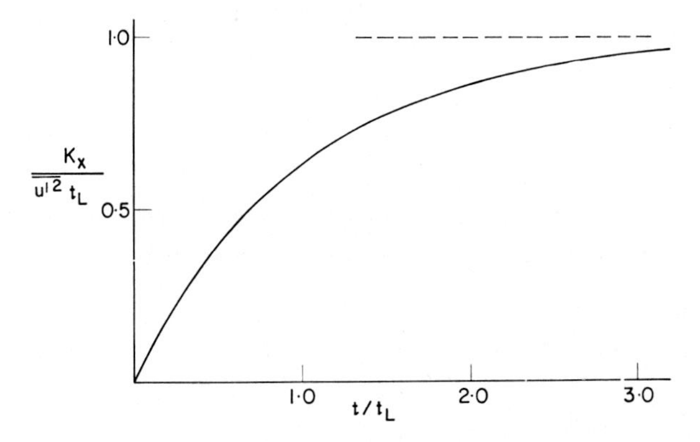

Fig. 3.4. Apparent eddy diffusivity as per Equation (3.26), using a correlation coefficient of the form given by Equation (3.3).

continous sources, for example, are simultaneously maintained in a turbulent field, one somewhat upstream of the other, their combined field may be obtained by simple linear superposition. At a point where both clouds contribute to concentrations, two different 'diffusivities' have to be employed to calculate the field. This led Taylor (1959) to label the notion of an apparent eddy diffusivity an 'illogical conception.' Undoubtedly, its indiscriminate use could lead to nonsensical conclusions. The constant asymptotic value of the apparent diffusivity, by contrast, is a genuine property of the turbulent field and may be used much as molecular diffusivity to describe concentration fields, once the diffusion time becomes large compared to the Lagrangian time scale.

Taking due note of these difficulties, it is still of considerable practical significance that a simple linear relationship exists between mean flux and mean concentration gradient (Equation (3.23) if x, y, z are principal axes, and its generalization otherwise) in the diffusion of an instantaneous point source, and of course in all its linear superpositions. We have arrived at this result by combining our knowledge of the 'P-field' with that of the 'χ-field,' without any questionable assumptions. It may be noted that a similar linear relationship between mean turbulent flux and mean concentration gradient may be derived by the well known mixing length argument (e.g., Schlichting, 1960), which however rests on several quite unsatisfactory assumptions. One would be led to think that a direct justification of the linear relationships (Equation (3.23)) should be possible on the basis of a more satisfactory statistical model than underlies the mixing-length argument. As we shall see later, many useful and realistic theoretical results may be derived by means of the concept of an eddy diffusivity.

We also note here that the approximate description of a continuous point-source cloud (Equation (3.14)) is a solution of the equation:

$$U \frac{\partial \chi}{\partial x} = K_y \frac{\partial^2 \chi}{\partial y^2} + K_z \frac{\partial^2 \chi}{\partial z^2} \tag{3.27}$$

which is a steady-state diffusion equation without diffusive flux in the x-direction, corresponding to the 'slender plume' approximation made in its derivation. The 'diffusivities' K_y and K_z are again defined as in Equation (3.13), with the substitution of diffusion time $t = x/U$:

$$K_y = \frac{U}{2} \frac{d\sigma_y^2}{dx}. \tag{3.28}$$

This apparent eddy diffusivity may thus be determined from observations on the spreading of point-source clouds. Its asymptotic behavior is

$$K_y = \frac{v_m^2}{U} x \quad (x \to 0) \tag{3.29}$$

$$K_y = v_m^2 t_L \quad (x \to \infty).$$

We note that observations on diffusing clouds may also be utilized to determine the relative turbulence intensity v_m/U and the Lagrangian time scale t_L. Similar results hold of course for K_z and w_m.

3.9. Application to Laboratory Experiments

It is not difficult to produce turbulent flow in the laboratory which is accurately stationary (statistically steady), while being also nearly homogeneous in space. Examples are steady flow in a channel or a pipe at high Reynolds numbers, if one confines attention to those regions of the flow sufficiently far removed from the walls. The mean concentration field of a continuous point source placed into such a flow may be investigated in detail and should conform to our theory developed in this chapter as long as the cloud does not approach the walls. Some illuminating experimental studies of this kind have been described, an early example in a channel (Kalinske and Pien, 1944) and a more recent one in a pipe (Becker *et al.*, 1966).

All of this work confirms the points made above namely that

(a) The concentration distribution across a plume diameter is Gaussian to the accuracy of the measurements, except when fairly obvious wall-influences appear.

(b) The standard deviation of the dispersion across the plume (σ_y or σ_z) behaves as predicted by Taylor's theorem, i.e., it grows linearly close to the source ($\sigma_y/x = v_m/U$, $\sigma_z/x = w_m/U$), while at large distances we find relations of the type

$$\sigma_y^2 = 2\frac{v_m^2}{U^2}\,x_0\,(x - x_0) \tag{3.30}$$

so that $K_y = (v_m{}^2/U^2)\,x_0$ is the asymptotic eddy diffusivity and x_0 is a kind of 'effective origin' of the diffusion process. By our earlier results $x_0 = Ut_L$, so that the Lagrangian time scale may be calculated from the measurements.

(c) The Lagrangian correlation coefficient, determined by a reverse application of Taylor's theorem (i.e., by double differentiation of the $\sigma_y^2(x)$ curve) is reasonably well represented by an exponential function

$$R = e^{-t/t_L}.$$

We note, however, that this method of determining the Lagrangian correlation coefficient is inherently rather inaccurate. This form of the correlation coefficient also implies $t_0 = t_1 = t_L$ (Equation (3.5b)), which is reflected in Equation (3.30) by the double appearance of $x_0 = Ut_L$.

Figure 3.5 shows lateral standard deviations σ_y observed by Kalinske and Pien (1944) demonstrating some of the above facts. The 'effective origin' x_0 is determined in Figure 3.5 by a backward extrapolation of the asymptotic σ_y^2 curve (which becomes a straight line).

The asymptotic eddy diffusivity $K_y = (v_m^2/U^2)x_0$ is a genuine property of the flow and should at large Reynolds numbers only depend on mean flow rate and channel or pipe dimensions (including roughness size). Thus at high Reynolds numbers, in

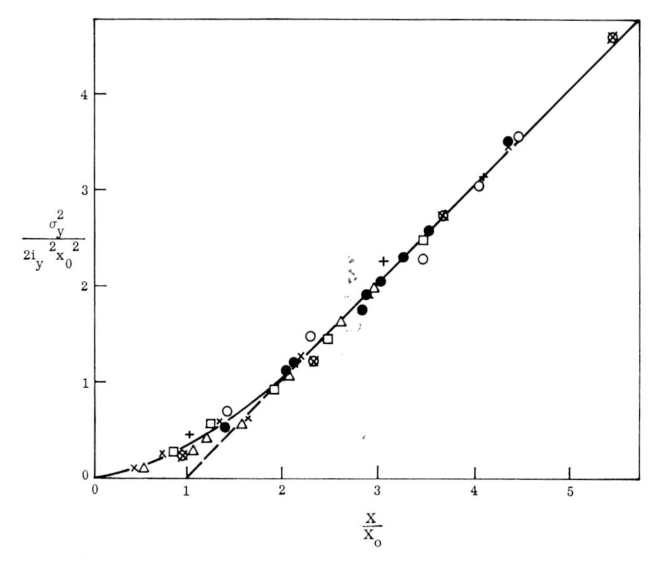

Fig. 3.5. Lateral diffusion in a two-dimensional channel, observed by Kalinske and Pien (1944). Illustration reproduced from Hinze (1969).

a given pipe or channel, the nondimensional number $(r_0 U_c / K_y)$ should be constant (r_0 = pipe radius, U_c center velocity).

Becker *et al.* (1966) show that this number is indeed constant, in their case equal to about 425; some other measurements put it as high as 600. The differences between individual measurements are no doubt due to differences in wall roughness, which influences the magnitude of the rms turbulent velocities. Some investigators have found small differences (order 10%) between Eulerian and Lagrangian rms velocities, but this is likely to have been due to the difficulties of measurement immediately behind a source, where the method of introduction of 'marked fluid' necessarily causes a disturbance. Apart from such small blemishes, the agreement with theory is good indeed. In particular, there is no doubt that the asymptotic regime described by Equation (3.30) is realized in such laboratory experiments.

3.10. Application to Atmospheric Diffusion

Beginning with the work of Sutton (1932, 1953), the elementary statistical theory of turbulent diffusion was widely utilized as a theoretical framework for the representation of data on atmospheric diffusion. However, particularly within the first 100 m height or so from the ground, this application is not straightforward because neither the wind speed nor the atmospheric turbulence are even approximately uniform. It is difficult enough to pick a 'mean' wind speed which would adequately represent the actual wind distribution in the homogeneous theoretical model, but an even greater difficulty is that the length-scale of eddies is approximately proportional to the height above ground level. Thus the spreading of a cloud tends to accelerate as it reaches into higher levels within the surface layer, a phenomenon particularly marked for clouds released at ground level. We shall discuss these complications in greater detail in Chapter V; here we shall use the elementary statistical theory with whatever empirical inputs are necessary to represent atmospheric observations. Quite successful predictions of atmospheric diffusion may be made on this basis, mainly because a good deal of empirical information may be absorbed in the functional form of the standard deviations $\sigma_y(t)$ and $\sigma_z(t)$.

Even if we regard the turbulence field homogeneous in a first approximation, the presence of the solid boundary at ground level must be allowed for to conserve mass. Usually, the ground is assumed to be a perfect reflector and its presence is represented by a mirror-image source placed below ground, much as we have done in Chapters I and II. By our fundamental theorem, this implies certain assumptions on particle-displacement probabilities, akin to those discussed in connection with random walk with a reflecting barrier. These assumptions apper ato be reasonable, and there is also experimental evidence (e.g., Csanady *et al.*, 1968) to show that the field of a 'real' plus that of a 'mirror image' source satisfactorily describes concentration distributions above ground level.

Thus following Sutton (1947) and others we may write for the mean concentration field due to a continuous elevated point source of strength q g s^{-1}:

$$\chi(x, y, z) = \frac{q}{2\pi U\sigma_y\sigma_z}\left[\exp\left\{-\frac{y^2}{2\sigma_y^2} - \frac{(z-h)^2}{2\sigma_z^2}\right\} + \right.$$
$$\left. + \exp\left\{-\frac{y^2}{2\sigma_y^2} - \frac{(z+h)^2}{2\sigma_z^2}\right\}\right], \tag{3.31}$$

where the coordinate origin is vertically below the source at ground level and the source is located at a height h. The appropriate value of the 'mean' wind speed U is in practice sometimes difficult to assign, while the specification of the functions $\sigma_y(x)$ and $\sigma_z(x)$ is a separate task of considerable complexity.

The main practical interest of the above formula lies in its prediction of ground level concentrations:

$$\chi(x, y, 0) = \frac{q}{\pi U\sigma_y\sigma_z}\exp\left\{-\frac{y^2}{2\sigma_y^2} - \frac{h^2}{2\sigma_z^2}\right\}. \tag{3.31a}$$

Specifically along the axis of the plume, $y=0$, one finds

$$\chi(x, 0, 0) = \frac{q}{\pi U\sigma_y\sigma_z}\exp\left(-\frac{h^2}{2\sigma_z^2}\right). \tag{3.31b}$$

The general appearance of this distribution is illustrated in Figure 3.6, with $a = \sigma_y\sigma_z =$ const.

A maximum occurs where the argument in the exponential is -1 (i.e., where $\sigma_z = h/\sqrt{2}$), while at large distances, where $\sigma_z \gg h$, the concentration varies as $(\sigma_y\sigma_z)^{-1}$. The latter is the same as the variation downwind of a ground-level point source, the formula for which may be obtained by setting $h=0$ in the above expressions.

Lines of constant mean concentration at ground level may also be calculated without difficulty from Equation (3.31a) and yield long cigar-shaped contours. Some of these are illustrated in Figure 3.7. Another illustration of the ground level mean concentration pattern is contained in Figure 3.8. Qualitatively, these illustrations are in agreement with observed patterns, but quantitative agreement depends on the correct prediction of the σ_y and σ_z functions to which we turn in the next section.

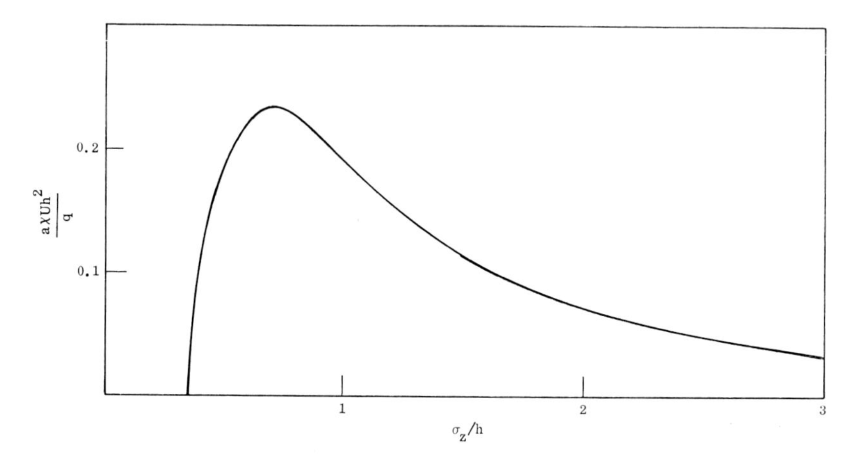

Fig. 3.6. Ground level concentration distribution below axis of horizontal plume.

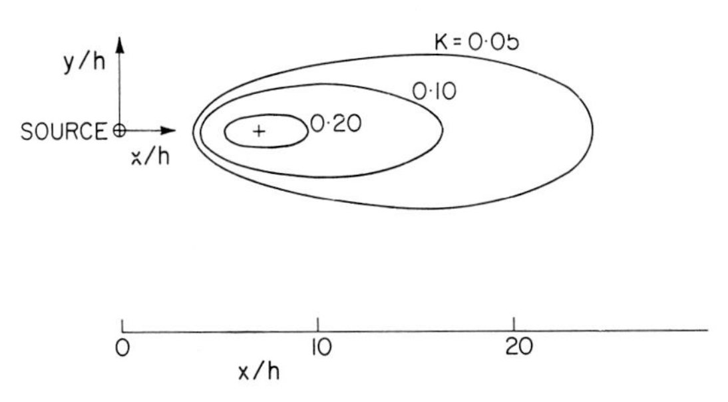

Fig. 3.7. Contours of constant concentration at ground level below horizontal plume, for $x = \sigma_z/i_z$ with $i_z = 0.1$, $K = 2xUh^2/q$.

3.11. Initial Phase of Continuous Plumes

For *elevated* sources it is reasonable to assume that the homogeneous-field theory predictions for *short* diffusion times (Equation (3.5a)) hold for a practically significant portion of the plume. This is not the case for *ground-level* sources, because the Lagrangian time scale becomes very short in the immediate vicinity of the ground (very small eddies). In describing diffusion from a chimney it is therefore resonable to set

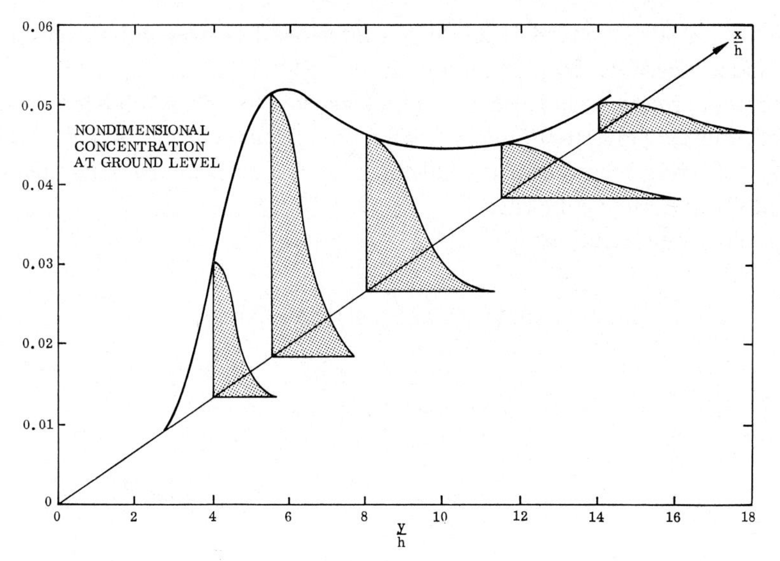

Fig. 3.8. Perspective illustration of ground level concentration downwind of elevated source.

for an initial phase

$$\begin{aligned} \sigma_y &= i_y x \\ \sigma_z &= i_z x, \end{aligned} \tag{3.32}$$

where we have used the nondimensional relative intensities defined in Equation (3.0).

The range of approximate validity of these simple relationships turns out to be quite appreciable: they often remain accurate enough up to the distance where the plume grows so large that it touches the ground, in which vicinity also the maximum ground level concentrations are observed. The only remaining practical question then concerns the 'climatology' of turbulence intensities i_y, i_z: with what frequency do low or high values of this intensity occur?

This question relates to the structure and properties of atmospheric turbulence, a complex topic in its own right, discussed e.g., in the monograph of Lumley and Panofsky (1964). We cannot do justice to this topic here and will merely summarize the most important points. Turbulence intensities in the atmosphere depend on

(1) the height of measurement,
(2) the roughness of the ground, and
(3) the stability of the atmosphere.

Under conditions of neutral stability, the rms turbulent velocities are approximately constant in the 'surface layer' (up to about 100 m height) above which they decay with height. The mean wind speed, by contrast, increases considerably (in the surface layer as the logarithm of the height) so that the *relative* turbulent intensity, being the ratio of rms turbulent velocity to mean wind speed, decreases with height. For elevated sources some kind of average turbulent intensity has to be used in our elementary theoretical model to characterize the air layer between source and ground level. Such an 'effective' value of the turbulent intensity therefore decreases with increasing source height, under otherwise identical conditions.

Over rough ground turbulence is more intense than over a smoother surface. Both this effect and the previously discussed influence of source height may be clearly seen from the following relationships, valid for a neutral surface layer (Panofsky, 1967) under ideally uniform conditions:

mean velocity distribution:

$$U = \frac{u^*}{k} \ln \frac{z}{z_0} \quad (\text{valid for } z \gg z_0, \text{ say } z > 100\, z_0)$$

rms turbulent velocity

$$\begin{aligned} \sqrt{\overline{v'^2}} &= 2.2\, u^* \\ \sqrt{\overline{w'^2}} &= 1.25\, u^*, \end{aligned} \tag{3.33}$$

where u^* is friction velocity ($\sqrt{\tau_0/\varrho}$, if τ_0 is shear stress at ground level, ϱ air density), k is 'Karman's constant', usually assumed to be about 0.4 and z_0 is roughness length. Combining Equations (3.32) and (3.33) one may write for neutral conditions

$$\begin{aligned} i_y &= \frac{0.88}{\ln z/z_0} \\ i_z &= \frac{0.5}{\ln z/z_0}. \end{aligned} \tag{3.33a}$$

The arithmetic average velocity between release height h and $z = z_0$ is attained at the level $z = h/e$, if the velocity distribution is logarithmic. It appears to be reasonable to use the value of i_y, i_z appropriate to this level in a representation of diffusion by means of a homogeneous field model. Thus under neutral conditions, for different release heights and surface roughnesses we find the estimates

$$\begin{aligned} i_y &= \frac{0.88}{\ln h/z_0 - 1} \\ i_z &= \frac{0.5}{\ln h/z_0 - 1}. \end{aligned} \tag{3.33b}$$

Table III.1 illustrates the variations in effective turbulence intensity predicted by Equation (3.33b). The direct relationship Equation (3.32) between observed turbulent intensity and standard deviations close to sources has been established by many atmospheric studies, notably those analyzed by Cramer (1959), but mostly for sources quite close to ground level. For atmospheric sources of appreciable elevation the experimental evidence is less plentiful but it also confirms the ground level data (see e.g., Csanady *et al.*, 1968) and the above reasoning in regard to an 'effective' turbulent intensity which varies with source height.

TABLE III.1

'Effective' vertical turbulence intensities i_z under neutral conditions (Equation (3.33b)) ($i_y = 1.76\ i_z$)

z_0 Roughness of ground, cm	h, Release height, m		
	20	50	100
1	0.076	0.067	0.061
30	0.156	0.121	0.104
100	(0.250)	0.171	0.139

(The bracketed entry refers to a h/z_0 ratio too small for the theory to be valid.)

Under non-neutral conditions the turbulent intensity also depends on the vertical heat flux, through the effects of buoyancy forces, and the relationship of turbulent intensity to the various external parameters becomes quite complex. We shall attempt to obtain some physical insight into these phenomena in Chapter VI. Here we merely note that large variations in turbulent intensities i_y, i_z have been observed to be associated with stability/instability, from zero turbulence in pronounced inversions (strong stability) to relative intensities of 0.5 and higher under strong instability.

In order to deal with the resulting wide variations in turbulent properties, meteorologists with a practical bent have introduced 'stability categories' into which atmospheric conditions may be classified (Singer and Smith, 1953; Pasquill, 1962). One such classification is shown in Table III.2. The correspondence of stability categories and 'typical' near-ground turbulence intensity over level prairie, as given by Cramer (1959) is listed in Table III.3. Over rough ground and at larger elevations these values have to be modified, much as we have seen above for neutral stability. For known heat flux to momentum flux ratios equations equivalent to Equation (3.33) may be written down for non-neutral conditions (using, e.g., data quoted by Panofsky, 1967) and the values of turbulent intensity calculated.

TABLE III.2

Pasquill stability categories (they correspond to categories in Figure 3.9 below)

Surface wind speed, m s^{-1}	Insolation			Night, mainly overcast or $\geqslant 4/8$ low cloud	$\leqslant 3/8$ Low cloud
	Strong	Moderate	Slight		
2	A	A–B	B	–	–
2–3	A–B	B	C	E	F
3–5	B	B–C	C	D	E
5–6	C	C–D	D	D	D
6	C	D	D	D	D

A – Extremely unstable, B – Moderately unstable, C – Slightly unstable D – Neutral, D – Slightly stable, F – Moderately stable.

TABLE III.3

'Typical' turbulent intensities near ground level (Cramer, 1959)

Thermal stratification	i_y	i_z
Extremely unstable	0.40–0.55	0.15–0.55
Moderately unstable	0.25–0.40	0.10–0.15
Near neutral	0.10–0.25	0.05–0.08
Moderately stable	0.08–0.25	0.03–0.07
Extremely stable	0.03–0.25	0.0 –0.03

3.12. Atmospheric Cloud Growth far from Concentrated Sources

The simple relationships of Equation (3.32) only depend on one atmospheric parameter, relative turbulent intensity. At larger distances from the source the elementary statistical theory predicts that another parameter, Lagrangian time scale, becomes important. In addition to this, when applying the theory to atmospheric diffusion, the inadequacies of the elementary theory become more troublesome and have to be compensated for by a functional form of the standard deviations $\sigma_y(x)$, $\sigma_z(x)$ which would not make sense for a field of stationary and homogeneous turbulence. Thus the theory becomes rather more complex and intellectually less satisfactory, but remains a practically useful tool, particularly if one understands the physical causes of the complexities (which we shall discuss further in Chapters V and VI).

An analysis of experimental data on the growth of clouds released at ground level has led Sutton (1932, 1953) to propose power law formulas for the standard deviations:

$$\begin{aligned} 2\sigma_y^2 &= C_y^2 x^{2-n} \\ 2\sigma_z^2 &= C_z^2 x^{2-n}, \end{aligned} \tag{3.34}$$

where n, C_y, and C_z are constants, their values depending on atmospheric stability and source height. Under neutral conditions Sutton found $n=0.25$, while over flat grassland near ground level he proposed $C_y=0.4\ \text{cm}^{1/8}$, $C_z=0.2\ \text{cm}^{1/8}$. The awkward physical dimensions of these constants (depending on the value of the exponent n) make this formulation unattractive and it is nowadays rarely used. However, at least under neutral conditions it represents observed facts adequately and we must discuss its implications.

By a reverse application of Taylor's theorem it follows that the Lagrangian correlation coefficient implied by Equation (3.34) is

$$R(\tau) = \frac{(2-n)(1-n)\, C_y^2 U^{2-n}}{4\overline{v'^2}}\, \tau^{-n} \tag{3.35}$$

which is of the form $R(\tau)=\text{const.}\ \tau^{-n}$ and has various physically undesirable properties. The integrals which we have used to define time scales t_0 and t_1 (Equation

(3.5b)) diverge for this form of $R(\tau)$. A stationary stochastic process with this type of correlation coefficient has an unrealistic spectrum: the spectral density becomes infinite at zero frequency. It is difficult to assign any physical meaning to such a process, but it would seem to imply the presence of infinitely large eddies.

Fortunately, a much more satisfactory physical explanation may be found in the fact that the field of turbulence near ground level is not uniform. We have already remarked that as a cloud released at ground level grows vertically, it becomes subject to the action of eddies of increasing size, so that its further growth accelerates. As we shall see in Chapter V, realistic diffusion models for such non-uniform layers yield cloud growth rates precisely as given by Sutton's formulas so that we may safely ascribe the peculiar behavior of the standard deviations to the non-uniformity of turbulence. In such a field, particle velocities do *not* constitute a stationary stochastic process, because the time scale of the 'steps' increases systematically. Our difficulties with $R(\tau)$ noted above therefore disappear: Equation (3.35) and its consequences become simply invalid deductions.

Sutton's power-law formulas have not been confirmed for non-neutral stability. In the course of more extensive experimental work on cloud growth under such more complex conditions some interesting facts came to light, which could not be represented by equations of the form of Equation (3.34). Diffusion data near ground level, over ideally flat terrain and under uniform conditions are most conveniently summarized in graphical form for the various stability categories we have described above. Charts of this kind have been published by Cramer (1959), Gifford (1961), Pasquill (1962), and others. Figure 3.9 shows a presentation taken from Gifford (1961).

When we attempt to relate the information presented in Figure 3.9 to our elementary theory, particularly to Taylor's theorem, several conflicts again come to light, much as in connection with Sutton's formula. Indeed the curves corresponding to neutral atmospheric stability more or less agree with Sutton's equations and the remarks made above apply to them. For non-neutral stability the first thing that strikes one is the qualitative difference between horizontal and vertical standard deviations, $\sigma_y(x)$ vs $\sigma_z(x)$. Horizontal standard deviations behave in much the same way under all stability conditions and the dispersion appears to be more or less proportional to turbulent intensity at all distances. Power laws of the type of Equation (3.34) could therefore represent σ_y more or less adequately, as long as the constant C_y is adjusted appropriately to take into account reduced turbulence in stable conditions, enhanced turbulence in an unstable atmosphere.

By contrast, vertical standard deviations approach a constant asymptote ($\sigma_z =$ const.) in strong stability and show an explosive behavior in strong instability. As we have pointed out before, we can reconcile a constant asymptote with Taylor's theorem, if we assume a Lagrangian correlation coefficient with some strong negative loops, but it requires a physical explanation why such a distinct departure from laboratory diffusion should occur. Furthermore, the explosive growth of σ_z under unstable conditions cannot be reconciled with any admissible form of the Lagrangian correlation coefficient: assuming $R = 1 =$ constant is the maximum theoretically possible value and

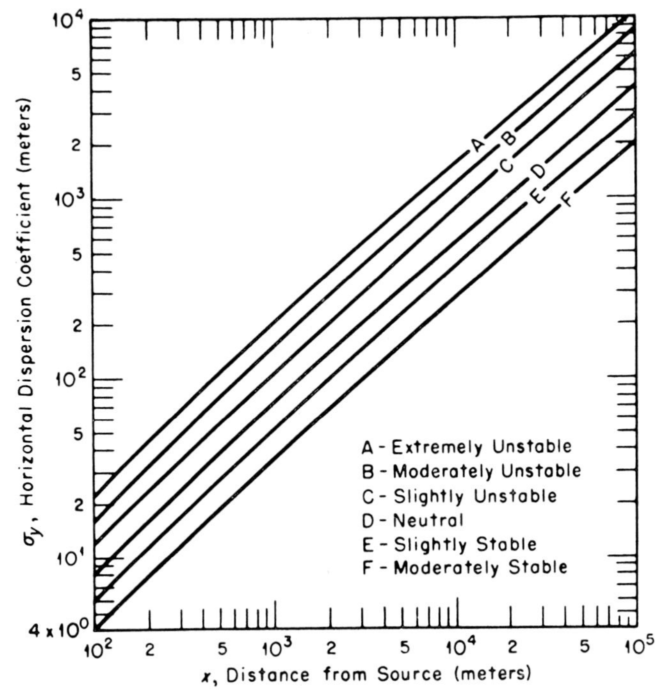

Fig. 3.9a. Horizontal dispersion coefficient as a function of distance from source (Gifford, 1961).

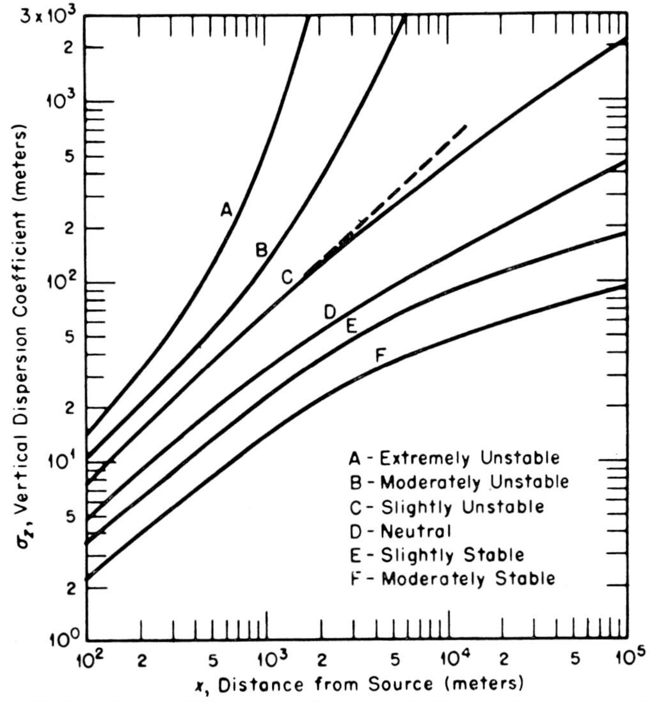

Fig. 3.9b. Vertical dispersion coefficient as a function of distance from source (Gifford, 1961).

the corresponding growth rate of σ_z (which is the initial behavior we already discussed) falls far short of observed data. In Chapter VI we shall find an explanation for these observed facts on σ_z: briefly, they are due to negative and positive buoyancy forces on diffusing elements of fluid.

The data summarized in Figure 3.9 may be put to direct practical use in air pollution estimates in connection with Equation (3.31). One must remember, however, that these data apply to flat, uniform terrain and ground-level release. For rough surfaces and elevated sources the effective turbulent intensity which may be expected in the different stability categories must be adjusted upward or downward, as discussed before. The standard deviations σ_y and σ_z must then be scaled up, in the lack of more detailed information simply in proportion to turbulence level. At the present time this appears to be the most accurate calculation possible for obtaining engineering estimates of ground level pollutant concentrations (*time-means* of course, for continuous sources) due to a pollutant plume centered at a constant height h.

3.13. The Non-Stationary Character of Atmospheric Turbulence

Up to now we have been proceeding on the assumption that the model of a *stationary* field of turbulence is adequate in atmospheric applications, even if there are some important effects connected with nonhomogeneity. It turns out, however, that also this is only a gross approximation and departures from it give rise to important discrepancies between experiment and theory.

The atmospheric analogue of statistically steady laboratory turbulent flow is when the 'mean wind' is constant in time. This is never the case for long, but one might expect it to be a reasonable approximation at certain times, for periods short compared to the typical passage time of weather cycles (of order 100 h). Indeed the wind frequently *appears* to be steady, but closer examination often reveals this to be an illusion. Perhaps the best way to describe the situation is to regard the wind as part of very large-scale geophysical turbulence, which contains all scales of motion, from weather cycles down to eddies of 1 cm diam or less. At any fixed site the large scale eddies cause what appears as an approximately steady wind for several hours, while eddies of a few meters in diameter mainly cause the diffusion of tracers or pollutants from continuous sources of the kind we have been discussing. If there were no intermediate size eddies, this separation would no doubt work well (as it does over ideally uniform terrain and under ideally 'steady' conditions), but in most actual situations a more or less continuous spectrum of eddy sizes is present. The typical eddy which is generated by shear flow near the ground has a Lagrangian time-scale of the order of 100 s: this is the atmospheric analogue of laboratory turbulence. The 'Eulerian' time scale of the same eddies (typical 'passage time' at a fixed point) is even less, of the order of 20 s. Thus in atmospheric diffusion experiments with continuous sampling, if averages are taken over periods of 3 min to 1 h it should be possible to approximate laboratory conditions, provided that eddies with the intermediate time scale of a few minutes to a few hours are relatively rare.

In a practical sense this approximation works, judging by the fact that we can quite successfully describe experimental results in the framework of the elementary theory. However, the actual presence of intermediate size eddies has two practical consequences:

(1) The observed 'mean' concentration profile (time-average over some chosen period of 3 to 60 min) often shows random irregularities. Intermediate-size 'eddies' (as all others) occur at random times and one is luckier in some experiments than others in escaping their influence. Experimentalists often reject data taken when wind fluctuations of such intermediate time scale were evidently present, saying that the 'mean wind had shifted.' The detailed record of velocity fluctuations observed with a bivane, for example, often shows clear cut examples of 'trends' or even complete fluctuation cycles over a (say) 10 min period. This shows up in the mean concentration field as a considerable departure from a Gaussian profile. To put it another way, the 3 to 60 min time-average concentration at a fixed point still remains a random variable. Our elementary theory applies to the idealized case of steady mean wind, which is infrequently realized at all accurately.

(2) The mean-square turbulent velocities, and with them the cross-wind dimensions of diffusing clouds depend on the period of averaging. This is because intermediate size eddies make a greater contribution to what appear as wind 'fluctuations' when the observation period is longer. It is easiest to see this in the obvious case where an intermediate size eddy causes a (nearly) linear trend in wind direction: the contribution of such an eddy to the rms lateral velocity goes up directly with increased observation period.

Some approximate theoretical arguments on the effects of sampling time on velocity variances have been advanced by Ogura (1959), Wipperman (1961), and others. These rely on assumptions regarding the Lagrangian velocity spectrum about which little experimental information exists. We do have, on the other hand, good experimental evidence directly on the effect of sampling time on the standard deviations σ_y and σ_z, summarized by Hino (1968) and shown here in Figure 3.10. In the practically important range both σ_y and σ_z grow approximately with the 4th root of sampling time. Maximum ground-level concentrations vary as $(\sigma_y\sigma_z)^{-1}$ and therefore decrease with the inverse square root of the sampling time. In making estimates of ground-level concentration levels it is therefore fairly important to be specific about the length of averaging time one has in mind. The data shown earlier in Figure 3.9 relate to a sampling time of 15 min.

3.14. The Hay-Pasquill Method of Cloud-Spread Prediction

In using Taylor's theorem for predicting cloud growth we have substituted the Eulerian mean-square turbulent velocities for the Lagrangian ones, a measure which worked very well in describing theoretically laboratory diffusion experiments. In applying the same method to atmospheric diffusion experiments, noting that atmospheric turbulence is not in fact stationary, one hesitates to assume the same equi-

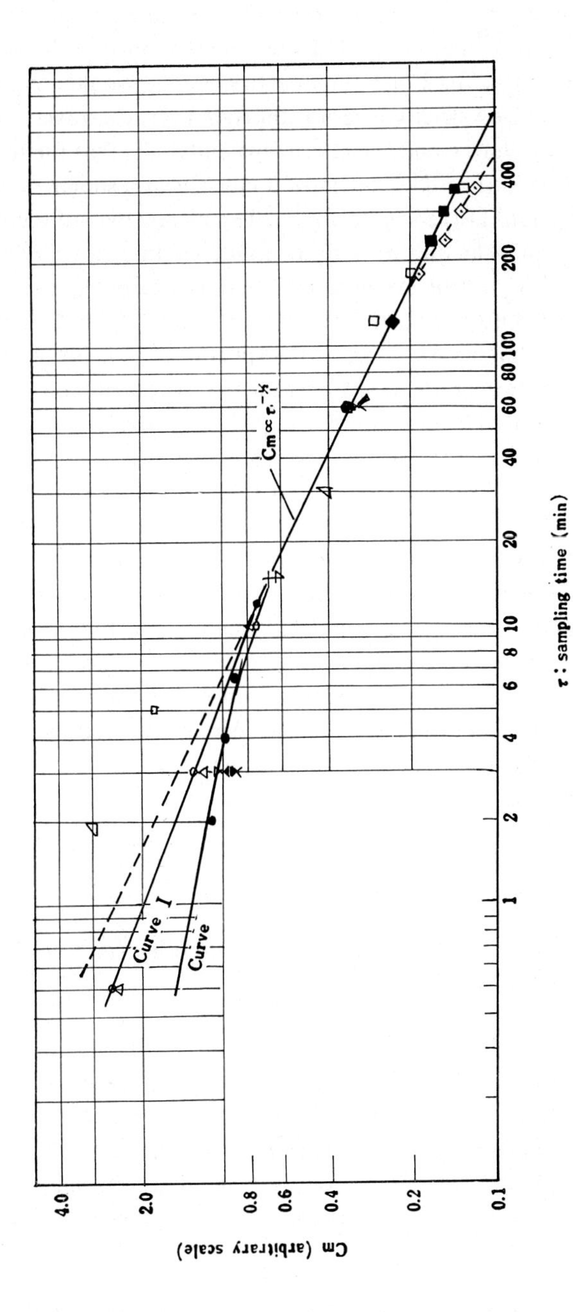

Fig. 3.10. Summarized data on the relationship between the maximum value of smoke concentration (or the axial concentration) and sampling time. Dots represent various experimental data (Hino, 1968).

valence without confirmatory evidence. In particular, the question arises: given that σ_y and σ_z are determined from concentrations averaged over a sampling period s, how is the 'relevant' value of $\overline{v'^2}$, $\overline{w'^2}$ to be determined as an Eulerian quantity, over what time period must v'^2 and w'^2 be averaged? Experiments show that the same sampling period s must be applied to the Eulerian velocities, as one would intuitively expect.

This equivalence is indeed strictly true for diffusion (particle travel) times so short that the Lagrangian correlation coefficient remains unity, i.e., that the particles retain their initial turbulent velocities. The spread of a cloud near a source and the fluctuations of a light vane placed at the source are in effect merely two different methods of measuring the turbulent velocities at a fixed point. Further downstream the equivalence is not so obvious, but the evidence on laboratory turbulence shows that fixed point instruments and diffusing particles sample the same 'population' of velocity fluctuations. Applied to the nonstationary atmospheric case, the same sampling time should allow the same fraction of the eddy spectrum to contribute to diffusion as is contributing to the measured fixed-point mean square velocity.

This idea can be carried a little further and made the basis of a practical diffusion prediction technique, as suggested by Hay and Pasquill (1959). The main difference between an 'Eulerian' (fixed-point) and a 'Lagrangian' (drifting particle) velocity-history is that at a fixed point the fluctuations appear rather more quickly, as eddies are convected past the measuring instrument. If one slowed down the fixed-point record by an appropriate factor, one would expect to obtain a velocity-history very similar to that of a drifting particle. Thus one could *simulate the displacement* of such a diffusing particle from a fixed-point velocity record by finding the average Eulerian velocity over an appropriately shortened period. Along the y-axis, for example, if a particle is released at time t_r, its simulated displacement during travel time t would be

$$y_T = \frac{t}{T} \int_{t_r}^{t_r+T} v'(t')\, dt', \tag{3.36}$$

where $v'(t')$ are Eulerian velocities, and the 'simulation time' T is some fraction of the diffusion or travel time t:

$$t = \beta T \tag{3.37}$$

with β=an empirical constant. In Equation (3.36) the averaged Eulerian velocity ($1/T$ times the integral) multiplies the travel time. Alternatively, the factor t/T may be regarded as providing the appropriate 'stretching' of the time axis.

The relationship between Eulerian velocity and simulated displacement (Equation (3.36)), is almost identical with Equation (2.6) and the argument leading to Taylor's theorem applies, so that

$$\overline{y_T^2} = 2\overline{v'^2}\beta^2 \int_0^T \int_0^{t'} R_E(\tau)\, d\tau\, dt', \tag{3.38}$$

where $R_E(\tau)$ is now the Eulerian velocity autocorrelation and overbars denote ensemble averages. The actual mean-square displacement of drifting particles is of course

$$\sigma_y^2 = \overline{y^2}(t) = 2\overline{v'^2} \int_0^t \int_0^{t'} R_L(\tau)\, d\tau\, dt' \tag{3.39}$$

with $R_L(\tau)$ the Lagrangian autocorrelation. The two expressions are identical for all values of t if, in addition to Equation (3.37) it is true that

$$R_E(\tau) = R_L(\beta\tau) \tag{3.40}$$

or that the shapes of the Eulerian and Lagrangian correlations are identical after appropriate rescaling of the time axis. We have already pointed out that this is *not* the case, although there is often a resemblance. Eulerian correlations often have negative loops because of constraints of continuity (see the early sections of Townsend, 1956) while no such constraints operate on the Lagrangian quantity. However, we have also seen before that the prediction of spread is not very sensitive to the exact shape of the correlation coefficient and therefore it is not surprising to discover that the simulation technique based on Eulerian velocities gives in practice good diffusion predictions.

In the actual application of the technique the ensemble averages are replaced by time-averages over the chosen sampling time s, i.e.,

$$\overline{y_T^2} \rightarrow \int_0^s y_T^2(t_r)\, dt_r, \tag{3.41}$$

where $y_T(t_r)$ is as defined in Equation (3.36). This procedure may be interpreted as forming a moving average velocity process

$$v_T'(t_r) = \frac{1}{T} \int_{t_r}^{t_r+T} v'(t')\, dt' \tag{3.36a}$$

the mean-square value of which over sampling time s then gives the dispersion estimate

$$y^2(t) = t^2 \overline{v'^2_{t/\beta, s}}, \tag{3.38a}$$

where it bears repetition that $\overline{v'^2_{t/\beta, s}}$ is the mean-square value of the moving-average velocity process (formed with averaging time $T = t/\beta$), evaluated over the sampling time s. The output of a hot-wire anemometer, proportional to the v' component, can, for example, easily be fed to an averaging circuit of time constant t/β, the signal then squared again and its mean value determined over averaging time s to give $\overline{v'^2_{t/\beta, s}}$.

When the instrument used to determine lateral velocity fluctuations is a light directional vane, its output is azimuth angle θ, which is, for small angles very nearly

$$\theta = \frac{v'}{U}, \tag{3.42}$$

where U is mean velocity. A signal proportional to azimuth angle may again be processed to form a moving average over period t/β, then squared and averaged over sampling time s to yield $\overline{\theta^2_{t/\beta,\,s}}$. The simulated mean-square displacement is then

$$y^2(t) = x^2 \overline{\theta^2_{t/\beta,\,s}}\,, \tag{3.43}$$

where $x = Ut$ is distance from the point of release. This may be written in a form identical with Equation (3.32):

$$\sigma_y = i_y \big|_{t/\beta,\,s} x\,, \tag{3.43a}$$

where $i_y\big|_{t/\beta,s} = \sqrt{\overline{\theta^2_{t/\beta,\,s}}}$ is the 'effective' value of lateral turbulent intensity for travel time t, sampling time s. An entirely analogous relationship holds for the vertical component:

$$\sigma_z = i_z \big|_{t/\beta,\,s} x\,, \tag{3.43b}$$

where i_z is of course evaluated from an Eulerian record of the vertical velocity fluctuations w'.

In spite of the theoretical weakness of its background, the Hay-Pasquill method gives good cloud-growth estimates in 'real-life' atmospheric diffusion problems, when the character of atmospheric turbulence is far from ideal. The appropriate value of the empirical constant β is somewhat uncertain, but the method is not very sensitive to the actual value of β adopted, and $\beta = 4$ has given good results in many different situations. The broader significance of this success is that it is more important to allow for the actual, nonstationary character of atmospheric turbulence, than to have accurate information on the Lagrangian correlation coefficient.

From an engineering point of view the agreement between diffusion data and turbulence measurements carried out during the same experiment is of relatively little value in, say, design estimates. A long series of Eulerian turbulence measurements, however, may be utilized to establish something like a diffusion climatology of a given site. Prior to building a chimney one could record turbulent velocities at e^{-1} times chimney height for a year or more, for example, and analyze the data to see with what frequency high or low values of σ_y and σ_z would occur. Instrumentation for the direct readout of these quantities is nowadays readily available and site studies of this kind play an important rôle in locating major power stations.

EXERCISE

1. Analyze the ground level SO_2 pollution caused by a chimney plume the axis of which is at a constant height of 120 m above ground in a 15 m s^{-1} wind. The emission rate of SO_2 is 1 m^3 s^{-1} at normal temperature and pressure (2.5 kg s^{-1}). Calculate the maximum ground level concentration and estimate the area over which the ground level concentration exceeds one half of the maximum. Carry out the calculations for several 'typical' atmospheric conditions.

References

Bartlett, M. S.: 1956, *An Introduction to Stochastic Processes*, Cambridge Univ. Press, London, 312 pp.
Batchelor, G. K.: 1949, *Australian J. Sci. Res.* **2**, 437.
Batchelor, G. K.: 1953, *Homogeneous Turbulence*, Cambridge Univ. Press, London, 197 pp.
Batchelor, G. K.: 1959, *J. Fluid Mech.* **5**, 113.
Becker, H. A., Rosensweig, R. E., and Gwozdz, J. R.: 1966, *Am. Inst. Chem. Eng.* **12**, 964.
Cramer, H. E.: 1959, *Am. Ind. Hyg. Assoc. J.* **20**, 183.
Csanady, G. T., Hilst, G. R., and Bowne, N. E.: 1968, *Atmos. Env.* **2**, 273.
Frenkiel, F. N.: 1953, *Adv. Appl. Mech.* **3**, 61.
Gifford, F. A.: 1961, *Nucl. Safety* **2**, 47.
Hay, J. S. and Pasquill, F.: 1959, *Adv. Geophys.* **6**, 345.
Hino, M.: 1968, *Atmos. Env.* **2**, 149.
Hinze, J. D.: 1959, *Turbulence*, McGraw-Hill Book Co., New York, 586 pp.
Kalinske, A. A. and Pien, C. L.: 1944, *Ind. Eng. Chem.* **36**, 220.
Leighton, P. A., Perkins, W. A., Grinnell, S. W., and Webster, F. X.: 1965, *J. Appl. Meteorol.* **4**, 334.
Lin, C. C.: 1959, 'Statistical Theories of Turbulence', *Turbulent Flow and Heat Transfer*, Princeton Univ. Press. p. 196.
Lumley, J. L.: 1962, 'The Mathematical Nature of the Problem of Relating Lagrangian and Eulerian Statistical Functions in Turbulence', in *Mécanique de La Turbulence*, Éd. du Centre Nat. de la Rech. Sci., Paris, pp. 17–26.
Lumley, J. L. and Panofsky, H. A.: 1964, *The Structure of Atmospheric Turbulence*, Interscience Publishers, New York, 239 pp.
Ogura, Y.: 1954, *J. Meteorol. Soc. Japan* **32**, 22.
Ogura, Y.: 1959, *Advan. Geophys.* **6**, 149.
Ozmidov, R. V.: 1965, *Izv. Atmos. and Oceanic Phys.* **1**, 439.
Panofsky, H. A.: 1967, *A Survey of Current Thought on Wind Properties Relevant for Diffusion in the Lowest 100 m*, *Symposium on Atmospheric Turbulence and Diffusion*, Albuquerque, N. M., pp. 47–58, Sandia Laboratories.
Pasquill, F.: 1962, *Atmospheric Diffusion*, D. van Nostrand Co. Ltd., New York, 297 pp.
Saffman, P. G.: 1960, *J. Fluid Mech.* **8**, 273.
Schlichting, H.: 1960, *Boundary Layer Theory*, McGraw-Hill Book Co., New York, 647 pp.
Schubauer, G. B and Tchen, C. M.: 1959, 'Turbulent Flow', in *Turbulent Flow and Heat Transfer*, Princeton Univ. Press, pp. 75–195.
Singer, I. A. and Smith, M. E.: 1953, *J. Meteorol.* **10**, 121.
Sutton, O. G.: 1932, *Proc. Roy. Soc. London* **A135**, 143.
Sutton, O. G.: 1947, *Quart. J. Roy. Meteorol. Soc.* **73**, 426.
Sutton, O. G.: 1953, *Micrometeorology*, McGraw-Hill Book Co., New York, 333 pp.
Taylor, G. I.: 1922, *Proc. London Math. Soc.* **A20**, 196.
Taylor, G. I.: 1959, *Advan. Geophys.* **6**, 101.
Townsend, A. A.: 1956, *The Structure of Turbulent Shear Flow*, Cambridge Univ. Press, London, 315 pp.
Wipperman, F.: 1961, *Int. J. Air Water Pollution* **4**, 1.

CHAPTER IV

'RELATIVE' DIFFUSION AND OCEANIC APPLICATIONS

4.1. Experimental Basis

We have briefly alluded above to the possibility of averaging concentrations at fixed distances from a diffusing cloud's center of gravity. This leads to the theory of 'relative' diffusion. The experimental basis of this theory is best explained by reference to some actual observations.

An approach based on relative diffusion theory is almost invariably adopted to evaluate observations on the diffusion of tracers or pollutants in oceans or lakes where fixed reference positions are difficult to establish. In experiments of this kind fluorescent dye is often released from a point source forming either a single 'puff' or a continuous 'plume.' The bodily motion of the puff or plume is irregular, the plume waving about in the current much as a smoke plume does in the wind. In one set of observations in Lake Huron, for example (Csanady, 1966) a number of concentration profiles at a given distance x (along the mean current) from a dye source were determined by crossing the continuous dye plume in a small boat between two marker buoys. The dye concentration was continually measured with the aid of an optical instrument which gives instantaneous readings. Thus a graph of concentration versus cross plume distance y resulted from each run; sample graphs of this kind are shown in Figure 4.1. In this particular experiment 25 runs were carried out resulting in 25 concentration profiles, all obtained at the same distance x from the source and at the same depth z.

As may be seen, the individual concentration profiles are irregular in the manner characteristic of turbulent phenomena. It is, however, possible to extract a much more regular concentration distribution from the data by averaging the 25 profiles, overlapping them so that their centers of gravity coincide. The average of 25 readings at the center of gravity, and at given distances y on one side or the other yields the distribution shown in Figure 4.2. This is already not too different from a Gaussian distribution and we may surmise that a larger number of runs would have resulted in a mean profile which is even closer. If we regard the diffusing dye elements as moving in a frame of reference attached to the center of gravity of the plume, we conclude that they have executed a random walk like process in the y-direction.

The absolute position of the center of gravity was not noted in these observations, but with some care it could have been related to the fixed marker buoys. However, it is known that the plume has gone through several cycles of back and forth bodily movements between the markers in the course of the 25 runs, with an amplitude equal to several times the observable instantaneous plume width. Thus had we averaged

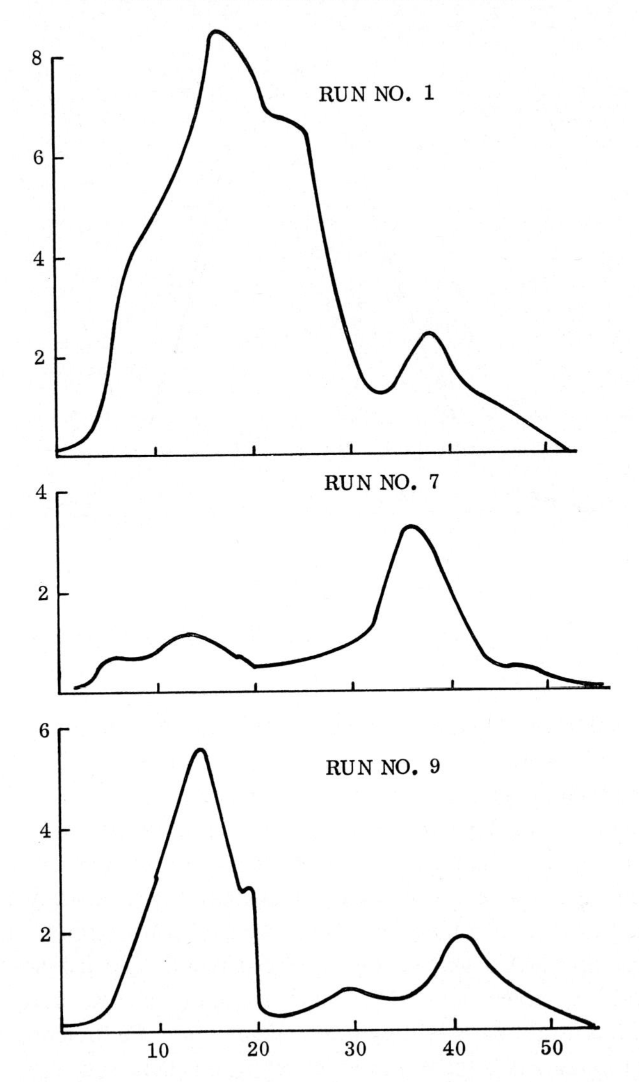

Fig. 4.1. Some observed concentration profiles across continuous plume of marked fluid (Csanady, 1966). The abcissa is distance in meters across a plume section and the ordinate is concentration in parts per hundred million.

concentrations at fixed spatial positions, we would have observed a mean distribution several times (perhaps 3 times) wider than the one shown in Figure 4. 2, with a peak value correspondingly lower by a factor of 3 or so. The profile averaged in the moving frame, shown in Figure 4.2, is rather closer to the irregular individual profiles in observable width and peak concentration than the fixed-frame average profile would be. We note, however, that the moving frame average profile is still a highly smoothed version of observed profiles, and that at given distances y from the center of gravity still quite appreciable 'fluctuations' in local concentration occur, comparable in mag-

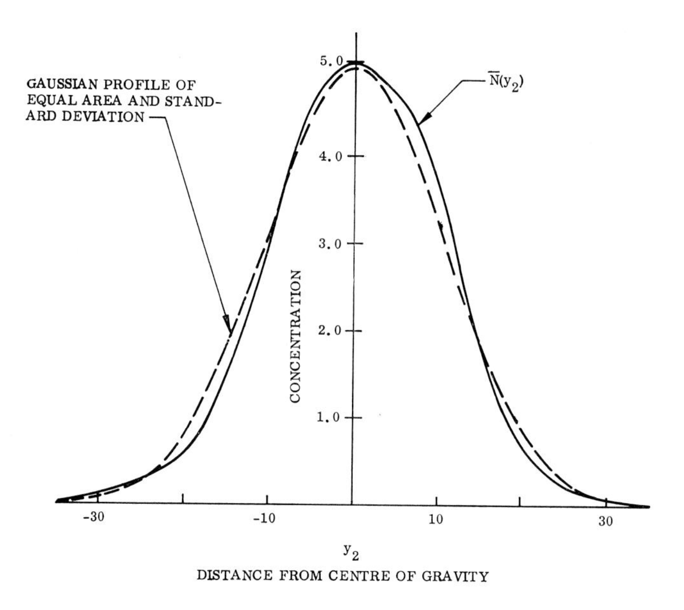

Fig. 4.2. Profile of concentrations averaged at fixed distances from center of gravity.

nitude to the local 'mean.' In a fixed frame the fluctuations are even greater relatively, because the 'mean' is lower.

The width of a plume grows of course with distance from the source. The increase in width of the plume regularized by averaging concentrations in a frame of reference moving with the center of gravity may be attributed to 'relative diffusion'. The irregular displacements of the center of gravity are usually described as 'meandering.' The growth in a fixed frame of reference (of an *averaged* concentration field) is due to 'absolute diffusion,' which is seen to consist of relative diffusion *and* meandering.

The practical advantage of moving-frame averaging is even more obvious when we deal with concentrations caused by the passage of an instantaneously released *puff* of tracer or pollutant. A regular mean concentration field subject to ascertainable physical laws can in this instance only be approximated by carrying out many experiments under identical environmental conditions and determining the stochastic average field. There is no direct physical measurement in this case, analogous to the time-average concentration at fixed points behind continuous sources, which results in anything resembling the fixed-frame ensemble average concentration field. In the usual puff experiment the field of total *dosage* is determined, caused by the passage of the entire puff. Because a 'puff' is relatively short, this dosage field differs little from an instantaneous cross section across the thickest part of the puff. A *moving-frame average* concentration or dosage field at least resembles the typical measured one in its width and peak concentration value to a tolerable degree. If the averaging of many puff-trials is carried out at *fixed points in space* on the other hand, a flattened-out distribution results which is more remote from individual trials. For these reason, many experimentalists have come to regard the theory of relative diffusion as relevant to the diffusion of a puff (or 'cluster', see e.g., Pasquill, 1962). It should be clear, however, that the theory is equally useful for describing continuous plumes, if one's interest is not confined to time-average concentrations at fixed points in space.

The reader may also observe that the experimental data shown in Figure 4.1 were collected at fixed distances from the source x, and at fixed depths z, and that only the cross-plume coordinate was in Figure 4.2 in fact referred to a 'moving' origin. However, the meandering of successive 'puffs' which combine to form a 'plume' along the x-axis is not likely to have interfered with the mean profile so obtained, on account of the 'slenderness' of the plume which results in little net diffusive transport along x. Also, the meandering of the plume was negligible along the vertical because of the proximity of the free surface. In the absence of meandering of course there is no difference between 'fixed' and 'moving' frames of reference, so that the observed mean profiles may in fact be regarded as having been referred to a frame of reference attached to the plume's center of gravity (meandering only horizontally). A similar approximation must frequently be made in experimental work on relative diffusion.

4.2. Mean Concentration Field in a Frame of Reference Attached to the Center of Gravity

Consider the release at time $t=0$ of marked fluid into a field of stationary and homogeneous turbulence. Let the observed concentration field at subsequent times in an *individual realization* of the experiment be $N(\mathbf{x},t)$. This field is subject to the continuity equation in differential and integral forms:

$$\begin{aligned} \partial N/\partial t &= -\nabla\cdot\mathbf{F} = -\nabla\cdot(\mathbf{u}N - D\nabla N) \\ Q &= \int N(\mathbf{x},t)\,\mathrm{d}\mathbf{x} = \text{const.}, \end{aligned} \tag{4.1}$$

where Q is the total amount of marked fluid released, $\mathbf{F}$ is flux of marked fluid due to bulk motion (velocity $\mathbf{u}$) as well as to molecular diffusion. The volume integral extends over all space. We shall here designate the components of the position vector $\mathbf{x}$ as (x_1, x_2, x_3), so that $\mathrm{d}\mathbf{x}$ is the volume element $\mathrm{d}x_1\,\mathrm{d}x_2\,\mathrm{d}x_3$. The first moment of the distribution $N(\mathbf{x},t)$ yields the position vector $\mathbf{c}$ of the center of gravity:

$$\mathbf{c} = 1/Q \int \mathbf{x}N(\mathbf{x},t)\,\mathrm{d}\mathbf{x}. \tag{4.2}$$

For an individual realization $\mathbf{c}$ is a random function of time. Its rate of change follows from differentiating Equation (4.2). After some reductions, noting in particular that $\partial N/\partial t$ equals the divergence of a vector, we find:

$$\mathrm{d}\mathbf{c}/\mathrm{d}t = 1/Q \int \mathbf{u}N(\mathbf{x},t)\,\mathrm{d}\mathbf{x}. \tag{4.3}$$

A 'relative' position vector may now be defined by

$$\mathbf{y} = \mathbf{x} - \mathbf{c} \tag{4.4}$$

with components (y_1, y_2, y_3). The observed concentration field may also be described in the 'relative' frame of reference, the center of which, $\mathbf{c}(t)$, moves about in a random manner: we have clearly $N(\mathbf{y},t) = N(\mathbf{x}-\mathbf{c},t)$. This only differs from the 'fixed' frame description in the trivial point of a different coordinate origin. Significant differences exist, however, between the *statistical properties* of N as observed at fixed $\mathbf{x}$ and fixed $\mathbf{y}$, respectively.

The zeroth and first moments of the concentration distribution in the $\mathbf{y}$-frame are clearly:

$$\begin{aligned} \int N(\mathbf{y}, t)\, \mathrm{d}\mathbf{y} &= Q \\ \int \mathbf{y} N(\mathbf{y}, t)\, \mathrm{d}\mathbf{y} &= 0 . \end{aligned} \tag{4.5}$$

The second moments of the concentration distribution in the $\mathbf{x}$ and $\mathbf{y}$ frames are also simply related. From the definition of a center of gravity the 'parallel axis' theorem of moments of inertia follows directly:

$$\begin{aligned} \int y_i y_j N(\mathbf{y}, t)\, \mathrm{d}\mathbf{y} &= \int (x_i - c_i)(x_j - c_j)\, N(\mathbf{x}, t)\, \mathrm{d}\mathbf{x} = \\ &= \int x_i x_j N(\mathbf{x}, t)\, \mathrm{d}\mathbf{x} - Q c_i c_j \end{aligned} \tag{4.6}$$

for any i and j $(=1, 2$ or $3)$.

As we have explained in some detail before, it is the statistical properties of the diffusion process which concern us, specifically in this chapter the ensemble-mean field at fixed $\mathbf{y}$, $\bar{N}(\mathbf{y},t)$. Therefore we next develop the consequences of the above kinematic relationships for ensemble average quantities. On repeating a given release a large number of times, an ensemble average value of $\mathrm{d}\mathbf{c}/\mathrm{d}t$ may be determined which is, from Equation (3.3):

$$\mathrm{d}\bar{\mathbf{c}}/\mathrm{d}t = 1/Q \int (\bar{\mathbf{u}}\bar{N} + \overline{\mathbf{u}'N'})\, \mathrm{d}\mathbf{x}, \tag{4.7}$$

where overbars denote ensemble averages and primes denote fluctuations (departures from the ensemble mean in an individual realization). We have identified before the mean product $\overline{\mathbf{u}'N'}$ as a local turbulent flux vector. In a homogeneous field, and provided that the cloud as released is symmetrical about the origin, the turbulent flux must be antisymmetrical, so that its space-integral is zero. The same result is obtained if we suppose the turbulent component of the flux to be proportional to the concentration gradient. Thus for symmetrical releases rigorously, for others at least approximately:

$$\mathrm{d}\bar{\mathbf{c}}/\mathrm{d}t = 1/Q \int \bar{\mathbf{u}}\bar{N}(\mathbf{x}, t)\, \mathrm{d}\mathbf{x}. \tag{4.7a}$$

In a homogeneous field the mean velocity of the diffusing particles $\mathbf{u}$ is constant: it may consist of a uniform bulk fluid motion and a constant drift velocity of particles relative to the fluid. In either case we may without loss of generality allow our 'fixed' coordinates to drift with the mean velocity $\bar{\mathbf{u}}$, i.e. choose our coordinates so as to make $\overline{\mathrm{d}\mathbf{c}}/\mathrm{d}t=0$. In the following we shall assume this (implying also that the space integral of the turbulent flux vanishes). The zeroth and first moments of the ensemble average fields are then, in 'fixed' and 'moving' frames:

$$Q = \int \bar{N}(\mathbf{x}, t)\,\mathrm{d}\mathbf{x} = \int \bar{N}(\mathbf{y}, t)\,\mathrm{d}\mathbf{y}$$
$$\bar{\mathbf{c}} = \int \mathbf{x}\bar{N}(\mathbf{x}, t)\,\mathrm{d}\mathbf{x} = \int \mathbf{y}\bar{N}(\mathbf{y}, t)\,\mathrm{d}\mathbf{y} = 0\,. \tag{4.8}$$

Any physically meaningful difference between 'fixed' and 'moving' frame ensemble average concentration fields $\bar{N}(\mathbf{x},t)$ and $\bar{N}(\mathbf{y},t)$ is therefore confined to their second and higher moments. To characterize second moments we define the following mean square dispersion tensors:

$$\Sigma_{ij} = 1/Q \int x_i x_j \bar{N}(\mathbf{x}, t)\,\mathrm{d}\mathbf{x}$$
$$S_{ij} = 1/Q \int y_i y_j \bar{N}(\mathbf{y}, t)\,\mathrm{d}\mathbf{y} \tag{4.9}$$
$$M_{ij} = \overline{c_i c_j}\,.$$

From Equation (4.6) we have then, after ensemble-averaging:

$$\Sigma_{ij} = S_{ij} + M_{ij}\,. \tag{4.10}$$

To exhibit the physical meaning of this last relationship a little more clearly consider a specific component of the dispersion tensors involved, say $i=j=1$, and write

$$\sigma_1^2 = \Sigma_{11}$$
$$s_1^2 = S_{11} \tag{4.11}$$
$$m_1^2 = M_{11}$$

so that σ_1, s_1 and m_1 are length-scales (standard deviations) of their respective distributions. We may call σ_1 standard deviation of the 'absolute' dispersion, s_1 that of 'relative' dispersion and m_1 that of 'meandering'. Equation (4.10) shows then that

$$\sigma_1^2 = s_1^2 + m_1^2\,. \tag{4.10a}$$

In other words absolute dispersion is the 'sum' of relative dispersion and meandering, in the sense that their variances are additive. Clearly, σ_1 is always greater than either s_1 or m_1. Similar relations naturally hold also for the coordinate directions x_2 and x_3. The nondiagonal components Σ_{12} etc. may be assumed to vanish on an appropriate choice of coordinates (principal axes).

4.3. Probability Distributions of Particle Displacements

In view of the insight we gained from our fundamental theorem (Equation (2.1)) into various aspects of absolute diffusion it seems worthwhile to explore whether a similar theorem also holds for relative diffusion. Indeed it turns out that we can make use of *two* such theorems in this context.

Consider first the release of a number of marked particles of total mass Q as a 'point'-cloud. On repeating the experiment many times, we may determine a mean concentration field at fixed distances $\mathbf{y}$ from the center of gravity, and at successive times t, $\bar{N}(\mathbf{y},t)$. The quantity $Q^{-1}\bar{N}(\mathbf{y},t)$ is also an experimental approximation to the probability that a marked particle will be found at $(\mathbf{y},t)$, hence it is also equal to the probability of *displacement* $\mathbf{y}$ (from the center of gravity) in time t for a diffusing particle. Denoting again the particle-displacement probability distribution by $P(\mathbf{y},t)$ we have the fundamental theorem for relative diffusion:

$$\bar{N}(\mathbf{y}, t) = QP(\mathbf{y}, t). \tag{4.12}$$

A complication arises in the extension of this theorem to initially finite size clouds. For reasons which should become clear in the next section, the displacement probability distribution of one element of such a larger cloud is *not* identical with $P(\mathbf{y},t)$ of a point-source release, and moreover it may depend on its position within the released cloud, $\mathbf{y}'$ say. The initial size and shape of the cloud and the concentration distribution within it at release, $N_0(\mathbf{y}',0)$, may significantly influence the probability of given displacements, so that a general statement of finite cloud history becomes quite complex. Taking the relatively simple case of a finite line-source of length b (perpendicular to the mean flow, this being the practically most important example) we define the displacement probability density relative to the center of gravity, for a single element located initially at $\mathbf{y}'$ by

$$P(\mathbf{y} - \mathbf{y}', t, \mathbf{y}', b)$$

so that the ensemble-average concentration field at subsequent times becomes:

$$\bar{N}(\mathbf{y}, t) = \int N_0(\mathbf{y}', 0)\, P(\mathbf{y} - \mathbf{y}', t, \mathbf{y}', b)\, d\mathbf{y}'. \tag{4.13}$$

After a sufficient time has elapsed we may expect diffusing particles to 'forget' their initial position in the cloud, so that the separate dependence of $P(\,)$ on $\mathbf{y}'$ in Equation (4.13) disappears, but not the dependence on initial size b. We shall find, however, that the finite cloud probability distribution becomes at some time indistinguishable from a point-cloud one, released at some earlier time:

$$P_{\text{finite}}(\mathbf{y} - \mathbf{y}', t, b) = P_{\text{point}}(\mathbf{y} - \mathbf{y}', t + t_0), \tag{4.14}$$

where $t = -t_0$ is an 'effective origin' in this case, the value of which depends on cloud size b.

As in the case of absolute diffusion, the one-particle fundamental theorem opens the door to deductions from the kinematics of particle movements, in this case in a frame of reference attached to the center of gravity of a diffusing cloud. Further insight is obtained by considering *two-particle* probabilities. It was perceived early by Richardson (1926) and exploited theoretically by Batchelor (1952) that 'relative diffusion' is closely linked to the rate at which two individual diffusing particles separate. To establish this relationship, let a point cloud be released at $t=0$ at the origin of a fixed frame, $\mathbf{x}=0$, and consider the mean product

$$(1/Q^2)\,\overline{N(\mathbf{x},t)\,N(\mathbf{x}',t)}\,\mathrm{d}\mathbf{x}\,\mathrm{d}\mathbf{x}'.$$

This product may be regarded as the experimental approximation to the joint probability of finding marked fluid particles both at $\mathbf{x}$ and $\mathbf{x}'$, hence it is also equal to the joint probability of particle *displacements* (for *two* diffusing particles) $\mathbf{x}$ and $\mathbf{x}'$, in time period t. Denoting the two-particle displacement probability density by $P(\mathbf{x},\mathbf{x}',t)$, (such that $P(\ )\mathrm{d}\mathbf{x}\mathrm{d}\mathbf{x}'$ is the probability of finding a particle at $\mathbf{x}$, another at $\mathbf{x}'$) we may also write

$$\overline{N(\mathbf{x},t)\,N(\mathbf{x}',t)} = Q^2 P(\mathbf{x},\mathbf{x}',t). \tag{4.15}$$

This last result may be called our 'two particle' fundamental theorem. An important deduction from it may be made by calculating the second moments of $P(\mathbf{x},\mathbf{x}',t)$ with respect to the separation vector $(\mathbf{x}'-\mathbf{x})$. Let

$$\Pi_{ij} = \int\int (x_i' - x_i)(x_j' - x_j)\,P(\mathbf{x},\mathbf{x}',t)\,\mathrm{d}\mathbf{x}'\,\mathrm{d}\mathbf{x}. \tag{4.16}$$

With the aid of the fundamental theorem (Equation (4.15)), the definition of a center of gravity, and the definitions of Σ_{ij}, etc, introduced before, we may now easily show that

$$\Pi_{ij} = 2\Sigma_{ij} - 2M_{ij} = 2S_{ij} \tag{4.17}$$

so that the mean-square separation of two diffusing particles is just twice their mean square distance from the center of gravity. Observations on drifting particle-pairs therefore provide experimental information relevant to the relative diffusion of clouds about their center of gravity. A cloud grows in relative diffusion, as particle-pairs increase their mean-square separation.

The probability density $P(\mathbf{x},\mathbf{x}',t)$ may also be regarded as specifying the probability of absolute displacement $\mathbf{x}$, and of a relative displacement $\boldsymbol{\eta}=\mathbf{x}'-\mathbf{x}$ of the two particles. Integrating over all displacements $\mathbf{x}$ we find a mean 'neighbor density' in function of distance $\boldsymbol{\eta}$:

$$F(\boldsymbol{\eta},t) = \int P(\mathbf{x},\boldsymbol{\eta},t)\,\mathrm{d}\mathbf{x} = 1/Q^2 \int \overline{N(\mathbf{x},t)\,N(\mathbf{x}+\boldsymbol{\eta},t)}\,\mathrm{d}\mathbf{x}. \tag{4.18}$$

The integral over the concentration product may indeed be determined for a single realization and supplies a somewhat smoothed picture of the distribution of particles within a cloud. The ensemble average neighbor density $F(\boldsymbol{\eta},t)$ constitutes a possible description of relative diffusion alternative to the mean concentration distribution in a 'moving' frame, $\bar{N}(\mathbf{y},t)$ and was originally suggested by Richardson (1926). Its practical usefulness, however, remains to be demonstrated. To give an illustration of the kind of information contained in $F(\boldsymbol{\eta},t)$ the reader may consider the value of this function for zero displacement $\boldsymbol{\eta}$:

$$F(0, t) = 1/Q^2 \int [\bar{N}^2(\mathbf{x}, t) + \overline{N'^2}(\mathbf{x}, t)] \, d\mathbf{x}. \tag{4.19}$$

This clearly depends on the field of mean square fluctuations, as well as on the mean concentration distribution, and it is difficult to attach a simple practical interpretation to it.

The main practical value of the two-particle fundamental theorem (Equation (4.15)), lies in some qualitative deductions one may make with its aid, following the ideas of Richardson (1926). With two particles separated at a given distance, turbulent eddies large compared to particle separation move them about together, not contributing much to their relative separation. On the other hand, eddies small compared to the separation distance perturb this distance relatively little. Most effective in increasing the separation distance are eddies of a size comparable to the separation distance itself. To illustrate this practically very important point quantitatively, let us model an 'eddy' by a single Fourier component of the spatial velocity distribution, along a given axis x:

$$u = a \sin kx, \tag{4.20}$$

where a is a velocity amplitude, a random variable of which the statistical properties depend on wave number k. If two particles are situated at x and $x+l$, respectively, their relative velocity along x is:

$$v = u(x + l) - u(x) = a [\sin k(x + l) - \sin kx]. \tag{4.21}$$

To estimate a mean square value of v, we square Equation (4.21), integrate over a full cycle $x=0$ to $x=2\pi/k$ and substitute the ensemble average value of the amplitude a:

$$\overline{v^2} = \overline{a^2}(1 - \cos kl). \tag{4.22}$$

With constant $\overline{a^2}$ this has a maximum at $k=\pi/l$ or a wavelength $2\pi/k$ of $2l$. If $\overline{a^2}$ varies as $k^{-5/3}$ (as in the inertial subrange) the maximum value of $\overline{v^2}$ occurs at $k=1.43/l$ or a wavelength of about $4.4\,l$. Much larger eddies ($k \to 0$) contribute little to separation because $\cos kl$ tends to unity. Much smaller eddies, $k \to \infty$ can give again a multiplier of order unity in the bracket in Equation (4.22), so that all smaller eddies effectively contribute to $\overline{v^2}$, but their amplitude $\overline{a^2}$ usually decreases rapidly with increasing k, at least as long as the potentially most effective eddy ($\overline{a^2}$ a maximum) does not exceed the size of the energy containing eddies.

When a large number of eddies (Fourier components) are present, the total mean square relative velocity is a sum of contributions from all eddy sizes, according to the above, down to a wave number not much smaller than π/l or so. As a cloud grows, the range of eddies effectively contributing to $\overline{v^2}$ in this manner increases, because the lower wave-number limit of the effective range reduces progressively, until the supply of eddies 'runs out', i.e., until $k=\pi/l$ becomes so low that the corresponding $\overline{a^2}$ is negligible.

When this condition is reached, $v^2 \to u^2$, and the relative motion of two particles begins to resemble the (absolute) motion of a single particle in a homogeneous field of turbulence, the mean square velocity of which is constant. Up to that point, however, the relative motion of two particles is an accelerating or 'evolutionary' process in turbulent flow, in the sense that the mean square velocity grows with separation distance, and therefore with time because a greater range of eddies contributes to it.

Velocities relative to the center of gravity of a diffusing puff are in effect relative velocities to all the other particles present and the above argument applies to them, at least insofar as mean square separations are concerned, as demonstrated by Equation (4.17).

4.4. Kinematics of Particle Movements in a Moving Frame

Let particle velocities referred to the fixed and moving frames be, respectively $\mathbf{u}(u_1, u_2, u_3)$ and $\mathbf{v}(v_1, v_2, v_3)$. From a differentiation of Equation (4.4) it follows that

$$\mathbf{u} = (\mathrm{d}\mathbf{c}/\mathrm{d}t) + \mathbf{v}. \tag{4.23}$$

From Equation (4.8) we also have that

$$\overline{\mathbf{v}} = \mathrm{d}\overline{\mathbf{y}}/\mathrm{d}t = 0. \tag{4.24}$$

Because the relative velocity and displacement of a diffusing particle are again related by

$$\mathbf{y} = \int_0^t \mathbf{v}\,\mathrm{d}t \tag{4.25}$$

the analogue to Taylor's theorem using relative velocities is easily derived. Along the $\mathbf{y}_1$-axis, for example, the mean square displacement varies as:

$$\frac{\mathrm{d}\overline{y_1^2}}{\mathrm{d}t} = 2\overline{y_1 \frac{\mathrm{d}y_1}{\mathrm{d}t}} = 2\int_0^t \overline{v_1(t)\, v_1(t')}\,\mathrm{d}t'. \tag{4.26}$$

As we have seen in the previous section, the relative velocity components $v_1(t)$ do *not* constitute a stationary process. For a 'point' source release the initial relative velocities are indeed molecular. As the cloud grows, the smallest turbulent eddies begin to contribute to $v(t)$, then increasingly larger ones until the limit of eddy size is reached and exceeded. Thus the velocity covariance $\overline{v_1(t)v_1(t')}$ is a function of diffusion time t as well as of the lag time $\tau = t - t'$. A modified Lagrangian correlation coefficient may be defined by

$$R(\tau, t) = \frac{\overline{v_1(t)\, v_1(t-\tau)}}{\overline{v_1^2}(t)} \tag{4.27}$$

which has a shape illustrated in Figure 4.3, for a given time t. The shaded area may

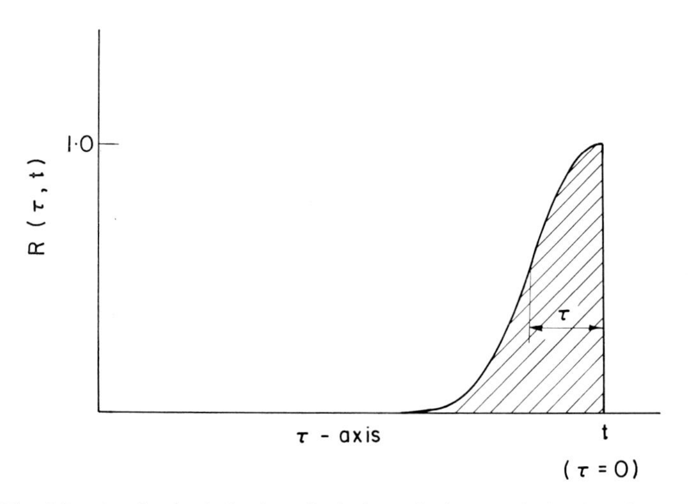

Fig. 4.3. Qualitative behavior of relative velocity correlation for given t.

again be regarded a Lagrangian time-scale for relative diffusion:

$$t_R(t) = \int_{t_0}^{t} R(\tau, t)\, d\tau, \tag{4.28}$$

where the lower limit t_0 is sufficiently far removed from t for the correlation to be zero at $\tau = t - t_0$. If the cloud has started as a point source such a lower limit $t_0 > 0$ may always be found because the persistence time of the initial (molecular) relative velocities is very short. The Lagrangian time-scale t_R is characteristic of all those eddies contributing to relative velocities, which we have seen to be mainly those comparable in size to typical particle separation in a cloud (i.e., also to cloud size). As the cloud grows, larger eddies begin to contribute, which, generally speaking, have longer 'lifetimes,' so that the Lagrangian time-scale t_R slowly increases. So does, of course, the mean-square relative velocity.

Using the definitions from Equation (4.27) and (4.28), Equation (4.26) may be written as

$$\overline{y_1^2} = 2 \int_0^t \overline{v_1^2}(t')\, t_R(t')\, dt'. \tag{4.29}$$

In virtue of the one-particle fundamental theorem, $\overline{y_1^2}$ is also equal to s_1^2, the dispersion along the y_1 axis of the ensemble-mean concentration field $\bar{N}(\mathbf{y},t)$

$$\overline{y_1^2} = s_1^2 = 1/Q \int y_1^2 \bar{N}(\mathbf{y}, t)\, d\mathbf{y} = 2 \int_0^t \overline{v_1^2}(t')\, t_R(t')\, dt' \tag{4.30}$$

with similar relations for the dispersion along the y_2 and y_3 axes.

4.5. Phases of Cloud Growth

Some general properties of s_1^2 (and of s_2^2 and s_3^2, of course: we are only using the y_1-axis as an example) now follow for certain phases of the relative diffusion process. While a point-source cloud is much smaller than the smallest turbulent eddies present, its relative diffusion proceeds by molecular movements only. Therefore

$$\frac{1}{2}\frac{ds_1^2}{dt} = D \quad \left(s_1 \ll \frac{\nu^{3/4}}{\varepsilon^{1/4}}\right), \tag{4.31}$$

where D is molecular diffusivity, ν is kinematic viscosity, ε is rate of turbulent energy dissipation (cm^2 s^{-3}). The quantity $\nu^{3/4}/\varepsilon^{1/4}$ is proportional to the typical size of the small (dissipative) eddies.

When relative diffusion is dominated by eddies in the inertial subrange (i.e., when s_1 is comparable in size to these eddies) the mean-square velocity and the Lagrangian time scale t_R are functions only of the energy dissipation rate ε, and of cloud size, of course. Dimensional considerations then show that (Batchelor, 1950, 1952):

$$\begin{aligned} \overline{v_1^2}(t) &= a_1 s_1^{2/3} \varepsilon^{2/3} \\ t_R(t) &= a_2 s_1^{2/3} \varepsilon^{-1/3}, \end{aligned} \tag{4.32}$$

where a_1, a_2 are constants of order unity. From Equations (4.30) we have then

$$\frac{1}{2}\frac{ds_1^2}{dt} = a_1 a_2 s_1^{4/3} \varepsilon^{1/3} \quad \left(\frac{\nu^{3/4}}{\varepsilon^{1/4}} \ll s_1 \ll L\right). \tag{4.33}$$

The quantity $(\frac{1}{2}ds_1^2/dt)$ may be interpreted as an eddy diffusivity in relative diffusion: Equation (4.33) shows this to increase as the $\frac{4}{3}$ power of cloud size. The stated conditions on s_1 express that the diffusion is dominated by inertial subrange eddies.

Equation (4.33) may be integrated and yields

$$s_1^{2/3} = \frac{2a_1 a_2}{3} \varepsilon^{1/3} (t - t_0), \tag{4.34}$$

where t_0 is an integration constant, having the physical significance of an effective time-origin for the inertial subrange-dominated phase of relative diffusion. In terms of time elapsed from such an effective origin, the cloud size grows as $(t-t_0)^{3/2}$ or rather faster than linearly with time. This is strikingly different from the $t^{1/2}$ growth of clouds in molecular diffusion or even from the linear growth in the initial phase of 'absolute' diffusion.

The energy dissipation rate ε is related to characteristic turbulent velocity u_m (a typical rms velocity) and length scale L by (e.g., Batchelor, 1953):

$$\varepsilon = A(u_m^3/L), \tag{4.35}$$

where A is constant, equal to 1.1 in isotropic turbulence. Substituting into Equation

(4.34) we derive

$$\frac{s_1}{L} = B\left[\frac{u_m(t-t_0)}{L}\right]^{3/2}, \tag{4.36}$$

where B is another constant of order unity $[B = A^{1/2}(2a_1a_2/3)^{3/2}]$. This is a particularly instructive formulation of the result and shows that the condition $s_1 \ll L$ can only hold as long as $(t-t_0)$ is small compared to L/u_m, the latter quantity being in order of magnitude equal to the Lagrangian time-scale t_L. Thus the total duration of this 'explosive' phase of cloud growth is rather short, shorter than the lifetime of the typical 'energy containing' eddy in the field.

Finally, when a cloud becomes large compared to the energy containing eddies, all eddies present contribute to its further diffusion and statistically the difference between 'absolute' and 'relative' velocities disappears, the center of gravity is no longer subject to further displacement by eddies larger than the cloud. Thus we have from Equation (4.30), just as in the final phase of absolute diffusion:

$$\frac{1}{2}\frac{ds_1^2}{dt} = \overline{u_1^2}t_L = \text{const.} \tag{4.37}$$

Integrating, we find again that s_1 grows with the square root of time as in molecular diffusion (although the eddy diffusivity $u_1^2 t_L$ is rather larger than the molecular values D normally are).

4.6. History of a Concentrated Puff

The mean concentration field, $\bar{N}(\mathbf{y},t)$, again appears to be Gaussian (in a homogeneous field) as we have already seen in Figure 4.2, although the experimental evidence is not extensive enough to establish this conclusively for all phases of growth. The theoretical justification remains as in absolute diffusion, the non-rigorous idea that many independent Fourier components combine to produce the total relative velocity.

In the initial (molecular) phase, a Gaussian distribution of course follows from the diffusion equation. For the initial condition describing a concentrated source:

$$\bar{N}(\mathbf{y}, 0) = Q\delta(\mathbf{y}), \tag{4.38}$$

where $\delta(\mathbf{y})$ is a three-dimensional delta function, the early phases of relative diffusion are described by

$$\bar{N}(\mathbf{y}, t) = \frac{Q}{8(\pi Dt)^{3/2}} \exp\left\{-\frac{|\mathbf{y}|^2}{4Dt}\right\}. \tag{4.39}$$

This holds as long as the puff is smaller than the dissipative eddies in the field, i.e.,

$$s = \sqrt{2Dt} \ll \frac{\nu^{3/4}}{\varepsilon^{1/4}}. \tag{4.40}$$

The relative diffusion in this phase is isotropic, $s_i^2 = s^2$ for all $i = 1, 2, 3$. From Equation (4.10) we find the connection with cloud spread in absolute diffusion,

$$\sigma_1^2 = m_1^2 + 2Dt. \tag{4.41}$$

If we now also note Equation (3.2), which takes into account molecular motions in absolute diffusion, we find that

$$\sigma_{1t}^2 = m_1^2 \tag{4.42}$$

or that the mean-square displacement of particles in a *fixed* frame due to the *turbulent* movements alone gives rise to the *meandering* of the center of gravity of such a small puff, while relative diffusion is due to molecular movements alone. As we have pointed out before, the molecular contribution to σ_1^2 is usually negligible, so that thus effectively *all* absolute 'diffusion' comes about by the random movements of the entire released puff. The initial behavior of the meandering-dispersion is therefore, in view of our results in the previous chapter:

$$m_1 = \sqrt{\overline{u_1'^2}}\, t \tag{4.43}$$

and similar linear relations for the x_2 and x_3 axes. The result throws further light on the nature of absolute diffusion in the initial phase.

As the point-cloud grows, turbulent eddies begin to influence its spread about its center of gravity, at least the larger ones of which are anisotropic. Thus the mean concentration field in the moving frame becomes, to the best of our present knowledge

$$\bar{N}(\mathbf{y}, t) = \frac{Q}{(2\pi)^{3/2}\, s_1 s_2 s_3} \exp\left\{-\frac{y_1^2}{2s_1^2} - \frac{y_2^2}{2s_2^2} - \frac{y_3^2}{2s_3^2}\right\}. \tag{4.44}$$

In the final phase of diffusion, when the cloud is large compared to the energy containing eddies, we have

$$\frac{1}{2}\frac{\mathrm{d}\sigma_1^2}{\mathrm{d}t} = \frac{1}{2}\frac{\mathrm{d}s_1^2}{\mathrm{d}t} = \overline{u'^2}\, t_L \tag{4.45}$$

from which we also find by Equation (4.10) that

$$\mathrm{d}m_1^2/\mathrm{d}t = 0 \quad (s_1 \gg L) \tag{4.46}$$

or that m_1 (and similarly m_2 and m_3) tends to a constant asymptotic value at large diffusion times (which should be of the order of the eddy length-scale L on dimensional grounds).

Figure 4.4 illustrates the behavior of the three standard deviations σ_1, s_1 and m_1, against diffusion time, for an initially concentrated puff. The duration of the 'molecular' phase depends on the diffusivity D and could be quite long for low D, but the practical significance of this is not very great because few releases in the atmosphere or ocean are truly 'point' sources in the sense of being small compared to the smallest eddy present. The duration of the steep-growth phase of s_1 is short, rather less than t_L, as we have pointed out before. This is a particularly important point to keep in mind in comparing the theory with experimental data. We should also remember that this figure summarizes qualitative theoretical arguments valid for the highly idealized case of a field of stationary and homogeneous turbulence and comparison with experimental data should be undertaken with appropriate caution. Some experimental data on meandering of dye plumes in the Great Lakes are shown in Figure 4.4b.

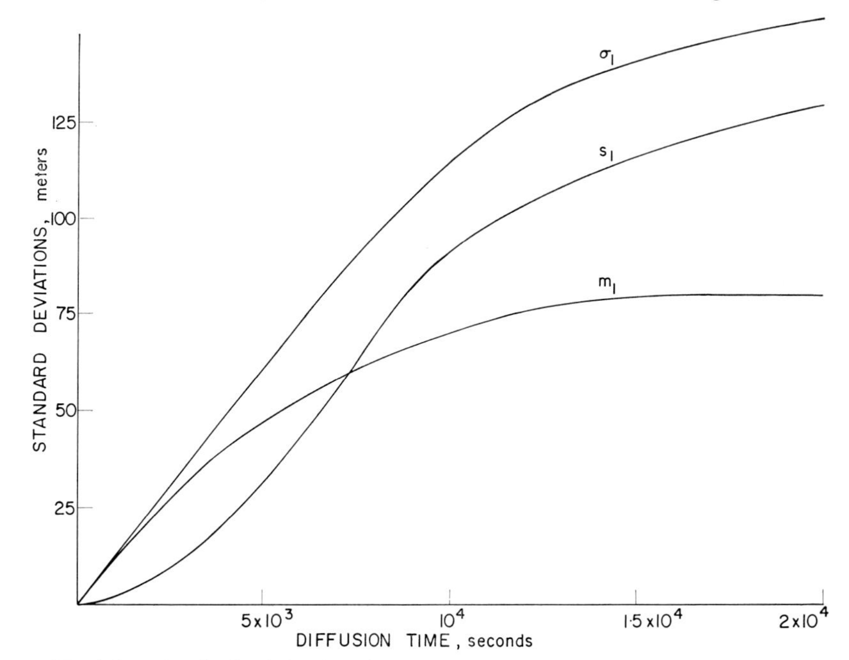

Fig. 4.4a. Qualitative behavior of standard deviations of 'absolute' and 'related' diffusion and meandering.

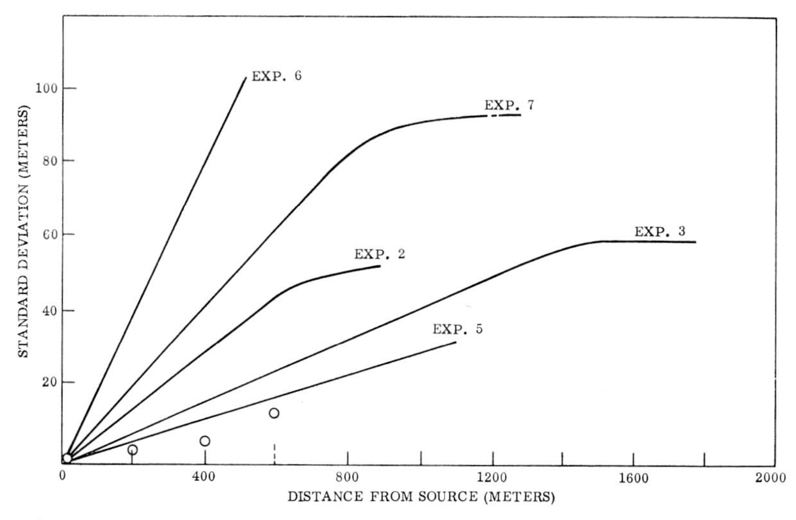

Fig. 4.4b. Standard deviation of meandering vs. distance from source observed in Lake Huron 1967.

4.7. Initially Finite Size Cloud

We have already pointed out in Section 4.3 that particle displacement probabilities from finite cloud releases depend on cloud size and possibly position in the cloud. The precise physical reason should now also be clear from the arguments on relative particle motions: a certain range of eddies is immediately active in dispersing a cloud of finite initial size. Also particles at the fringes of a cloud are further from their neighbors than average and tend to separate at a faster rate.

As before, we can follow cloud growth by focusing on the second moments S_{ij} of the concentration distribution in a moving frame (these were defined in Equation (4.9)). From Equation (4.13) we have then

$$S_{ij} = 1/Q \int N_0(\mathbf{y}', 0)\,\mathrm{d}\mathbf{y}' \int y_i y_j P(\mathbf{y} - \mathbf{y}', t, \mathbf{y}', b)\,\mathrm{d}\mathbf{y}, \tag{4.47}$$

where we have assumed for concreteness a finite line source of length b. The second integral in this expression represents the mean value of $y_i y_j$ for a *sub-ensemble* of particles released at a constant distance $\mathbf{y}'$ from the initial center of gravity. This sub-ensemble average may be related to the relative velocity-history of the same element by an equation identical to Equation (4.26) above (Taylor's theorem, for this particular case). If we use an overbar to denote such a sub-ensemble average, and $\langle\rangle$ brackets to denote 'cloud average', which is expressed by the first integral in Equation (4.47), we may write Taylor's theorem for this case:

$$\mathrm{d}S_{ij}/\mathrm{d}t = 2 \int_0^t \overline{\langle v_i(t)\, v_j(t')\rangle}\,\mathrm{d}t'. \tag{4.48}$$

An equation equivalent to this result has been derived by Smith and Hay (1961). Formally this is similar to Taylor's theorem in its original form, but the velocity covariance it contains consists of sub-ensemble average *relative* velocity products, averaged again over an entire diffusing cloud. At *short* diffusion times the sub-ensemble average products remain close to their initial values and we have

$$S_{ij} = S_{ij}\big|_0 + \langle v_i(0)\, v_j(0)\rangle\, t^2 \quad (t \to 0), \tag{4.49}$$

where $S_{ij}|_0$ is the initial mean-square dispersion (for the finite line source of Figure 4.5 if the initial concentration is constant, we have $S_{22}|_0 = b^2/12$).

After a sufficiently long period the particles in the cloud must 'forget' their initial positions. Also, the initial concentration distribution (whether it is constant over a source volume or not) is obliterated by turbulent mixing and a Gaussian distribution should be approached. In other words, at a sufficiently long time after release the cloud will behave as if it had been discharged from a point source further upstream, as already stated in Equation (4.14). Figure 4.6 illustrates the growth of such an initially finite size cloud: at some time its history joins that of a point-source cloud. The details of this curve prior to coalescing with the point-source cloud history are difficult to

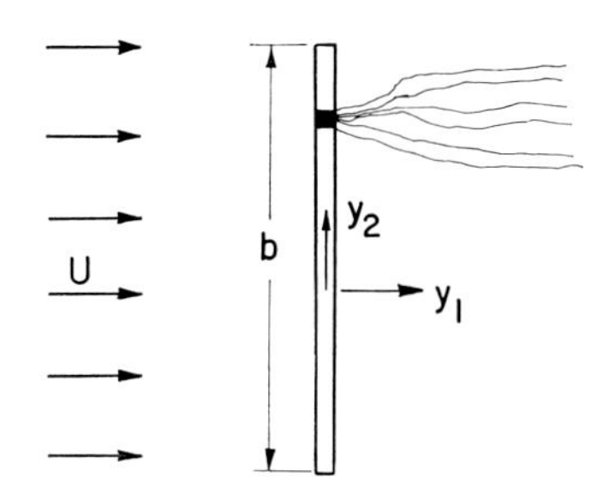

Fig. 4.5. Finite line source in uniform stream.

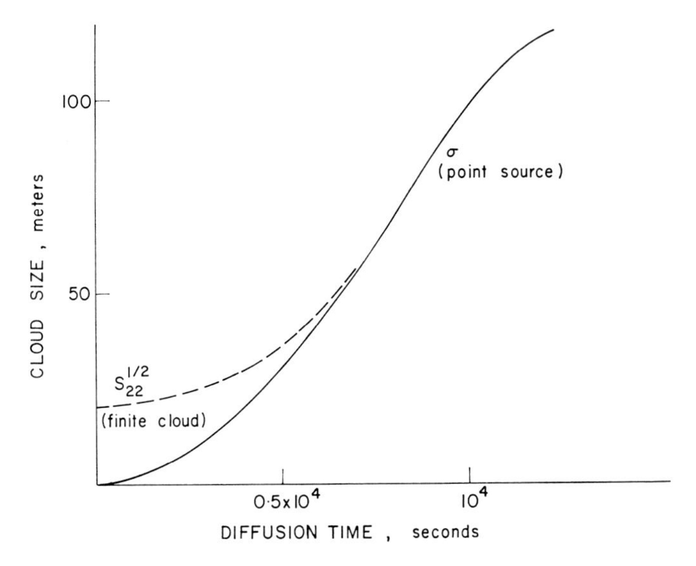

Fig. 4.6. Initial behavior of finite cloud.

elucidate quantitatively, because they depend in a complex manner on both Eulerian and Lagrangian properties of turbulence. Approximate theoretical calculations (such as those of Smith and Hay, 1961) demonstrate this difficulty by being forced to rely on several crude assumptions. Similar results can therefore only be regarded as quite rough predictions; Smith and Hay (1961) predict for example, that the maximum slope of the curve in Figure 4.6 should be proportional to turbulence intensity squared, i_y^2. This is highly doubtful and it conflicts with some dimensional deductions of Batchelor (1950), (our Equation (4.34) above).

4.8. Use of the Diffusion Equation

The equation of continuity may also be exploited to throw further light on the relative diffusion problem. For any stage of an individual realization, referring coordinates to the center of gravity we may write the diffusion equation:

$$\partial N(\mathbf{y}, t)/\partial t = -\nabla \cdot [\mathbf{v}N(\mathbf{y}, t) - D\nabla N(\mathbf{y}, t)]. \tag{4.50}$$

The divergence of the flux on the right-hand side consists of terms involving bulk motion and molecular diffusion. Taking ensemble averages at fixed $(\mathbf{y}, t)$ and noting that $\bar{\mathbf{v}}=0$ we find

$$\partial \bar{N}/\partial t = -\nabla \cdot [\mathbf{F}_t - D\nabla \bar{N}], \tag{4.51}$$

where the turbulent flux component is

$$\mathbf{F}_t = \overline{\mathbf{v}N'} = \overline{\mathbf{v}(N - \bar{N})}. \tag{4.52}$$

For a *point-source* release we have already seen that the mean concentration distribution $\bar{N}(\mathbf{y}, t)$ probably remains Gaussian at all times t. By the argument of Section 3.8 it follows then that components of the flux are proportional to the concentration gradients. For the y_1-direction, for example

$$F_{t1} = -K_1 \frac{\partial \bar{N}}{\partial y_1} \quad \text{where} \quad K_1 = \frac{1}{2}\frac{\mathrm{d}s_1^2}{\mathrm{d}t} \tag{4.53}$$

in complete analogy with our results on absolute diffusion. Similar relationships may be written down for the y_2- and y_3-axes.

In the initial phase of relative diffusion of a *finite* cloud the concentration distribution is not in general Gaussian and the above identification cannot be made. If we did insist on defining a diffusivity, this would have to be a function not only of initial cloud size but also of position within the cloud, and the whole exercise would become rather pointless.

A rather crude approximation to the initial behavior of a finite cloud may be obtained by using the diffusion equation notwithstanding the above objections, with a diffusivity appropriate to a point source cloud at a stage of its history where S_{ij} equals the released cloud's moments $S_{ij}|_0$. The subsequent growth of S_{ij} may then be assumed to proceed according to Figure 4.4, or some equivalent empirical function appropriate to the situation treated. This type of approach has been proposed by Brooks (1960) for oceanic pollution calculation, and Csanady (1970) for modeling effluent plumes in the Great Lakes.

One great advantage of using the equation of continuity is that one may model non-conservative effects, such as die-off of bacteria (in a field of diffusing sewage) or flocculation, radioactive disintegration, etc. In many such cases the rate of loss of the diffusing substance (bacteria, for instance) is proportional to their concentration, the factor of proportionality being a 'decay constant' of dimensions s^{-1}. Let this constant by denoted by k: then the diffusion equation in the moving frame becomes

$$\frac{\partial \bar{N}}{\partial t} + k\bar{N} = \frac{\partial}{\partial y_1}\left(K_1 \frac{\partial \bar{N}}{\partial y_1}\right) + \frac{\partial}{\partial y_2}\left(K_2 \frac{\partial \bar{N}}{\partial y_2}\right) + \frac{\partial}{\partial y_3}\left(K_3 \frac{\partial \bar{N}}{\partial y_3}\right). \tag{4.54}$$

In applying this equation to continuous plumes (but always measuring lateral and vertical distances from the center of gravity of the plume) we replace the time-dependence by dependence on distance along the plume, $x = Ut$ (ignoring the curvature of the plume center line) and neglect diffusion along the usually 'slender' plume. This gives:

$$U \frac{\partial \bar{N}}{\partial y_1} + k\bar{N} = \frac{\partial}{\partial y_2}\left(K_2 \frac{\partial \bar{N}}{\partial y_2}\right) + \frac{\partial}{\partial y_3}\left(K_3 \frac{\partial \bar{N}}{\partial y_3}\right). \tag{4.55}$$

As in the application of absolute diffusion theory to air pollution prediction, this equation may be used to arrive at engineering decisions in sewage dispersal, etc., provided that reliable empirical information is available on the diffusivities $K_2(y_1)$ and $K_3(y_1)$ and the decay constant k.

4.9. Horizontal Diffusion in the Ocean and Large Lakes

In the introductory section of this chapter we have already indicated that the theory of 'relative' diffusion is the appropriate framework for the consideration of oceanic diffusion of pollutants. The problem arises mainly in the context of oceanic (or lacustrine) disposal of 'sewage' or of 'sludge', i.e., of domestic or industrial waste liquids. The usual method of disposal is to convey the liquid waste through a submarine pipe to a point some distance offshore in a large lake, sea or ocean and release it there through a system of 'diffuser ports.' The ports are designed to provide considerable initial mixing with sea or lake water. The net result is that 'marked fluid' is distributed in some initial concentration χ_0 over a considerable volume, which then drifts along with the lake or ocean current and mixes with its surroundings under the influence of the turbulence naturally present. The theory we have developed above may be exploited to predict the efficiency of this further dilution process.

In many such applications *vertical* diffusion makes no appreciable contribution to the dilution of pollutants. Such is the case, for example, when the initial cloud extends from the sea or lake bottom to the free surface. Even if the depth of water increases in the direction in which the pollutant cloud drifts, so that its vertical size increases, this does not imply vertical mixing (the streamlines in a horizontal plane must then converge to compensate for vertical divergence). Often the waste liquid has some buoyancy and 'boils' to the surface, effectively distributing itself over the available vertical depth during the initial mixing phase. In other situations a stable density interface (thermocline) stops downward or upward vertical mixing beyond a short initial phase, much as an inversion lid does in the atmosphere (cf. Chapter VI). In a later section we shall briefly consider the process of vertical mixing in the sea; here we first discuss the practically more important *horizontal* spread of pollutants through oceanic turbulence.

If the release of marked fluid takes place at some distance from shore, the field of turbulence surrounding the effluent in a large lake or ocean may be expected to be horizontally homogeneous to a good approximation. Our theoretical results should therefore apply to such releases: in particular, if the initial cloud size is small compared to the typical horizontal eddy, a phase of 'accelerated' relative diffusion should be observable during which the cloud size grows faster than linearly with time. The duration of this phase should however be relatively short, and be followed by a phase in which the cloud grows as in molecular diffusion, i.e. with the square root of time.

There are many difficulties of observation in the sea or in large lakes which make it difficult to compare theory and experiment. One problem is that, particularly in large-scale diffusion experiments, it is rarely possible to observe an *ensemble* of clouds discharged under identical conditions. Like it or not, one is often forced to make deductions from what can be observed in *individual* realizations of diffusion experiments. Any finer test of theory is under such circumstances out of the question and the most one can achieve is to deduce general trends from many similar (but far from identical) experiments.

Nevertheless, some systematic ensemble averages are available from experiments on small point-source plumes of Rhodamine B dye (artificially generated near the surface) in the Great Lakes (Csanady, 1966, 1970) and off the California coast (Foxworthy *et al.*, 1966). We have already seen above that these observations suggested a Gaussian distribution of ensemble mean concentration across a diffusing plume. The rate of growth of the plume was as typified in Figure 4.7 for the Great Lakes experiments: an 'explosive' phase followed by a more gentle rate of diffusion, the latter more or less as would follow from the assumption of constant eddy diffusivity. The same value of the diffusivity as determined from pilot dye-plume observations was found to describe quite well the diffusion of initially moderately large (order 100 m diam.) sewage clouds off the California coast (Foxworthy *et al.*, 1966). The numerical values of diffusivity

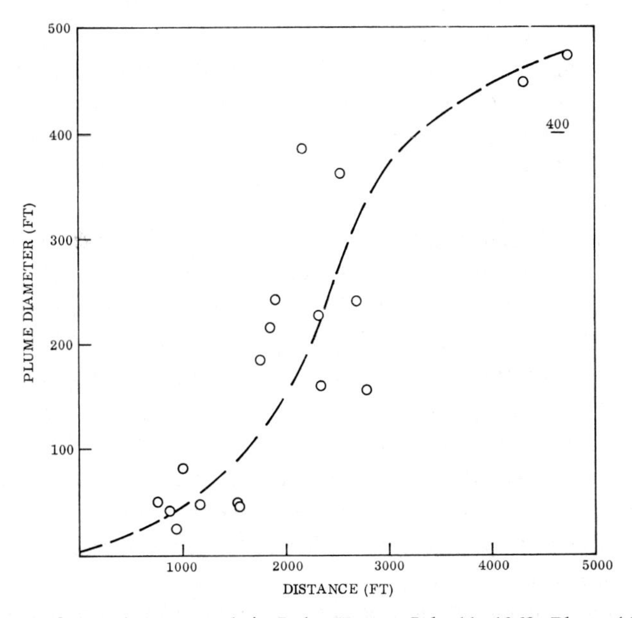

Fig. 4.7. Example of dye-plume growth in Lake Huron, July 11, 1962. Plume 'diameter' is 10% width of cross-section at surface.

determined from such experiments in the Great Lakes were almost identical with those determined in the Pacific Ocean off California, as long as the size of the cloud was of the same order of magnitude, i.e. about 100 m. The horizontal eddy size implied by these eddy diffusivities was about 20 m, which is more or less equal to the depth of water at the location of the California experiments and to the depth of the thermocline or of the bottom in the Lake Huron experiments. Thus these data are in every way consistent with the theory developed earlier in this chapter, and the known properties of turbulence in this kind of flow.

Some rather puzzling facts come to light, however, when we examine experimental data from much larger-scale observations. When the dye patch becomes of the order of 1 km and larger, its rate of diffusion increases again, almost as if the theoretical $\frac{4}{3}$ power law were valid over a much broader range of eddies. Indeed earlier data (Stommel, 1949) indicated a generally good proportionality of eddy diffusivity to the $\frac{4}{3}$ power of cloud size. A recent summary of more accurate data by Okubo (1971) shows, however, a somewhat less rapid increase of effective diffusivity. Okubo's curves are reproduced here in Figures 4.8 and 4.9 on account of their immediate practical usefulness. The definition of the symbols used in this diagram is

$$\begin{aligned}\sigma_{rc}^2 &= 2\sigma_x\sigma_y\\ K_a &= \sigma_{rc}^2/4t\\ l &= 3\sigma_{rc},\end{aligned}$$

where t is diffusion time and σ_x, σ_y are horizontal standard deviations of a diffusing patch along its two principal axes. The distance σ_{rc} is also the mean square *radius* of diffusing particles from the center of the patch. The small scale diffusion data of Foxworthy *et al.* (referred to above) are contained in the left-hand corners of Okubo's diagrams. As remarked before, diffusivities observed on small plumes in the Great Lakes are identical with those values. Murthy (1969) has shown that for larger scale dye patches in Lake Ontario the diffusivity increases with cloud size in much the same way as in the ocean.

The discrepancy between simple statistical theory and observation is of much the same kind as we encountered in the atmosphere: the standard deviations do not behave as would follow from a homogeneous-field model. From our present vantage point we must conclude that larger and larger 'eddies' come into play as a diffusing cloud grows, even though a constant diffusivity 'plateau' may be intermittently reached and then surpassed. Whether the larger scale flow structures, which must be present to cause an increase in apparent diffusivity, should be called 'eddies' is perhaps mainly a question of terminology. It would at the same time, certainly require a broadening of the definition of the term 'eddy' if we regarded, say, the flow pattern of the entire North Atlantic as one large eddy (note that the points in Okubo's diagrams extend to a cloud length-scale of 200 km).

If we regard the larger scale flow structures as parts of the 'mean flow' field, we must attribute the increased cloud growth rates to the non-uniformity of this field. In

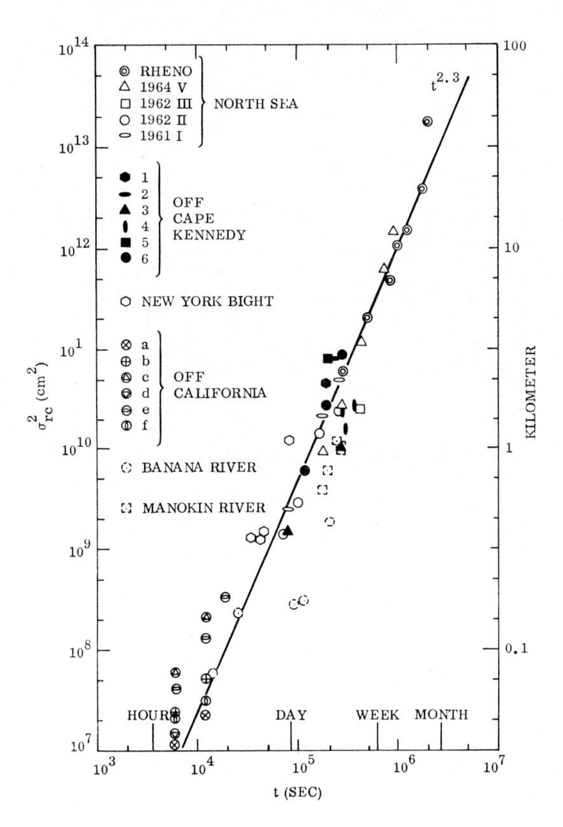

Fig. 4.8. A diffusion diagram for variance vs. diffusion time (from Okubo, 1971).

Chapter V we shall discuss in detail how such nonuniformity may lead to precisely the kinds of effects shown by large-scale oceanic diffusion experiments. Such a view appears physically more satisfactory. As a practical measure in dealing with oceanic diffusion problems, however, we may retain the homogeneous field model as an approximation and use empirical data on cloud growth to predict maximum concentrations, exactly as we have done in atmospheric diffusion problems.

4.10. Application to Diffusion of Sewage Plumes

A useful theoretical model of oceanic diffusion was developed by Brooks (1960), arising apparently out of work on sewage disposal in the Los Angeles area. At its point of introduction the sewage 'boils' to the surface, after having been mixed with sea water by a diffuser of some 300 m length lying at sea bottom. The effluent is forced through many ports of the diffuser, forming a number of jets which mix with the ambient fluid until their momentum is dissipated. The initial sewage field is thus fairly large horizontally and is well mixed from surface to bottom.

Because the currents near the shores usually follow the depth-contours, an adequate

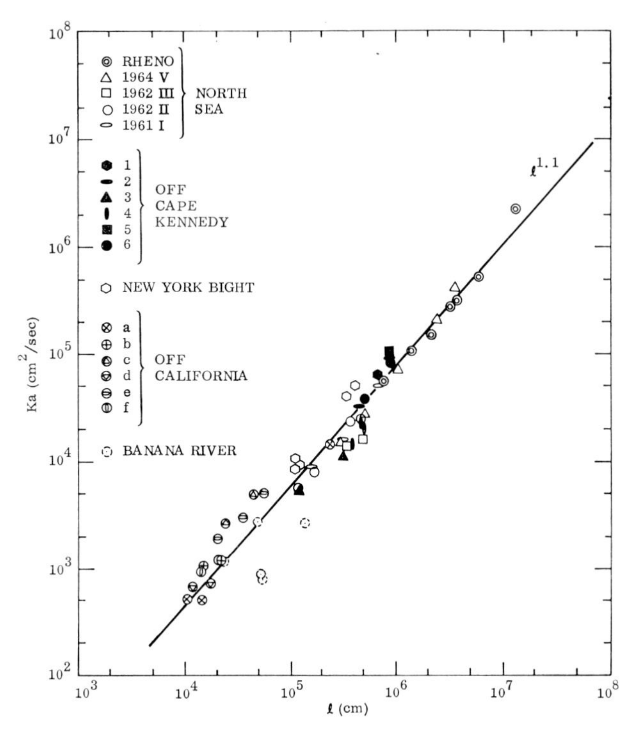

Fig. 4.9. A diffusion diagram for apparent diffusivity vs. scale of diffusion (from Okubo, 1971).

model is to assume that the depth of the sewage plume remains constant and that its further dilution is caused by lateral diffusion alone. The relevant form of the diffusion equation is then

$$U\frac{\partial \chi}{\partial x} + k\chi = \frac{\partial}{\partial y}\left(K_y \frac{\partial \chi}{\partial y}\right), \tag{4.56}$$

where we have returned to our earlier notation used in applications of the theory, writing $\chi(x, y)$ in place of $\bar{N}(\mathbf{y}, t)$, although it is understood that we are dealing with *relative* diffusion, i.e., measuring lateral distances y from the centerline of the plume. The (x, y, z) coordinate system with z vertical replaces the suffix notation now that we consider practical applications. Because diffusion along the axis of the sewage plume is of no consequence, the discharge region may be represented as a 'line source' of a certain length b, perpendicular to the mean current. The resulting theoretical model of the sewage field is essentially as already illustrated in Figure 4.5.

According to our discussion earlier in this chapter, the eddy diffusivity, defined for a plume by

$$K_y = \frac{U}{2}\frac{\mathrm{d}s_y^2}{\mathrm{d}x} \tag{4.57}$$

is a function of cloud size (i.e. of s_y, which takes the place of s_2 with the change of notation) in a homogeneous field. Over the extent of a sewage plume the flow field in the ocean may be assumed to be horizontally homogeneous to a satisfactory approximation, but diffusion on a larger scale must certainly be affected by non-homogeneities. In analogy with atmospheric applications, we may expect that the effects of such departures from the theoretical model may be taken into account by using empirical information on K_y (or equivalently, on s_y^2). We have already discussed the details in the previous section. The time-dependence of K_y or σ_y is replaced in a continuous plume by dependence on $x=Ut$.

A solution of Equation (4.56) is now easily found, appropriate to the model illustrated in Figure 4.5. The die-off term may be eliminated by the change of variable:

$$\chi = \phi \, e^{-kx/U}. \tag{4.58}$$

Substituted into Equation (4.56) this yields, noting also that K_y is a function of x and not of y:

$$U \frac{\partial \phi}{\partial x} = K_y \frac{\partial^2 \phi}{\partial y^2} \tag{4.59}$$

which is a simple diffusion equation for the field $\phi(x, y)$. Once we determine the 'conservative' field ϕ, the actual concentration (with die-off) may be found from Equation (4.58).

Let K_0 be the value of the diffusivity appropriate to the initial size of the cloud. Then we define a new distance variable

$$\xi = \int_0^x \frac{K_y}{K_0} \, dx. \tag{4.60}$$

This amounts to a stretching of the abscissa, in proportion to the growth of the eddy diffusivity. In terms of ξ, Equation (4.59) becomes:

$$U \frac{\partial \phi}{\partial \xi} = K_0 \frac{\partial^2 \phi}{\partial y^2} \tag{4.61}$$

which is a simple form of the classical diffusion equation with constant diffusivity. Its solution for a finite line-source is easily written down; after returning to our original variables and using Equation (4.57) we find, if χ_0 is the (constant) initial concentration:

$$\chi(x, y) = \frac{\chi_0 \, e^{-kx/U}}{\sqrt{2\pi} \, s_y} \int_{-b/2}^{b/2} \exp\left[-\frac{(y-y')^2}{2s_y^2}\right] dy' =$$

$$= \tfrac{1}{2}\chi_0 \, e^{-kx/U} \left[\operatorname{erf}\left(\frac{b/2 + y}{\sqrt{2} \, s_y}\right) + \operatorname{erf}\left(\frac{b/2 - y}{\sqrt{2} \, s_y}\right)\right]. \tag{4.62}$$

This is physically equivalent to arranging many infinitesimal point sources along b, all of strength $\chi_0 dy'$, and simply adding their fields. As we have noted before, this is a crude approximation in relative diffusion, even if we use a value of K_0 appropriate to the initial size of the cloud. The main physical features of the problem are, however, probably reflected reasonably faithfully by this solution. As we have already seen in Chapter I, the peak concentration remains equal to the initial value χ_0 until the point-source standard deviation s_y becomes about equal to $b/6$ (see Equation (1.33)). At rather larger distances the plume behaves as a point-source cloud, the maximum concentration in which decreases as s_y^{-1}. This is illustrated in Figure 4.10.

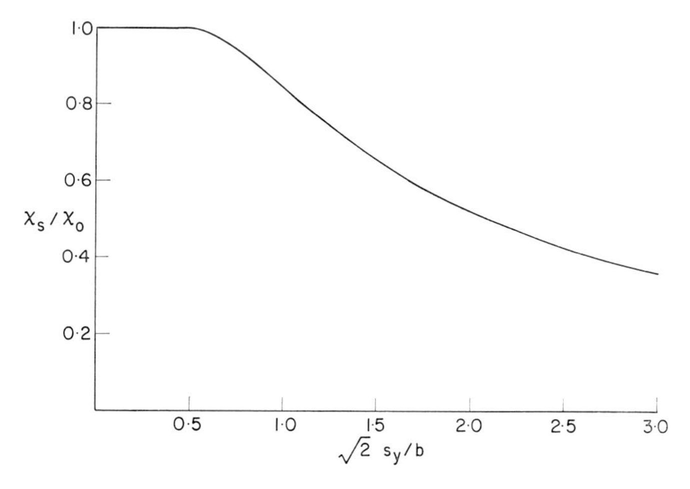

Fig. 4.10. Decay of maximum concentration with distance, for initially large cloud.

Practical estimates of maximum mean concentration may therefore be arrived at if we have data on the decay constant k (which we shall take for granted) and on the point-source standard deviation s_y. In large-scale oceanic diffusion problems we may use the Okubo diagrams (Figures 4.8 and 4.9) to estimate the behavior of s_y. However, as we have already pointed out, these variations are due to large-scale flow inhomogeneities, which may be absent in a given application. In a nearly homogeneous field plume growth is rather slower than suggested by Figures 4.8 and 4.9 and may be calculated to a satisfactory approximation using a constant eddy diffusivity. Experimental evidence on the behavior of sewage plumes in their early dispersal phase (which is practically the most important part) suggests an eddy diffusivity of order 10^3 cm^2 s^{-1}. It would seem to be safer to make such an assumption the basis of engineering estimates of oceanic dispersal, rather than using the more optimistic projections which may be taken from Figures 4.8 and 4.9.

4.11. Vertical Diffusion in Lakes and Oceans

The subject of vertical oceanic diffusion has not been explored very well so far, perhaps because its engineering importance is not as great as that of horizontal mixing, as we have just seen. We shall only give a very brief review of the subject here, basing our approach on experimental evidence from the Great Lakes (Csanady, 1970).

If some marked fluid is released into the topmost layers of a lake or sea, its initial vertical spread is somewhat similar to its horizontal spread, except that the field is vertically far from homogeneous. The speed, direction and turbulence level of a current usually change with depth. Therefore the growth rate of vertical cloud size (which we shall designate as s_z, a point-source standard deviation) is more complicated than that of s_y, but the process is basically similar, unless a stable layer is encountered by the diffusing cloud. A stable layer acts as a diffusion 'floor' and stops further vertical growth. Figure 4.11 illustrates vertical growth with and without a diffusion floor.

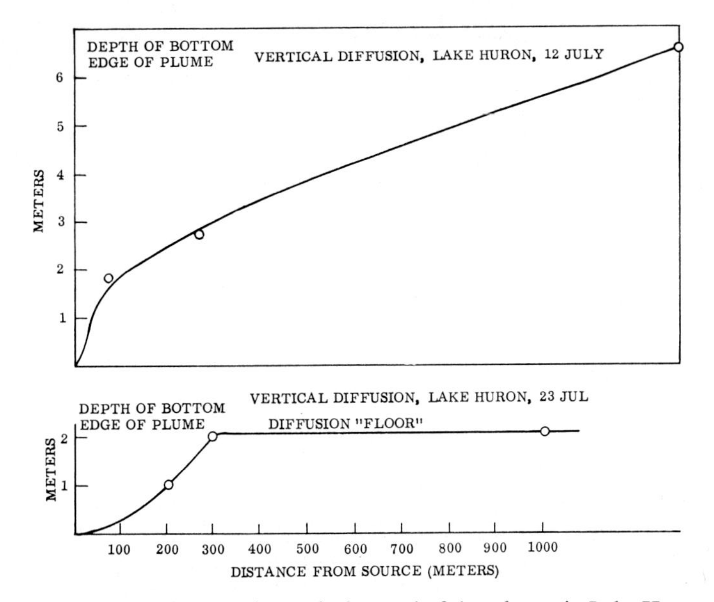

Fig. 4.11. Some observations on the vertical spread of dye plumes in Lake Huron, 1964.

A vertical diffusivity may again be defined by

$$K_z = \frac{U}{2}\frac{\mathrm{d}s_z^2}{\mathrm{d}x}.$$

This quantity varies with the properties of diffusion, reaching typically a value of 30 cm^2 s^{-1} when the cloud is of order 30 cm deep, and reducing to some 5 cm^2 s^{-1} as the cloud grows further to a depth of several meters. Except when there is pronounced surface cooling, a diffusion floor is invariably encountered at a depth of at most 30 m.

The slowness of vertical diffusion compared to horizontal spread is noteworthy. The difference is much greater than in the atmosphere: as we have seen in Chapter III, the ratio σ_z/σ_y is usually of order 0.5 in the atmosphere. In the ocean or large lakes the ratio is more like 1:30. To illustrate the point, Figure 4.12 shows a rather strongly smoothed cross-plume concentration profile, first with the vertical scale exaggerated, then with the same scale in the vertical as in the horizontal.

The above remarks apply only when the wind above the lake or ocean is relatively light. In stronger winds, and in particular when the surface is strongly cooled, large-scale flow structures become established which reach from the free surface down to the first stable sheet. These so-called 'Langmuir circulations' (after Langmuir, 1938) are similar to large horizontal roll-vortices with their axes parallel to the wind, neighboring vortices having opposite signs (Figure 4.13 shows a highly idealized illustration). On the surface, the presence of such Langmuir circulations is noticeable on account of the

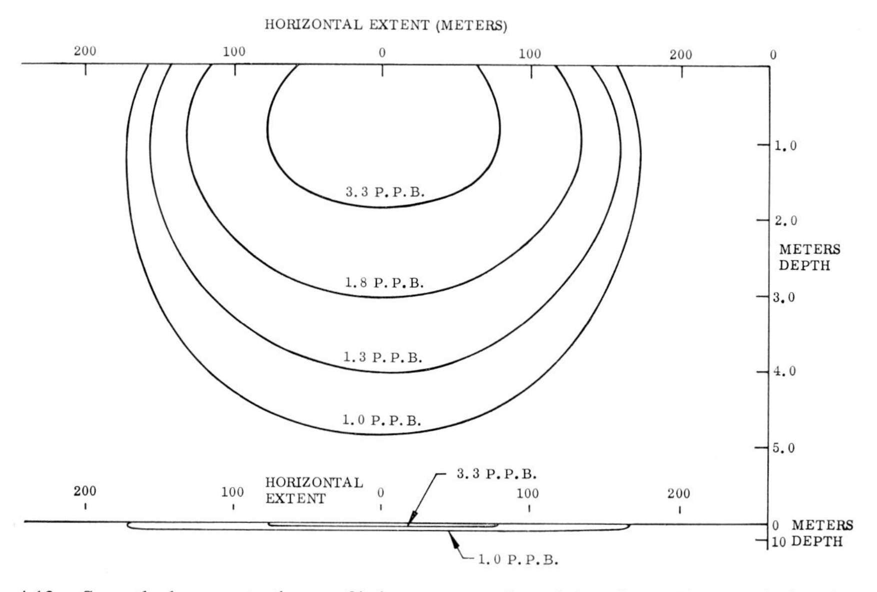

Fig. 4.12. Smoothed concentration profile in a cross section of dye plume. Top: vertical scale exaggerated by a factor of 50; bottom: correct vertical scale. Lake Huron, July 12, 1964.

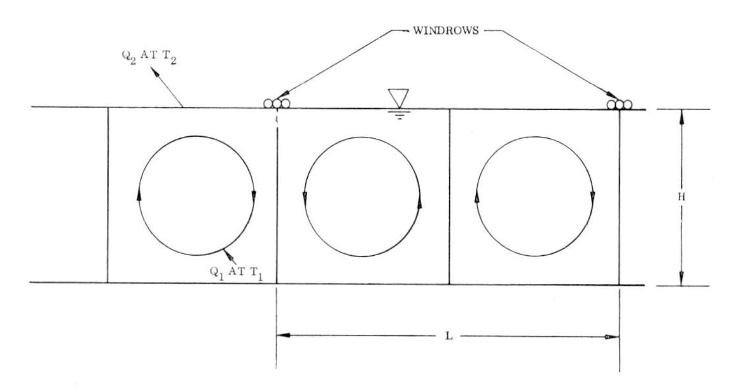

Fig. 4.13. Schematic illustration of Langmuir circulations extending to depth H and giving rise to windrows of spacing L.

accumulation of foam and debris at the lines of confluence between neighboring roll-vortices. The exact physical mechanism of these Langmuir circulations is not known. Laminar stability theory (Faller, 1966) suggests that they may be produced by the shear instability of the surface Ekman layer. In the presence of surface cooling the Rayleigh-Benard type instability (Chandrasekhar, 1961) may also establish a system of roll-vortices. Several other explanations are possible (e.g., Scott *et al.*, 1969) but some experimental facts are inconsistent with all of them.

Whatever their physical mechanism, once Langmuir circulations are present (and they invariably are above a wind speed of some 5 m s^{-1}) they rapidly distribute any marked fluid over their entire vertical range, i.e., from the free surface to the first stable sheet. This mode of transport is definitely not of the gradient type and a vertical diffusivity cannot usefully be associated with it.

Stable sheets or diffusion floors are a consequence of turbulence suppression by the force of gravity, when the density increases with depth in a fluid. The physical aspects of this phenomenon are discussed further in Chapter VI. Why or how a stable sheet becomes established at a specific depth (particularly in the presence of Langmuir circulations) is a difficult question to which no very satisfactory answer exists today. Theories of thermocline formation are described in oceanographic texts (e.g. Defant, 1961) but they can hardly be regarded conclusive.

Mixing across stable interfaces is not completely absent, but is sporadic and is believed to be due to the 'breaking' of internal waves (Mortimer, 1961). There is not enough quantitative information on the topic for us to pursue it here further.

EXERCISE

Analize the decay of maximum mean concentration in a sewage field of initial dimensions $b = 300$ m (well mixed from top to bottom) in a current of speed 15 cm s^{-1}. At what distance from the source is the concentration 10 times less than immediately after the initial (diffuser) mixing, if the decay constant k is (1) zero, (2) equal to 1/10 h?

References

Batchelor, G. K.: 1950, *Quart. J. Roy. Meteorol. Soc.* **76**, 133.
Batchelor, G. K.: 1952, *Proc. Cambridge Phil. Soc.* **48**, 345.
Batchelor, G. K.: 1953, *Homogeneous Turbulence*, Cambridge Univ. Press, 197 pp.
Brooks, N. H.: 1960, in *Proc. 1st Int. Conf. on Waste Disposal in the Marine Environment*, Pergamon Press, p. 246.
Chandrasekhar, S.: 1961, *Hydrodynamic and Hydromagnetic Stability*, Oxford Univ. Press, 652 pp.
Csanady, G. T.: 1966, Publ. No. 15, Great Lakes Div. Univ. Michigan, *Proc. 9th Conf. on Great Lakes Res.*, p. 283.
Csanady, G. T.: 1970, *Water Res.* **4**, 79.
Defant, A.: 1961, *Physical Oceanography*, Vol. I, Pergamon Press, 729 pp.
Faller, A. J.: 1966, in *Proc. 5th U.S. National Congress of Applied Mechanics*, p. 651.
Foxworthy, J. E., Tibby, R. G., and Barsom, G. M.: 1966, *J. Water Pollution Control Federation* **38**, 1170.
Langmuir, I.: 1938, *Science* **87**, 119.
Mortimer, C. H.: 1961, *Verh. Intern. Verein. Limnol.* **14**, 79.

Murthy, C. R.: 1969, in *Proc. 12th Conf. on Great Lakes Res.*, Int. Assoc. Great Lakes Res., Ann Arbor, Mich., p. 635.
Okubo, A.: 1971, *Deep-Sea Res.* **18**, 789.
Pasquill, F.: 1962, *Atmospheric Diffusion*, D. van Nostrand Co. Ltd., London.
Richardson, L. F.: 1926, *Proc. Roy. Soc. London* **A110**, 709.
Scott, J. T., Myer, G. E., Stewart, R., and Walther, E. G.: 1969, *Limnology-Oceanography* **14**, 493.
Smith, F. B. and Hay, J. S.: 1961, *Quart. J. Roy. Meteorol. Soc.* **87**, 82.
Stommel, H.: 1949, *J. Marine Res.* **8**, 199.

CHAPTER V

DISPERSION IN SHEAR FLOW

5.1. Introduction

The assumption that atmospheric or oceanic pollutants diffuse in a field of stationary and homogeneous turbulence is in many ways an overidealization. Even under quite simple conditions (e.g. over flat terrain) the wind or current velocity and turbulence intensity change markedly with at least the vertical coordinate, as does the length scale of turbulence. Such changes are certain to influence the movement and spread of any marked fluid introduced into the field. In the present chapter we shall concern ourselves with similar influences, all being basically attributable to mean velocity differences, e.g., to turbulent *shear flow*.

A simple argument in the spirit of the last chapter will illustrate one of the two main physical effects of shear flow on diffusion. Consider the release of a small cloud of tracer over flat terrain at *ground level* into the atmosphere. Given a steady wind and near-neutral conditions the mean velocity distribution within the first 50 m or so from the ground is very much as in the 'wall' or 'inner' layer portion of a two-dimensional boundary layer over a flat plate (Townsend, 1956; Lumley and Panofsky, 1964). Associated with this velocity distribution is an eddy length scale distribution which is linear with height to a very good approximation, as if eddies were all reaching down to ground level. Therefore as our diffusing cloud grows, it comes under the influence of larger and larger eddies, much as in relative diffusion in a homogeneous field. We already know that such a process will lead to 'accelerated' diffusion in the sense that the effective diffusivity increases in time. This results in fixed-frame cloud growth rather faster than would be the case in a homogeneous field.

A second important effect on diffusion due to a nonuniform velocity distribution is more subtle and will be discussed in Section 5.10: this is also an accelerating effect, but the role of increasing eddies is taken over by increasing mean velocity differences.

'Accelerated' diffusion in a boundary layer in the above sense is mainly of importance in connection with *atmospheric* releases of pollutants at or very near ground level. As already remarked, a simplified model of the low-level wind and turbulence distribution is provided by the inner portion of a two-dimensional turbulent boundary layer over a flat plate. In the next few sections we shall discuss turbulent diffusion within such an idealized 'wall layer,' known in the atmosphere as the 'surface layer.' We shall begin with a brief description of the flow in the entire planetary boundary layer under ideally simple conditions. For greater detail on the planetary boundary layer the reader may wish to consult the review article of Monin (1970) or the monograph of Plate (1971).

5.2. Properties of the Planetary Boundary Layer

Let us assume that the surface of the earth is ideally flat, that there is no heat transfer between the atmosphere and this surface, and that the flow outside the planetary boundary layer is steady, horizontally uniform, and proceeds along straight horizontal streamlines, while the temperature of the air is also constant horizontally. Under such conditions geostrophic balance is known to hold outside the boundary layer, meaning that the pressure gradient is balanced by the Coriolis force. Inside the boundary layer also shear stresses play a role, but in the equilibrium case no local or convective accelerations are present and the flow is strictly parallel to the surface, i.e., no slow thickening of the boundary layer in the direction of the flow takes place. The above idealizations also imply that the horizontal pressure gradient is independent of height. The equations of motion for the two horizontal components of the mean velocity, (u, v) then become:

$$\begin{aligned} -fv &= -\frac{1}{\varrho}\frac{\partial p}{\partial x} + \frac{1}{\varrho}\frac{\partial \tau_x}{\partial z} \\ fu &= -\frac{1}{\varrho}\frac{\partial p}{\partial y} + \frac{1}{\varrho}\frac{\partial \tau_y}{\partial z}, \end{aligned} \tag{5.1}$$

where $f = 2\Omega \sin\phi$ is the Coriolis parameter (ϕ = latitude, Ω = angular speed of earth), p is hydrostatic pressure, and τ_x, τ_y are components of the shear stress in horizontal planes.

Outside the boundary layer the shear stress vanishes and (u, v) tend to the components of the geostrophic velocity, (u_g, v_g):

$$v_g = \frac{1}{\varrho f}\frac{\partial p}{\partial x} \qquad u_g = -\frac{1}{\varrho f}\frac{\partial p}{\partial y}. \tag{5.2}$$

Let the x-coordinate coincide with the direction of the shear stress at ground level, i.e.

$$\tau_x = \tau_0 \qquad \tau_y = 0 \qquad (z = 0). \tag{5.3}$$

Integrating Equations (5.1) across the entire depth of the boundary layer we find

$$\begin{aligned} \tau_0 &= -\varrho f \int_0^\infty (v_g - v)\,\mathrm{d}z \\ 0 &= \varrho f \int_0^\infty (u_g - u)\,\mathrm{d}z. \end{aligned} \tag{5.4}$$

Observing now that $(u, v) \to 0$ as $z \to 0$, the first of these relationships shows that the

y-component of the velocity is non-zero at least somewhere in the boundary layer, i.e., that there is 'cross-flow.' The second relationship shows that the x-component overshoots u_g before asymptotically becoming equal to it. In other words, the structure of the outer part at least of the Ekman layer is likely to be quite complex.

Closer to the surface we may expect to find somewhat simpler conditions. The Coriolis force vanishes with the velocity, so that according to Equation (5.1) the pressure gradient is balanced there by the stress gradient, much as in other kinds of boundary layers. Indeed there is good evidence from many different kinds of turbulent flow to show that this balance becomes more or less incidental and that out to a certain distance from the solid surface the magnitude of the shear stress and the constraint of the surface dominate the structure of the flow. In other words, there is a 'law of the wall' to which all turbulent flow adjacent to a solid surface conforms, provided only that the shear stress is not too small. This proves to be true also above the earth's surface.

The 'law of the wall' is most simply stated in terms of the vertical velocity gradient:

$$du/dz = u^*/kz, \tag{5.5}$$

where k is Karman's constant, the value of which is usually quoted as 0.4, although recent evidence suggests 0.35 (Businger *et al.*, 1971). The definition of u^* is

$$u^* = \sqrt{\frac{\tau_0}{\varrho}} \tag{5.6}$$

and is known as the 'friction velocity'. As stated, Equation (5.5) holds for both 'smooth' and 'rough' surfaces, respectively, outside of a viscous sublayer, or a layer several times the depth of roughness elements within which the flow is irregular. Equation (5.5) follows from dimensional argument, if we postulate that the flow structure depends only on shear stress τ_0 and distance from the wall z. These postulates are reasonable outside the (viscous or rough) sublayer, and provided the total fractional change of shear stress is little, i.e., out to a distance of the order of $z=\tau_0/|\nabla p|$ (the latter may be deduced from Equation (5.1)). Because of the uncertainty regarding the value of the velocity at the outer edge of the sublayer, the law of the wall has to be formulated for further velocity *changes*, i.e. conveniently for the velocity gradient.

Integrating the law of the wall we arrive at the well-known logarithmic velocity distribution:

$$u = \frac{u^*}{k} \ln \frac{z}{z_0}, \tag{5.7}$$

where z_0 is an integration constant of the physical dimension of length. This 'constant' depends on what happens in the sublayer, and for a smooth surface it is proportional to v/u^*. For a rough surface there is an experimentally determinable value of z_0 (some 30 times less than the physical height of roughness elements) which is unaffected by flow conditions, so that z_0 is a genuine 'roughness parameter,' or 'roughness length.' For an appreciation of the physical role of the roughness length the reader should consult Schlichting (1960) and Clauser (1956).

Experimentally, the law of the wall may be verified in the 'inner' or 'surface' portion of the planetary boundary layer, typically up to 30 to 100 m height. Above this level the logarithmic law begins to break down and there is also a change of wind direction with height, as the influence of the Coriolis force begins to be felt.

The above results are valid for an adiabatic surface, i.e., a boundary layer in 'neutral' hydrostatic equilibrium. When there is upward or downward heat flux at the ground, the resulting small density changes of the air, acted upon by the force of gravity, produce important dynamical effects and modify the flow. Such dynamical factors are discussed in greater detail in Chapter VI. Here we confine ourselves to some elementary dimensional arguments. The dynamically important new variable in the presence of heating or cooling of the air by the ground is the flux of positive or negative buoyancy, which is

$$F_0 = \frac{g\beta}{\varrho c_p} H_0 \quad [\mathrm{m^2\, s^{-3}}], \tag{5.8}$$

where H_0 is ground level heat flux in kcal $\mathrm{m^{-2} s^{-1}}$, c_p is specific heat in kcal $\mathrm{kg^{-1}\,°C^{-1}}$, ϱ is density in kg $\mathrm{m^{-3}}$, β is expansion coefficient of the fluid in $\mathrm{°C^{-1}}$, and g is acceleration of gravity in m $\mathrm{s^{-2}}$. For a perfect gas $\beta = T^{-1}$, with T = absolute temperature. This may be assumed in the case of atmospheric air.

A convenient length scale may be formed from the buoyancy flux and the momentum flux $\tau_0 = \varrho u^{*2}$ at ground level. This is known as the Obukhov length:

$$L^* = u^{*3}/F_0 \quad [\mathrm{m}]. \tag{5.9}$$

Conventionally, the Obukhov length is regarded positive if both heat and momentum flow downward, i.e., when the surface cools the air in contact with it, negative otherwise. Surface cooling leads to a stable arrangement of air layers and tends to suppress turbulence. With negative L^*, on the other hand, an unstable distribution of density and enhanced turbulence is associated.

Heating or cooling of the ground has a quite profound influence on the structure of the entire planetary boundary layer, which is so far rather incompletely understood. The extent of this influence is much clearer for the *surface layer* than for the outer portion of the boundary layer. Dimensional reasoning supplies the following modified form of the law of the wall for the diabatic case:

$$\frac{\mathrm{d}u}{\mathrm{d}z} = \frac{u^*}{z} \phi\left(\frac{z}{L^*}\right), \tag{5.10}$$

where $\phi(\lambda)$ is a function known from accurate experimental work for the ideal case of uniform flat terrain. The mean velocity distribution follows from integration and may be regarded as empirically known. It is a distribution similar in character to the logarithmic law, the velocity gradient being, however, somewhat less in unstable than in stable conditions.

The complexity of the factors involved in boundary layer mechanics has prompted many investigators to express the mean velocity distribution in terms of a simple power law

$$u = U_h (z/h)^\alpha, \tag{5.11}$$

where h is a suitable reference height, U_h the velocity at $z = h$ and α is an exponent which depends on the choice of h, the stability of the atmosphere and on the roughness of the ground. Such a law can be fitted to the logarithmic distribution, Equation (5.7) as well as to its analogue under non-neutral conditions. The value of α determined for a best fit ranges between 0 and 1.0, a frequent range being 0.15 to 0.25, corresponding to near-neutral conditions and low to moderate roughness. In many diffusion calculations such an approximation is adequate because the mean velocity distribution plays a relatively subordinate role in determining the distribution of pollutants across wind. In some other diffusion phenomena the mean velocity distribution is crucial and a crude approximation to $u(z)$ leads to a similarly crude approximation in predicting the mean concentration distribution.

So far we have been focusing our discussion on the mean velocity distribution in the planetary boundary layer, and particularly its surface layer portion. At least equally important, from the point of view of diffusion, are the turbulence parameters intensity and scale. For the rms velocities u_m, v_m, w_m, dimensional arguments yield, in the surface layer

$$\frac{u_m}{u^*}, \quad \frac{v_m}{u^*}, \quad \frac{w_m}{u^*} = \text{func}\left(\frac{z}{L^*}\right). \tag{5.12}$$

Under neutral conditions, in particular, u_m, etc., are equal to a constant factor times u^*, independently of height. The eddy length scale L_t is proportional to z under neutral conditions, otherwise

$$L_t/z = \text{func}(z/L^*). \tag{5.13}$$

While all the above predictions are verified by observations over ideally flat terrain and under strictly steady conditions, (Panofsky, 1968) we must remember that in almost any practical application atmospheric conditions are far from these ideals.

5.3. Particle Displacements in a Wall Layer

The fundamental theorem connecting the field of particle displacement probabilities to the mean concentration field remains valid in shear flow (although both now depend on source location) so that useful insight also into shear flow diffusion may be gained by analyzing the motion of individual diffusing elements. We shall at first discuss such motions in the parallel shear flow of a neutral wall layer.

Figure 5.1 illustrates the release of marked fluid from a source a distance h above ground into a layer with varying mean velocity $u(z)$ and varying eddy length scale $L_t(z)$. A very short distance downwind of the source the cloud is still narrow and the non-uniformity of $L_t(z)$ and $u(z)$ cannot have a significant effect, the mean concentration remains nearly Gaussian as in a uniform field. Farther downwind however, particles in the half-space above the source are subject to the action of larger eddies and are stretched out over a longer portion of the *z-axis* than those below the source level. Also, the former particles are transported with a higher velocity and therefore distributed over a somewhat longer piece of the *x-axis* than the latter. Concentrations are reduced on both counts above the source, relative to those below, and a concentration profile results which is illustrated qualitatively in the second diagram downwind of the source in Figure 5.1.

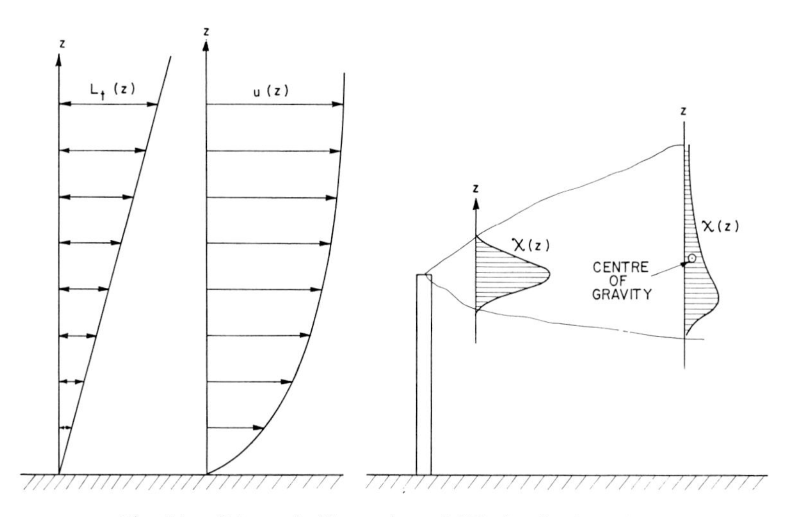

Fig. 5.1. Schematic illustration of diffusion in shear flow.

Because of the upward stretching of the cloud, the center of gravity moves vertically up, i. e. the mean vertical particle displacement becomes positive:

$$\bar{z} = \int \bar{w}\, \mathrm{d}t + h > h\,. \tag{5.14}$$

The Lagrangian mean velocity $\bar{w}$ in this case clearly does not equal the Eulerian one, which is zero everywhere by hypothesis. The dispersion of the cloud about its center of gravity is

$$\overline{z'^2} = \overline{(z - \bar{z})^2} = \int_0^t \int_0^t \overline{[w(t') - \bar{w}(t')]\,[w(t'') - \bar{w}(t'')]}\, \mathrm{d}t'\, \mathrm{d}t''$$

$$= 2 \int_0^t \int_0^{t'} \overline{w'^2}(t')\, R(\tau, t')\, \mathrm{d}\tau\, \mathrm{d}t'\,, \tag{5.15}$$

where

$$w' = w - \bar{w}$$

and

$$R(\tau, t') = \frac{\overline{w'(t')\, w'(t' - \tau)}}{\overline{w'^2(t')}}.$$

If the diffusion time t is long enough for the released particles to 'forget' their initial velocity, i.e., for $R(\tau, t)$ to be zero for an initial piece of the τ-axis, one of the integrations may be carried out to yield a Lagrangian time scale (much as in Chapter IV, Section 4.4)

$$t_L(t) = \int_0^t R(\tau, t)\, d\tau. \tag{5.16}$$

In terms of this time scale, the rate of change of mean square dispersion may be written:

$$\frac{1}{2}\frac{d\overline{z'^2}}{dt} = \overline{w'^2(t)}\, t_L(t) \quad (t \text{ large}). \tag{5.17}$$

This shear-flow analogue of Taylor's theorem shows that the asymptotic 'diffusivity' (defined by the left-hand side of Equation (5.17)) does not necessarily remain a constant. Even if $\overline{w'^2(t)}$ should be constant in a wall layer (where the Eulerian mean square turbulent velocity *is* constant), the Lagrangian time scale is likely to increase with the size of eddies acting on the cloud.

Some further progress may be made by noting the extreme simplicity of the turbulent flow in a neutral wall layer: as we have seen in the previous section, all such wall layers are similar, their portion outside the (viscous or rough) sublayer differing from another wall layer only to the extent that the friction velocity u^* is different. Eulerian turbulence properties therefore behave in a particularly simple way: $\overline{u'^2}$, etc., are proportional to u^{*2}, while the length scales L_t are (at given levels) the same in all wall layers. Lagrangian properties of the turbulence must likewise be determined by the external parameters of the flow and should be subject to equally simple laws. This notion of 'Lagrangian similarity' was apparently first enunciated by Batchelor (1959) and developed further by others, notably Cermak (1963) and his collaborators.

The lengths $\bar{z}$ and $(\overline{z'^2})^{1/2}$ are integral properties of the full particle displacement probability distribution $P(\mathbf{x}, t, h)$. In accordance with the remarks above they should only depend on diffusion time t, source height h and friction velocity u^* (once a cloud is large compared to the thickness of the viscous or rough sublayer):

$$\bar{z}, (\overline{z'^2})^{1/2} = \text{func}(u^*, t, h). \tag{5.18}$$

Specifically for ground-level release, $h=0$, we have the simple results

$$\begin{aligned} \bar{z} &= b_1 u^* t \\ (\overline{\bar{z}'^2})^{1/2} &= b_2 u^* t \end{aligned} \quad (h = 0), \tag{5.19}$$

where b_1, b_2 are constants of order unity. Similar results hold for length scales defined with higher moments of the distribution $P(\mathbf{x}, t)$. All these lengths are in a constant ratio (at successive times t), so that the distribution $P(\mathbf{x}, t)$, and with it the concentration distribution in a diffusing cloud, $\chi(\mathbf{x}, t)$ is 'self-similar', i.e., has invariant shape, with a length scale increasing linearly in time.

Returning now to our earlier arguments regarding displacement statistics we observe that the 'effective diffusivity' for ground-level release is

$$\frac{1}{2}\frac{\mathrm{d}\overline{z'^2}}{\mathrm{d}t} = b_2^2 u^{*2} t \quad (h = 0). \tag{5.20}$$

Comparing this result with Equation (5.17) we note that the Lagrangian mean square velocity $\overline{w'^2}(t)$ must itself be proportional to u^{*2}, so that the Lagrangian time-scale is proportional to diffusion time t, and therefore by Equation (5.19) also to the size of the diffusing cloud. Physically this is quite reasonable, because the eddy length scale L_t is proportional to distance from the ground and the 'effective eddy' acting on a cloud at a given stage of its growth may therefore be expected to be proportional to cloud size. Observe, however, that this is only the case for a cloud released at ground level.

For elevated source releases we find

$$z/(u^* t) = \mathrm{func}\,(h/(u^* t)) \tag{5.21}$$

and similar relationships for other characteristic lengths of the distribution $P(\mathbf{x}, t, h)$. The function of $(h/u^* t)$ may well be different for different characteristic lengths and therefore self-similarity is not likely to hold, nor is the time dependence of the length scales as simple as for ground level release. However, at sufficiently long diffusion times the argument of the unknown function in Equation (5.21) tends to zero, and the previous results are recovered. We have already seen in another context that sufficiently far downwind of an elevated source the maximum concentration shifts to ground level and the cloud behaves further as if it had been discharged at the ground. In practice, however, this is not very useful because maximum ground-level concentrations occur well before Lagrangian similarity is established. The similarity arguments are thus only useful in describing the field of a *ground-level* source and we shall proceed to a further discussion of this specific case.

5.4. Continuous Ground-Level Line Source

The above results may be exploited in discussing the time mean concentration field of the practically important *continuous* ground-level sources. Several aspects of continuous plumes have already been described in Chapters I and III and here we need only dwell upon peculiarities due to the structure of a wall layer. We shall consider first a line source coinciding with the y-axis, while x-axis is at ground level, along the mean wind. The field of the continuous source may again be thought of as built up of many successive puffs, discharged at successive times. Diffusion along x is again relatively slow and the mean concentration distribution across the line source plume, $\chi(x, z)$, may be assumed to be proportional to particle displacement probabilities $P(x, z, t)$ in a puff at an appropriate diffusion time $t = t_c$. Thus the growth of the plume along its axis is described by the relationships, for $\bar{z}$, etc. derived in the last section, but we have to replace the time variable by the distance variable x. The question is, what distance x does the puff cover in time $t = t_c$.

If the strength of the line source per unit length is q, the same amount of material has to pass each downwind cross section. The integral expressing this total transport may be used to define a convection velocity U_c:

$$q = \int_0^\infty u(z)\,\chi(x, z)\,\mathrm{d}z = U_c \int_0^\infty \chi(x, z)\,\mathrm{d}z. \tag{5.22}$$

$U_c(x)$ is clearly the bulk velocity of the material at a section x, so that the time t_c taken by a puff to reach a given downwind station x is

$$t_c = \int_0^x \frac{\mathrm{d}x}{U_c}. \tag{5.23}$$

We may now exploit our earlier results on puff-growth.

For a length scale L_c of the concentration distribution we may use the radius of inertia, or the position of the center of gravity $\bar{z}$, or another length proportional to these two. For the time being we leave the precise definition open, but note that the similarity argument of the last section requires for any specific choice, that

$$\mathrm{d}L_c/\mathrm{d}t = bu^*, \tag{5.24}$$

where $b = \text{const}$. Identifying the diffusion time t with t_c defined by Equation (5.23) we also have

$$\frac{\mathrm{d}L_c}{\mathrm{d}x} = b\,\frac{u^*}{U_c}. \tag{5.25}$$

We observe that U_c increases slowly with x, as the cloud grows, so that the slope $\mathrm{d}L_c/\mathrm{d}x$ must also slowly decrease with distance from the source.

The self-similarity of the concentration distribution may be expressed by writing

$$\chi(x, z) = \chi_0(x)\,f(z/L_c), \tag{5.26}$$

where $f(\zeta)$ (with $\zeta = z/L_c$) is conveniently defined so as to make $f(0) = 1$, in which case $\chi_0(x)$ is the ground-level concentration. Substitution of the logarithmic velocity distribution (Equation (5.7)) into Equation (5.22) causes some difficulties: the mean velocity extrapolates to zero at $z = z_0$, but the profile itself does not apply below z of order $30\,z_0$. We shall indicate the lower limit of integration as $z = 0$, merely noting that the presence of the viscous or rough sublayer introduces a certain inaccuracy into the results below, neglected terms being of the order of z_0/L_c times those retained. Where the logarithm of z/L_c enters explicitly into a result, we shall extrapolate it only to

z_0/L_c. It may be noted that similar difficulties arise with the sublayer in calculating the resistance coefficient of pipes or two-dimensional boundary layers (see e.g., Schlichting, 1960), but the resulting inaccuracies do not prove to be practically serious.

The substitution of a self-similar concentration distribution into Equation (5.22) supplies the relationship

$$U_c L_c \chi_0 = q/a_1 \quad (q \text{ in g s}^{-1}\text{ cm}^{-1}), \tag{5.27}$$

where $a_1 = \int_0^\infty f(\zeta)\mathrm{d}\zeta = \text{const.}$ Writing out the logarithmic velocity distribution explicitly we arrive at, from the same equation:

$$\begin{aligned} U_c &= \frac{1}{a_1}\int_0^\infty \left(\frac{u^*}{k}\ln\frac{z}{z_0}\right) f(\zeta)\,\mathrm{d}\zeta = \\ &= \frac{u^*}{k}\left(\ln\frac{L_c}{z_0} + \frac{a_2}{a_1}\right), \end{aligned} \tag{5.28}$$

where $a_2 = \int_0^\infty \ln\zeta f(\zeta)\mathrm{d}\zeta = \text{const.}$, if we neglect the problems arising from the presence of a sublayer. The value of a_2 may be positive or negative, depending on the exact choice of L_c: for $z < L_c$ the contributions to the integral defining a_2 are negative, for $z > L_c$ positive. For a certain specific choice of L_c we have exactly $a_2 = 0$. In the present section we shall assume this particular choice of L_c for simplicity and write

$$U_c = \frac{u^*}{k}\ln\frac{L_c}{z_0} \tag{5.29}$$

so that U_c is the mean velocity at the level L_c.

The variation of L_c with distance x may now be determined from Equations (5.25) and (5.29):

$$L_c \ln\frac{L_c}{ez_0} = kbx. \tag{5.30}$$

This is a functional relationship between L_c/z_0 and kbx/z_0 and may be regarded a universal 'law of growth' for a ground level source in a wall layer. We should once more note that in arriving at this result we have ignored the sublayer. Equation (5.3) is shown in Figure 5.2 (a) and (b) on both a linear and a logarithmic scale.

In experimental work the variation $L_c(x)$ is usually approximated by a power law:

$$L_c(x) = \text{const.}\, x^\gamma. \tag{5.31}$$

The exponent γ may be determined from a plot of $\log L_c$ vs. $\log x$. From our above results, Equations (5.25) and (5.30), we may show that

$$\begin{aligned} \gamma &= \frac{\mathrm{d}\log L_c}{\mathrm{d}\log x} = \frac{x}{L_c}\frac{\mathrm{d}L_c}{\mathrm{d}x} = \left(1 + \frac{L_c}{kbx}\right)^{-1} = \\ &= 1 - \frac{1}{\ln(L_c/z_0)}. \end{aligned} \tag{5.32}$$

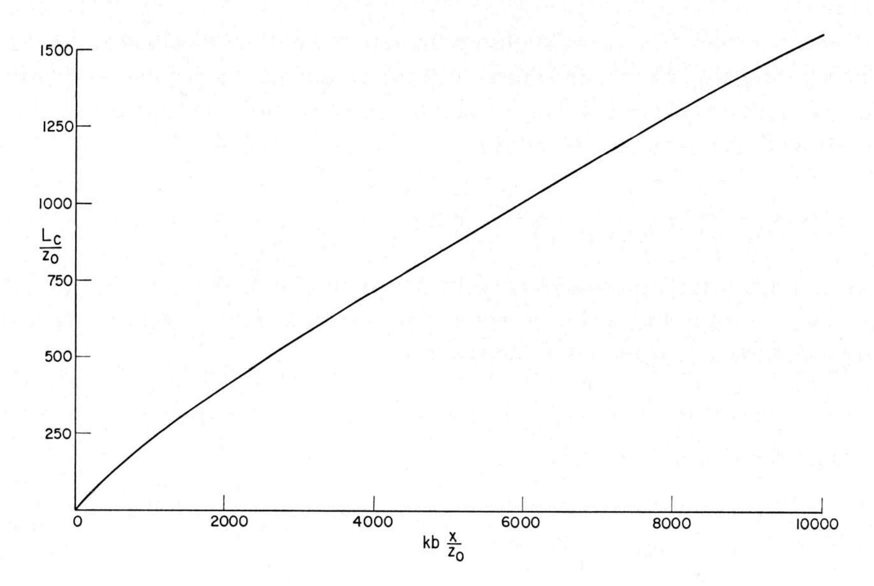

Fig. 5.2a. Universal growth law of ground level source, linear scale (10% height$\equiv$3.5L_c).

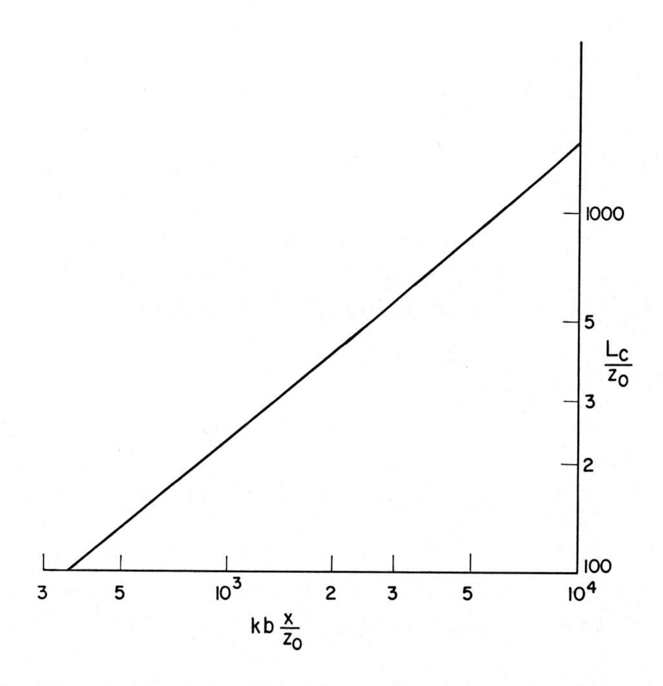

Fig. 5.2b. Universal growth law of ground level source: logarithmic scale.

In order for the wall layer relationships to apply to a diffusing cloud L_c has to be much larger than z_0. Therefore γ differs little from 1 and increases quite slowly with x to an asymptotic value of 1.0. The variation of ground level concentration χ_0 with x is now from Equations (5.27, 29 and 30):

$$\chi_0 = \frac{q}{a_1 u^* (bx + L_c/k)} = \frac{q\gamma}{a_1 b u^* x}, \tag{5.33}$$

where γ is again the exponent defined by Equation (5.32). We note that χ_0 varies nearly as x^{-1}, except that a slow increase of γ reduces the rate of decrease slightly. If we try to represent χ_0 again as a power law in x:

$$\chi_0 = \text{const.}\, x^{-\mu} \tag{5.34}$$

we find for the apparent exponent

$$\mu = -\frac{\mathrm{d}\log\chi_0}{\mathrm{d}\log x} = \frac{x}{U_c L_c}\frac{\mathrm{d}}{\mathrm{d}x}(U_c L_c) = \gamma(2-\gamma). \tag{5.35}$$

In terms of the small quantity $\Delta = 1/\ln(L_c/z_0)$ we have therefore $\gamma = 1-\Delta$, $\mu = 1-\Delta^2$.

5.5. Flux and Eddy Diffusivity

As in earlier chapters, we may add to the insight so far gained by focusing next directly on the χ-field of a continuous line source, which is subject to the constraint of continuity in differential as well as integral form (the latter was embodied in Equation (5.27)). Because we have already neglected diffusion along x, for the two-dimensional problem of line source diffusion the equation of continuity becomes

$$u\frac{\partial\chi}{\partial x} = -\frac{\partial F_z}{\partial z}, \tag{5.36}$$

where F_z is the mean (vertical) flux of marked fluid at any (x, z) due to molecular and turbulent transport. Using the postulate of self-similarity, Equation (5.26), and some other results above we can calculate F_z from Equation (5.36):

$$F_z = \frac{q}{a_1 L_c}\frac{\mathrm{d}L_c}{\mathrm{d}x}\psi(\zeta), \tag{5.37}$$

where

$$\psi(\zeta) = \zeta f(\zeta) + \Delta\zeta\ln\zeta f(\zeta) + \Delta^2\int_{z_0/L_c}^{\zeta} f(\zeta)\ln\zeta\,\mathrm{d}\zeta \tag{5.37a}$$

with $\Delta = [\ln(L_c/z_0)]^{-1}$, as before. According to our previous remarks, the logarithmic term should only be extrapolated downward to z_0/L_c and even there the expressions are inaccurate.

For self-similarity the flux, as well as the mean concentration distribution, should be the product of an x-dependent scale factor and a function of ζ. Equation (5.37a) shows that this is *not* the case, the function $\psi(\zeta)$ being also dependent on the factor Δ in a fairly complex way (and even more so if we follow through its derivation, consistently retaining the lower integration limit $\zeta_0 = z_0/L_c$ in terms involving the velocity distribution). We are forced to conclude that *strict* self-similarity at least is inconsistent with continuity in a logarithmic wall layer, bounded by a thin sublayer below. The observed self-similarity of concentration profiles (which we shall discuss below) must therefore be approximate in such a layer. Analogous cases of *approximate* self-similarity of velocity profiles in the turbulent wake and boundary layer are well known, and have been discussed, e.g., by Townsend (1956). In our case strict similarity of the concentration and flux distribution only holds for a given value of L_c/z_0, although the actual distributions apparently change only quite slowly with this variable.

These difficulties due to the presence of a sublayer and with the logarithmic velocity profile at low z/z_0, encourage the replacement of the latter by the power-law formula, Equation (5.11), for the purposes of our further discussions. A reasonably accurate correspondence between the two distributions may be obtained by an appropriate choice of the adjustable parameters entering the power law (particularly of the exponent α). The values of these parameters also depend on the height-range over which the power law is expected to simulate the logarithmic distribution. Because we are interested in the effects of the mean velocity field on a diffusing cloud of a height of order L_c, the choice of α, etc., must be such as to optimize the agreement at z = order L_c. It is then not difficult to see that α must be a function of L_c/z_0. For example, if we equate the velocity and velocity gradient at $z = h = L_c$ between power law and logarithmic formulas, we find $\alpha = (\ln L_c/z_0)^{-1}$. A less arbitrary relationship will be established below.

From the power-law profile and the hypothesis of self-similarity the convection velocity may be calculated:

$$U_c = \frac{\int_0^\infty u\chi \, dz}{\int_0^\infty \chi \, dz} = \frac{a_3}{a_1} U_h \left(\frac{L_c}{h}\right)^\alpha, \tag{5.38}$$

where

$$a_1 = \int_0^\infty f(\zeta) \, d\zeta$$

$$a_3 = \int_0^\infty \zeta^\alpha f(\zeta) \, d\zeta .$$

Note that we have not presupposed a specific choice of L_c to simplify the resulting expression for U_c, unlike in Equation (5.29). The step analogous to the earlier choice would be to assume a length scale to make $a_3 = a_1$. The choice to be used below is more convenient in dealing with experimental data.

On substituting into Equation (5.25) now and integrating we find the growth law:

$$\frac{L_c}{h} = A_1 \left(\frac{x}{h}\right)^{1/(1+\alpha)}, \tag{5.39}$$

where

$$A_1 = \left(\frac{b a_1 u^*}{(\alpha + 1)\, a_3 U_h}\right)^{1/(1+\alpha)} = \text{const}.$$

The factor u^*/U_h appearing in the constant is the square root of a local friction coefficient:

$$\frac{u^*}{U_h} = \sqrt{\frac{\tau_0}{\varrho U_h^2}} = \sqrt{c_f} \tag{5.40}$$

which is again a slow function of relative roughness, z_0/L_c.

It seems now reasonable to require for maximum correspondence with the logarithmic velocity profile model that the growth law (Equation (5.39)) should be locally similar to the more accurate Equation (5.30), i.e., that the apparent exponent γ calculated in Equation (5.32) should equal $(1+\alpha)^{-1}$. This leads to the identification

$$\alpha = \frac{1}{\ln \dfrac{L_c}{z_0} - 1} \tag{5.41}$$

which is illustrated in Figure 5.3. The ground level concentration may be expressed with the aid of Equation (5.27):

$$\chi_0 = \frac{q}{a_1 U_c L_c} = \frac{q}{b a_1 (\alpha + 1)\, u^* x}. \tag{5.42}$$

For constant α, this varies as x^{-1}, but the slow decrease of α with increasing x reduces the numerical value of the effective (negative) exponent below 1.0.

Consider now the vertical flux, which is easily shown to be

$$F_z = -\frac{\partial}{\partial x} \int_0^z u\chi \, \mathrm{d}z = -\frac{q}{a_3} \frac{\partial}{\partial x} \int_0^\zeta \zeta^\alpha f(\zeta)\, \mathrm{d}\zeta = \frac{q}{a_3 L_c} \frac{\mathrm{d}L_c}{\mathrm{d}x} \zeta^{1+\alpha} f(\zeta). \tag{5.43}$$

This result is of very much the same form as Equation (5.37). The function replacing $\psi(\zeta)$ is self-similar for constant α, but is subject to slow change as L_c/z_0 increases. We observe also in passing that a power-law velocity distribution is consistent with *strict* self-similarity.

It is further instructive to calculate the vertical eddy diffusivity

$$K_z = -\frac{F_z}{\partial\chi/\partial z} = -\frac{q}{a_3 \chi_0} \frac{\mathrm{d}L_c}{\mathrm{d}x} \zeta^{1+\alpha} \frac{f}{f'} = -\frac{a_1 b}{a_3} u^* L_c \left(\zeta^{1+\alpha} \frac{f}{f'}\right). \tag{5.44}$$

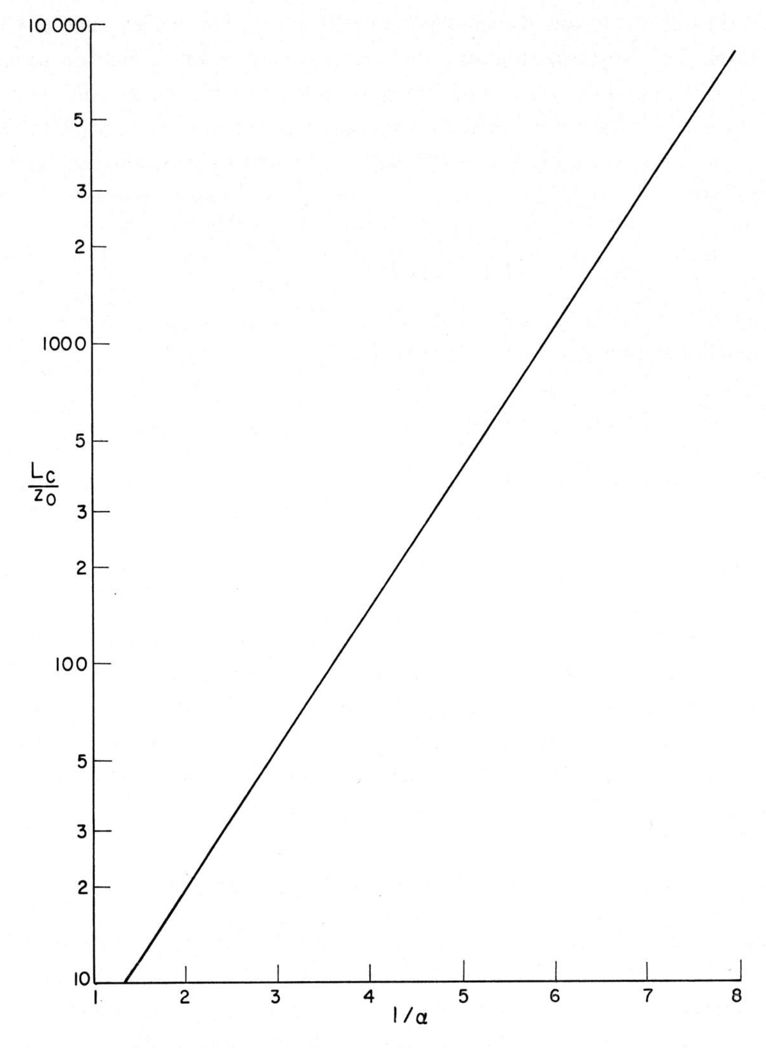

Fig. 5.3. Apparent power law exponent α.

This is again a product of an x-dependent diffusivity scale and a self-similar distribution (function of ζ). The x-dependent part may be taken to be

$$K_1(x) = bu^* L_c = \frac{1}{2}\frac{dL_c^2}{dt} \tag{5.45}$$

having used Equation (5.24). Therefore this diffusivity-scale is related to cloud size exactly as in a homogeneous field. The effective diffusion velocity is clearly proportional to u^*, the eddy size to L_c, quite consistently with our results in Section 5.3.

As to the variation with the z-coordinate of the vertical eddy diffusivity K_z, it is reasonable to expect this to be much as in an area-source diffusion problem, when marked fluid diffuses in one dimension, to or from an infinite area sink or source at ground level. The Reynolds analogy may be invoked between such an area-source diffusion problem and the vertical transport of momentum to the ground, because the boundary conditions are the same. For vertical momentum transport a momentum exchange coefficient (or eddy viscosity) may be defined by, assuming constant stress in the wall layer:

$$K_m = \frac{u^{*2}}{\mathrm{d}u/\mathrm{d}z} = \frac{u^{*2} h^{\alpha} L_c^{1-\alpha}}{\alpha U_h} \left(\frac{z}{L_c}\right)^{1-\alpha}. \tag{5.46}$$

The momentum exchange coefficient will have the same ζ-dependence as the corresponding part of K_z in Equation (5.44) provided that

$$\zeta^{1-\alpha} = -\zeta^{1+\alpha} \frac{f}{f'} \cdot \text{const}. \tag{5.47}$$

On integration we find now that

$$f(\zeta) = \exp\{-a_4 \zeta^{1+2\alpha}\} \tag{5.48}$$

with $a_4 = \text{const}$.

This already satisfies the previously imposed condition $f(0) = 1.0$. The value of a_4 depends on the exact choice of L_c. A convenient choice from the point of view of comparison with experiment is to prescribe $f(1) = 0.5$, which yields $a_4 = 0.693$, independently of α. With any other choice, a_4 also becomes a function of α. L_c is thus the 'half-height' of the cloud, the height where the concentration drops to one-half the maximum. For our previous choice of length scale made following Equation (5.28) we shall henceforth write $L_c^*(x)$.

5.6. Comparison with Experiment

The above theoretical predictions based on the postulate of Lagrangian similarity and applying to diffusion from ground-level sources differ quite markedly from the results of homogeneous-field statistical theory. With ground level release the persistence of the initial velocity should be negligible and the asymptotic stage of diffusion represented by Equation (5.17) should prevail throughout. In a homogeneous field this would imply a growth rate of $L_c = \text{const.}\ t^{1/2}$. Given the nonuniformity of the velocity distribution, this should lead to a dependence of L_c on x with a power exponent γ something less than 1/2. The same result follows from the classical diffusion equation with constant diffusivity, which was used in early attempts to represent atmospheric diffusion.

Early experimental data on atmospheric diffusion showed that the growth rate was more nearly linear, with an apparent exponent γ of something like 0.85. Sutton (1932) rationalized this finding in terms of the homogeneous field theory by introducing a rather arbitrary form of the Lagrangian autocorrelation function $R(\tau)$ which led to the correct functional form of the standard deviations $\sigma_y(x)$, $\sigma_z(x)$. As we have seen already in Chapter III, the approach has met with practical success and is still the basis of most engineering diffusion models. However, interpreting the experimental facts in the framework of Lagrangian similarity theory leads to physically rather more satisfactory conclusions: Equation (5.39) shows that the cloud should grow precisely as $x^{0.85}$ if $\alpha=0.18$ or so, which is a reasonable 'typical' exponent of the velocity distribution. Other experimental facts on diffusion from ground-level sources recorded by Sutton (1953) in his book are equally well in accord with our above results.

Field investigations are relatively difficult and more intricate features of the problem (such as the details of the vertical concentration distribution) remained essentially unknown until laboratory studies of wall-layer diffusion were carried out by Wieghardt (1948), and more systematically by Cermak and his collaborators (Poreh and Cermak, 1964; Malhotra and Cermak, 1964; Davar and Cermak, 1964). Cermak (1963) has presented a fairly detailed comparison of observed facts with Lagrangian similarity theory, including in his material atmospheric evidence, much of which was also summarized by Pasquill (1962). From this evidence one may distill the following conclusions:

(1) The concentration distribution behind a continuous, ground level line source in a wall layer is in fact self-similar within the accuracy of the observations. Some measured vertical profiles behind line sources (taken from Poreh and Cermak, 1964) are shown in Figure 5.4.

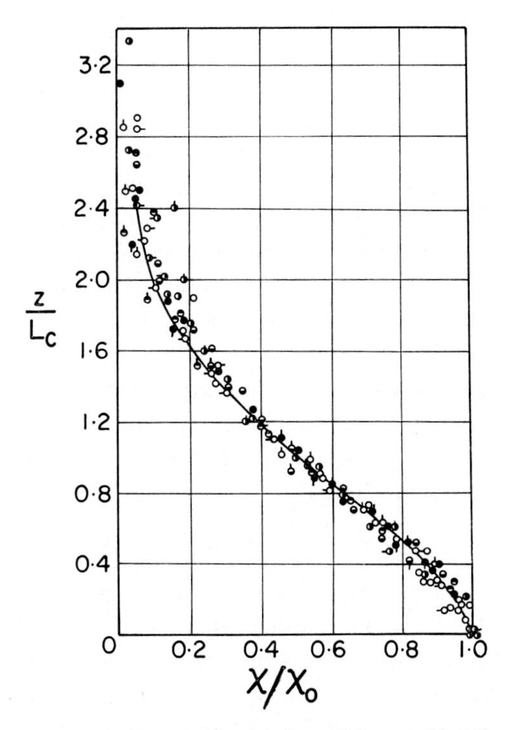

Fig. 5.4. Concentration distribution in wall layer behind line source Poreh and Cermak, 1964).

(2) The universal growth law of Figure 5.2 realistically depicts the relationship $L_c^*(x)$. This particular length scale is approximately 3.5 times less than the 10% height of the cloud, or it is some 63% of the half-height L_c (which, as we have seen, is an alternative convenient definition of a length scale). The parameter kb has an approximate value of 0.08. All these numerical data, particularly the length ratios, should be regarded as only approximate, however, because such ratios change slowly with relative roughness L_c/z_0 (in consequence of the approximate nature of Lagrangian similarity).

(3) Apparent power law exponents γ and μ have values pretty much as predicted by the theory. Specifically, their variation from one experiment to another may be satisfactorily related to differences in relative roughness L_c/z_0. One must add, however, that the exact values of the exponents γ and μ determined in one experiment are not always consistent with both Equations (5.32) and (5.35). These small discrepancies may legitimately be attributed to the experimental departures from the theoretical ideals: it is in practice difficult to carry out observations in a neutral wall layer on clouds which are large compared to the roughness elements, yet small enough to remain within the wall layer for a sufficient distance to yield accurate power-law exponents.

(4) The functional form of the self-similar distribution $f(\zeta)$ given by Equation (5.48) is precisely what was proposed empirically by Wieghardt (1948), Malhotra and Cermak (1964), and others. Malhotra and Cermak give $\alpha = 1/5.5$ and $f(\zeta) = \exp(-0.693\,\zeta^{1.4})$, in good agreement with Equation (5.48). The agreement between the empirically determined values of $(1+2\alpha)$ and the exponent of ζ in the exponential is not always so good, but exponential curves of the form of Equation (5.48) may be fitted to the observed self-similar profiles within a range of the exponent $(1+2\alpha)$ without serious loss of accuracy. One may say that Equation (5.48) fits the data within their accuracy. Therefore it is sufficiently accurate to assume that the *height dependent part* of eddy diffusivity behaves exactly as in area-source diffusion, the source coinciding with ground level. This conclusion has apparently first been arrived at by Morkovin (1965) and is a physically very illuminating one. It also justifies to some extent the use of the diffusion equation in more complex shear flow diffusion problems to which we turn in a later section.

(5) The experimental data extend to clouds generated by continuous *point* sources. These show apparent power-law exponents in accordance with the theoretical results of the next section. The concentration distribution is self-similar in two dimensions, well described by the universal function:

$$\phi(\eta, \zeta) = \exp\{-0.693(\zeta^{1+2\alpha} + a_5\eta^2)\} \tag{5.49}$$

with $\eta = y/L_c$, $\zeta = z/L_c$ and $a_5 = \text{const}$. Clearly, both height and width of the cloud are proportional to L_c: if the two were scaled differently, no self-similarity would exist. The function $\phi(\eta, \zeta)$ may be seen to be a product of the line-source universal distribution $f(\zeta)$ and a Gaussian distribution in the cross wind direction. Plots of $\phi(\eta, \zeta)$ at $\zeta = \text{const.}$ show in fact accurately similar distributions of the same width (Figure 5.5). In a sense, diffusion along y and z appears to proceed independently.

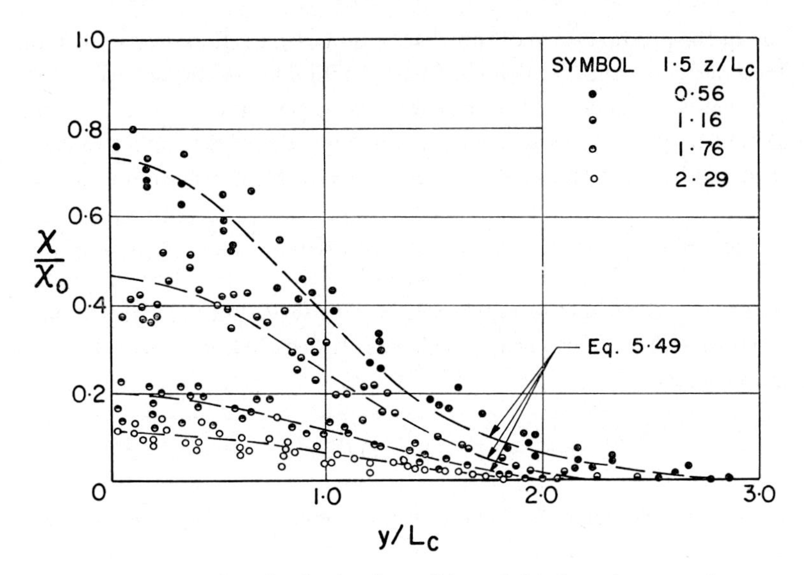

Fig. 5.5. Concentration distribution in wall layer behind continuous point source (Malhotra and Cermak, 1964).

5.7. Continuous Point Source at Ground Level

The arguments in favor of Lagrangian similarity also apply to the field of a continuous *point* source at ground level, in which case the scale width of the distribution at given diffusion time t should also be proportional to u^*t and therefore to any conveniently chosen length scale L_c of the vertical distribution. That self-similarity is in fact observed in such a situation has just been pointed out. The concentration distribution is therefore of the form

$$\chi(x, y, z) = \chi_a \phi(\eta, \zeta), \tag{5.50}$$

where $\chi_a(x)$ is the ('axial') concentration along the x-axis, and $\phi(\eta, \zeta)$ has already been defined in Equation (5.49). The integral expression of continuity becomes, in analogy with Equation (5.27)

$$q = a_6 U_c L_c^2 \chi_a \quad (q \text{ in g s}^{-1}), \tag{5.51}$$

where

$$a_6 = \iint \phi(\eta, \zeta)\, \mathrm{d}\eta\, \mathrm{d}\zeta = \text{const}.$$

The convection velocity must accordingly be defined as

$$U_c(x) = \frac{\iint u(z)\, \chi(x, y, z)\, \mathrm{d}y\, \mathrm{d}z}{\iint \chi(x, y, z)\, \mathrm{d}y\, \mathrm{d}z}. \tag{5.52}$$

In virtue of the product form of the distribution $\phi(\eta, \zeta)$ (Equation (5.49)) this yields exactly the same convection velocity as we found for a line-source cloud.

The arguments in Section 5.4 in regard to the variation of L_c with distance therefore remain exactly as for the line source, and Figure 5.2 also applies to point-source clouds. The axial concentration χ_a may then be seen from Equation (5.51) to vary as

$$\chi_a = \frac{q}{a_6 k b^2 u^* x^2} \frac{\gamma^2}{1-\gamma}, \tag{5.53}$$

where γ is again as defined in Equation (5.32). This shows that χ_a varies almost as x^{-2}. A more accurate apparent power law is also easily deduced: if $\chi_a = \text{const.}\ x^{-\nu}$, we have from our above results

$$\nu = -\frac{\mathrm{d}\ln\chi_a}{\mathrm{d}\ln x} = \gamma(3-\gamma) = (1-\Delta)(2+\Delta) = 2 - \Delta - \Delta^2, \tag{5.54}$$

where again we have written $\Delta = 1/\ln(L_c^*/z_0)$. As remarked in the previous section, this result again agrees with observation. (Sutton (1953), for example, quotes $\nu = 1.76$, corresponding to $\Delta = 0.2$, $L_c^*/z_0 = 150$, $\alpha = 0.25$, all fairly consistently with other information on Sutton's data).

Some interesting conclusions follow from a calculation of the vertical and horizontal components of flux, F_z and F_y, with the aid of the continuity equation and the empirically determined form of $\phi(\eta, \zeta)$. For the three-dimensional problem of point-source diffusion, neglecting however diffusion along x, the continuity equation is

$$\frac{\partial}{\partial x}(u\chi) = -\frac{\partial F_y}{\partial y} - \frac{\partial F_z}{\partial z}. \tag{5.55}$$

From Equations (5.49) to (5.51) we may express $(u\chi)$ as a product:

$$(u\chi) = \lambda_1(y, x)\,\lambda_2(z, x), \tag{5.56}$$

where

$$\lambda_1(y, x) = \frac{a_1}{a_6 L_c} \mathrm{e}^{-a_5\eta^2}$$

$$\lambda_2(z, x) = \frac{q}{a_1 L_c U_c} u f(\zeta)$$

and η, ζ and $f(\zeta)$ are as previously defined. The function $\lambda_2(z, x)$ is exactly what appears in the continuity equation for a line source diffusion problem; hence, if $\psi_2(z, x)$ is the vertical flux F_z for *line*-source diffusion:

$$\frac{\partial\lambda_2}{\partial x} = -\frac{\partial\psi_2}{\partial z}. \tag{5.57}$$

It is now apparent that Equation (5.55) is satisfied if the flux components are

$$\begin{aligned} F_z &= \lambda_1(y, x)\,\psi_2(z, x) \\ F_y &= \lambda_2(z, x)\,\psi_1(y, x) \end{aligned} \tag{5.58}$$

provided that

$$\frac{\partial \lambda_1}{\partial x} = -\frac{\partial \psi_1}{\partial y}. \tag{5.59}$$

On substituting the definition of $\lambda_1(y, x)$ into Equation (5.59) and integrating we arrive at

$$\psi_1(y, x) = \frac{1}{L_c}\frac{dL_c}{dx}\frac{a_1}{a_6}\,\eta\, e^{-a_5\eta^2}. \tag{5.60}$$

Therefore the distribution along z of the vertical flux component F_z is exactly as in line source diffusion, multiplied only by a scale factor and the Gaussian distribution along the y-axis. The vertical diffusivity K_z remains exactly as we have calculated in Section 5.5. On the other hand, the horizontal flux component and eddy diffusivity become

$$\begin{aligned} F_y &= u(z)\,\chi_a \frac{dL_c}{dx} f(\zeta)\,\eta\, e^{-a_5\eta^2} \\ K_y &= -\frac{F_y}{\partial\chi/\partial y} = \frac{1}{2a_5} L_c \frac{dL_c}{dx} u(z) = \frac{1}{2a_5 a_3} U_c L_c \frac{dL_c}{dx} \zeta^{\alpha}. \end{aligned} \tag{5.61}$$

The diffusivity K_y is seen to be the product of an x-dependent scale factor and a self-similar distribution, the former being of course identical with the similar scale factor multiplying K_z (Equation (4.44)). The dependence on ζ is, on the other hand, different, ζ^{α} vs. $\zeta^{1-\alpha}$. This again underlines the proposition that an eddy diffusivity is not a property of the turbulent medium in which the diffusion takes place, but of the specific diffusion problem. One should also add that the x-dependent scale factors in both K_y and K_z are more important than the ζ-dependence. If we replaced the ζ-dependent parts by constants, reasonably realistic solutions $\chi(x, y, z)$ could be calculated from the diffusion equation (e.g., $\exp(-\zeta^2)$ in place of $\exp(-\zeta^{1+2\alpha})$, no very striking change). On the other hand, ignoring the x-dependence of the scale factor yields a completely unrealistic $\chi_a(x)$ distribution.

This concludes our discussion of Lagrangian similarity. The argument that yielded the simple relations in Equation (5.19) leads to a more complicated result if another length scale is involved, such as a release height h or an Obukhov length L^* in a non-neutral wall layer. In the latter case approximate self-similarity of the concentration distribution with ground-level sources apparently remains a fact. The non-neutral velocity distribution may also be expressed by a power-law formula and we have already seen that self-similarity is consistent with such a distribution. The growth law of Figure 5.2 is, however, replaced by a more complex function, the local rate of growth depending also on L_c/L^*. Gifford (1962) has discussed this case theoretically and arrived at results quite analogous to those obtained for the neutral layer, also in reasonable agreement with observation. The empirical input required for definitive conclusion becomes rather heavy in this further extension of Lagrangian similarity theory to yield much further physical insight. For this reason we shall not pursue this topic here any further.

5.8. Use of the Diffusion Equation

The classical diffusion equation with appropriately chosen space-dependent eddy diffusivities has long been used to provide theoretical models of atmospheric and oceanic diffusion phenomena. Sutton (1953) has given a fairly detailed account of these developments, although he also shows that many erroneous conclusions may be reached by uncritical use of the diffusion equation. The basic difficulty is of course that, as we have seen repeatedly above, an eddy diffusivity is not a property of the turbulent medium, but of the diffusion process. Choice of a $K_z(z)$ or $K_y(z)$ distribution in a given situation amounts to a guess on the properties of the concentration field and may not always be an inspired one.

The fairly detailed and physically satisfactory description of diffusion in a neutral wall layer which we have obtained above may be utilized to highlight particularily clearly the limitations of all such so-called 'K-theories.' Consider again the vertical diffusivity K_z we obtained in Equation (5.44) and substitute for the ζ-dependence the consequences of Equation (5.48):

$$\begin{aligned} -\zeta^{1+\alpha}\frac{f}{f'} &= \frac{1}{a_4(1+2\alpha)}\left(\frac{z}{L_c}\right)^{1-\alpha} \\ K_z &= \frac{a_1 b}{a_3 a_4(1+2\alpha)}\,u^* L_c\left(\frac{z}{L_c}\right)^{1-\alpha}. \end{aligned} \tag{5.62}$$

As we have seen, K_z varies with z in the same manner as the momentum exchange coefficient K_m of Equation (5.46), the precise relationship being

$$K_z = A_2\left(\frac{L_c}{h}\right)^{\alpha} K_m, \tag{5.63}$$

where

$$A_2 = \frac{a_1 b\alpha}{a_3 a_4(1+2\alpha)}\cdot\frac{U_h}{u^*}.$$

The equation of continuity for the two-dimensional problem of line-source diffusion may therefore be written

$$\frac{\partial}{\partial x}(u\chi) = A_2\left(\frac{L_c}{h}\right)^{\alpha}\frac{\partial}{\partial z}\left(K_m\frac{\partial\chi}{\partial z}\right). \tag{5.64}$$

By contrast, the usual K-theory approach is to postulate $K_z = K_m$ and calculate the field of a line source from Equation (5.64) without the factor $A_2(L_c/h)^a$. The substitution of K_m for K_z is justified by an appeal to the Reynolds analogy, which is of course inapplicable between line-source mass diffusion and area-sink momentum diffusion. The usual form of the K-theory equation is recovered, however, if we formulate the problem in a (ξ, z) coordinate system, where ξ is a stretched distance variable

$$\xi = A_2\int_0^x\left(\frac{L_c}{h}\right)^{\alpha}\mathrm{d}x = A_3 h\left(\frac{x}{h}\right)^{(1+2\alpha)/(1+\alpha)}, \tag{5.65}$$

where

$$A_3 = \frac{1+\alpha}{1+2\alpha} A_1^{\alpha}$$

A_1 having been defined in Equation (5.39). In terms of this variable we have

$$\frac{\partial (u\chi)}{\partial \xi} = \frac{\partial}{\partial z}\left(K_{\mathrm{m}} \frac{\partial \chi}{\partial z}\right). \tag{5.66}$$

The transformation expressed by Equation (5.65) implies a distortion of the x-axis of relatively modest proportions (because α is usually a small quantity), so that the K-theory solution of the problem may be regarded as moderately relevant. Specifically, the calculated concentration distribution with z is correct in functional form (Sutton, 1953, p. 281, Equation (8.12)) although of course the dependence of χ on ξ is slightly different from its dependence on x. In experimental work such differences are particularly difficult to establish on account of uncertainties in the velocity profile, and in the shear stress (which affects the 'right' value of K_{m}).

In light of the above we may now propose some reasonable assumptions on the K_z-distribution which may apply to line-source diffusion behind an *elevated* source. At short distances downwind the results of homogeneous-field theory may be used, which predicts

$$K_z(x) = \overline{w'^2} t_{\mathrm{L}} (1 - \mathrm{e}^{-x/U_{\mathrm{h}} t_{\mathrm{L}}}), \tag{5.67}$$

where the Lagrangian time scale t_{L}, the mean square vertical turbulent velocity $\overline{w'^2}$ and the velocity U_{h} all relate to the release height h. It is also easy to see that $\overline{w'^2} t_{\mathrm{L}}$ is proportional to K_{m} at $z=h$. As the cloud becomes larger, the variation of K_z with z should become important and a reasonable hypothesis regarding an 'effective' K_z distribution would be

$$K_z = A_4 (1 - \mathrm{e}^{-x/U_{\mathrm{h}} t_{\mathrm{L}}}) K_{\mathrm{m}}(z), \tag{5.68}$$

where A_4=constant of order unity. The resulting concentration field could be obtained by a solution of Equation (5.66), with the stretched distance variable:

$$\xi = A_4 \int_0^x (1 - \mathrm{e}^{-x/U_{\mathrm{h}} t_{\mathrm{L}}})\, \mathrm{d}x = A_4 [x - U t_{\mathrm{L}} (1 - \mathrm{e}^{-x/U t_{\mathrm{L}}})]. \tag{5.69}$$

At large x the stretching becomes unimportant, but close to the source a concentration distribution is obtained which is much more in accordance with experiment than the solution of the diffusion equation with $\xi = x$. At large x on the other hand, the initial elevation of the source should become unimportant, and the ground-level solution should be approached. This implies the asymptotic relevance of the coordinate stretching given by Equation (5.65).

Turning now to the horizontal diffusivity K_y, we note that this varies according to Equation (5.61) exactly the same way with x as K_z *at fixed* ζ, but not at fixed z. To use the diffusion equation for three-dimensional problems (point source release, specifically) would therefore be unduly complex. On the other hand, it is also unnecessary because a Gaussian concentration distribution is observed in the cross wind horizontal direction (a direction in which the turbulent field is uniform) just as in a homogeneous field. There is no point in trying to calculate some minor perturbation of the Gaussian profile from K-theory, which is in itself some sort of an ad-hoc model of turbulent diffusion, of probably lesser accuracy than a Gaussian distribution.

We conclude that solutions of the two-dimensional diffusion equation with $K_z = K_m$ are likely to provide some insight into certain effects of shear flow on diffusion, if we regard our x-coordinate as freely rescaleable. However, the method is inherently an approximation and only the use of the simplest possible mathematical models should have practical merit.

5.9. Elevated Sources

As an example for the use of the diffusion equation we shall discuss the case of an elevated line source in a wall layer. We take the diffusion equation as

$$u(z)\frac{\partial \chi}{\partial x} = \frac{\partial}{\partial z}\left(K_z \frac{\partial \chi}{\partial z}\right) \tag{5.70}$$

with $u = U_h(z/h)^\alpha$, $K_z = K_h(z/h)^{1-\alpha} = (c_f/\alpha)U_h h$ (c_f = friction coefficient $= u^{*2}/U_h^2$).

The source is situated at $x = 0$, $z = h$. Introduce the nondimensional variables

$$\begin{aligned} \zeta &= \frac{z}{h} \qquad \xi = \frac{K_h x}{U_h h^2} = \frac{c_f}{\alpha}\frac{x}{h} \\ C &= \frac{\chi U_h h}{q} \qquad (q = \text{source strength, g cm}^{-1}\,\text{s}^{-1}). \end{aligned} \tag{5.71}$$

(q = source strength, g cm^{-1}s^{-1}.)

The solution of the equation may be obtained by transform methods (Smith, 1957). It is:

$$C = \frac{\zeta^{\alpha/2}}{(2\alpha+1)\,\xi}\exp\left\{-\frac{1+\zeta^{1+2}}{(2\alpha+1)^2\,\xi}\right\} I_{-\alpha/(2\alpha+1)}\left\{\frac{2\zeta^{(2\alpha+1)/2}}{(2\alpha+1)^2\,\xi}\right\}, \tag{5.72}$$

where $I_\nu(\,)$ is the Bessel function of imaginary argument of fractional order ν. As $\zeta \to 0$ (i.e., at ground level) this solution reduces to

$$\begin{aligned} C_0 &= \frac{(1+2\alpha)^{-1/(1+2\alpha)}}{\Gamma\left(\dfrac{1+\alpha}{1+2\alpha}\right)\xi^{(1+\alpha)/(1+2\alpha)}}\exp\left\{-\frac{1}{(1+2\alpha)^2\,\xi}\right\} = \\ &= \frac{a}{\xi^m}\,e^{-b/\xi} \quad (\text{where } a,\ b,\ m = \text{const.}). \end{aligned} \tag{5.73}$$

Because of the uncertainties affecting the x-coordinate, the main interest of the solution lies in its prediction of a vertical concentration distribution. Figure 5.6 shows this for different values of ξ. The locus of the maximum descends rapidly to ground level, as we have already seen in simpler models.

It is of interest to note that these concentration distributions differ little from one predicted by the homogeneous field theory ('Sutton model', Gaussian concentration distribution with mirror image term, see Section 3). Figure 5.7 shows a comparison of Equation (5.72) (labeled 'Smith model') and the Sutton model with experimental data. Both are clearly equally good, but we may note that the fitting of the first set of points requires a rather lower source height in the Sutton model than in the Smith model. The effect may simply be an aberration, but if it is real, it arises presumably from a shear-flow induced depression of the locus of maximum χ, which has to be allowed for empirically in the Sutton model. Apart from this minor point one may indeed conclude that the variations of wind speed and eddy size with height have no really important effects on diffusion, except insofar as they affect the rate of cloud growth (which would appear as some kind of distortion of the ξ-axis).

Close to the source the distortion of the ξ-axis is certain to be serious. As an approxi-

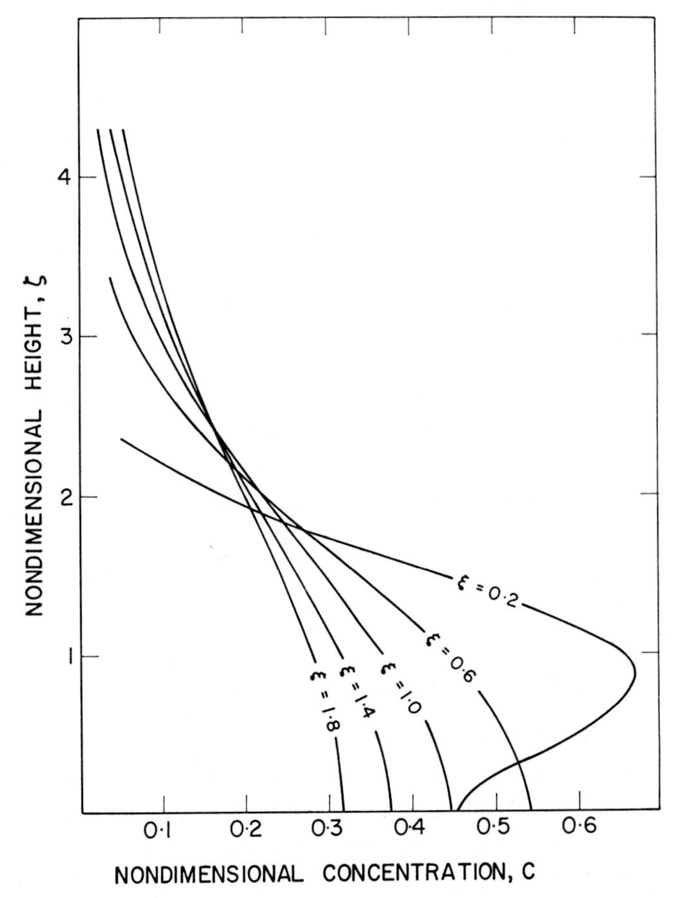

Fig. 5.6. Development of concentration profile behind elevated source (Csanady *et al.*, 1968).

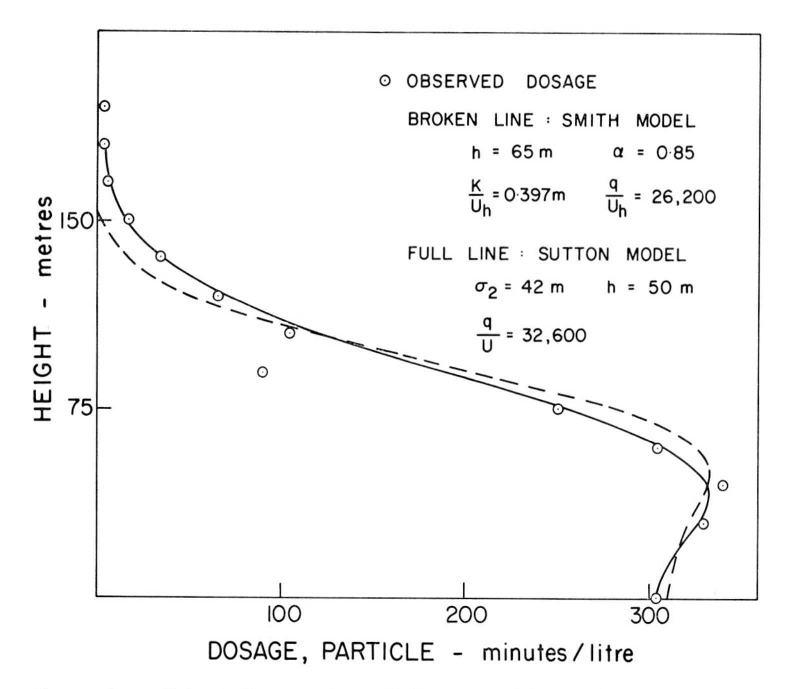

Fig. 5.7a. Comparison of elevated source theoretical models with experimental data (dosage caused by passage of instantaneous line source cloud) (Csanady *et al.*, 1968).

mation, we may use the transformation indicated in Equation (5.69). An easily observable feature of plume behavior is the locus z_m of the maximum concentration at a fixed x, calculated from $\partial\chi/\partial z = 0$. Figure 5.8 illustrates the locus $z_m(\xi)$ calculated from the diffusion equation and its transformation by an equation similar to Equation (5.69). The resultant modified path $z_m(x)$ is precisely of the character observed in laboratory experiments by Davar and Cermak (1964). An example of these is shown in Figure 5.9. We may indeed conclude that judicious use of the diffusion equation can lead to a fairly accurate description of the concentration distribution. Furthermore our results justify the approximate use of homogeneous field theory already described in Chapter III and supplies the physical insight for the understanding of observed cloud growth rates.

5.10. Longitudinal Dispersion in Shear Flow

By concentrating on diffusion from continuous sources we have so far avoided the need to analyze the x-wise growth of a cloud in a field where the x-component of the velocity was nonuniform. To rectify this omission, consider the growth along x of a puff in parallel shear flow, already illustrated in Figure 5.1. The nonuniformity of the velocity field distorts the shape of the puff, the top portion of it traveling further in a given time than the bottom. Turbulent eddies continually mix marked fluid vertically,

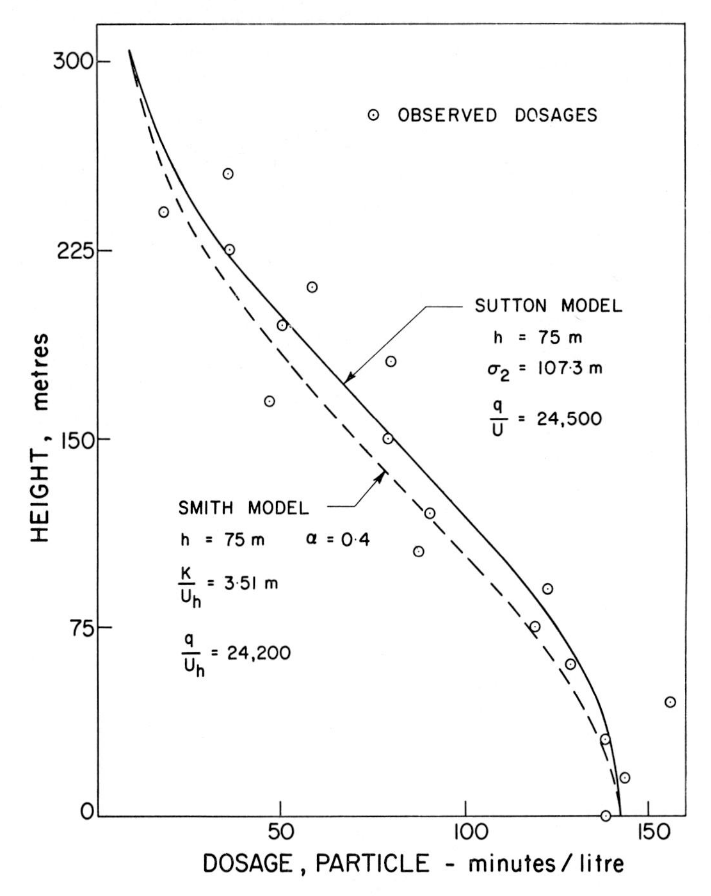

Fig. 5.7b. As Figure 5.7a, another experiment.

producing a *longer* cloud at any z than would be the case in a flow field of uniform velocity. We conclude that a combination of shear flow and cross-flow turbulent diffusion leads to a new form of dispersal (sometimes called turbulent 'dispersion' rather than 'diffusion'). This combined effect was first discussed by Taylor (1953) for dispersion in laminar flow in a pipe, i.e., a combination of shear flow and *molecular* diffusion.

The corresponding phenomenon in turbulent flow may also be treated theoretically with the aid of the diffusion equation, if due note is taken of the approximate nature of this approach. Indeed this appears to be one of the principal benefits to be derived from 'K-theory.' We shall discuss the shear-flow dispersion problem below with the aid of the so-called 'concentration moment' method due to Aris (1956), following mainly an article by Saffman (1962).

We assume that the spreading of marked fluid from an instantaneous point or line source is adequately described by the diffusion equation with z-dependent velocity

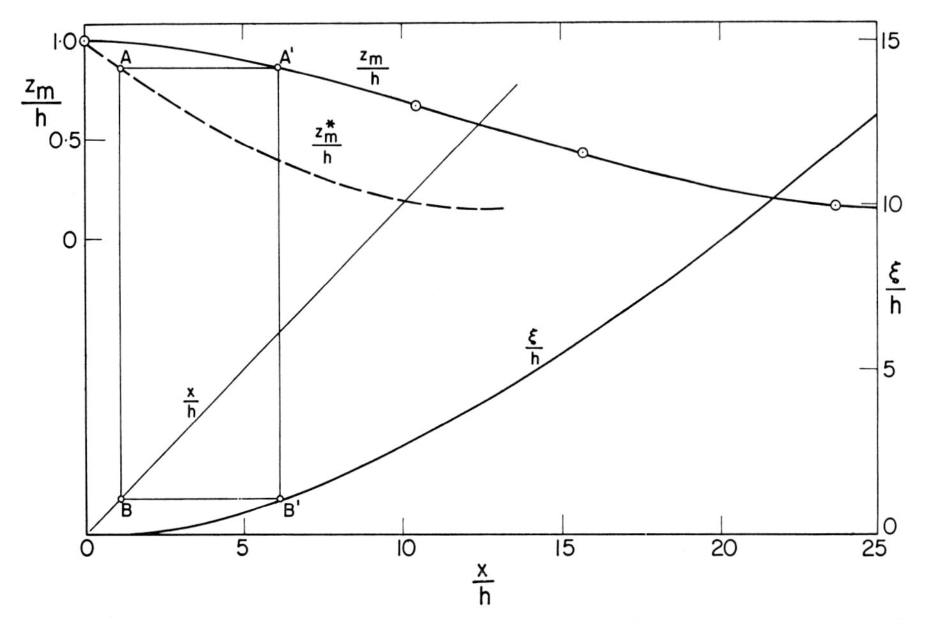

Fig. 5.8. Transformation of K-theory results in the neighborhood of a concentrated source: $z_m{}^*/h =$ locus of maximum concentration given by K-theory; $z_m/h =$ transformed locus. The transformation is indicated by the curve ξ/h, the steps being illustrated by the construction of point A′ on the transformed curve from a given point A on the K-theory curve.

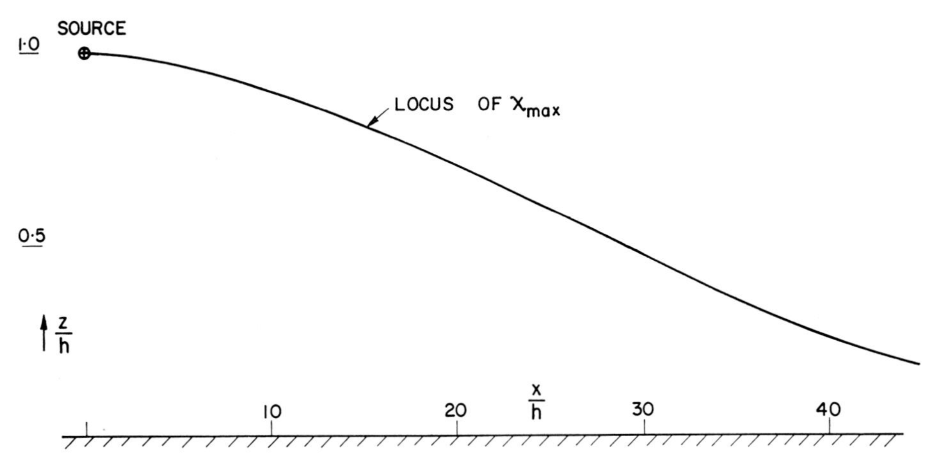

Fig. 5.9. Locus of maximum concentration for elevated source plume in the laboratory (Davar and Cermak, 1964). The boundary layer was $3h$ thick in this experiment, so that the plume was diffusing mainly (but not entirely) in the wall layer.

$u(z)$ and eddy diffusivities K_x, K_y, K_z:

$$\frac{\partial \chi}{\partial t} + u\frac{\partial \chi}{\partial x} = \frac{\partial}{\partial x}\left(K_x \frac{\partial \chi}{\partial x}\right) + \frac{\partial}{\partial y}\left(K_y \frac{\partial \chi}{\partial y}\right) + \frac{\partial}{\partial z}\left(K_z \frac{\partial \chi}{\partial z}\right). \tag{5.74}$$

For arbitrary distributions $u(z)$, $K_x(z)$, etc., it is not possible to find analytical solutions of this equation. However, use may be made of the fact that these quantities are functions of the z-coordinate alone, by investigating integral moments of the χ-distribution defined as follows:

$$\theta_{\mathrm{nm}}(z, t) = \int_{-\infty}^{\infty} \int_{-\infty}^{\infty} x^n y^m \chi(x, y, z, t)\, \mathrm{d}x\, \mathrm{d}y \tag{5.75}$$

with $n, m = 0, 1, 2 \ldots$. A differential equation for θ_{00} may be obtained by integrating Equation (5.74) over all x, y:

$$\frac{\partial \theta_{00}}{\partial t} = \frac{\partial}{\partial z}\left(K_z \frac{\partial \theta_{00}}{\partial z}\right). \tag{5.76}$$

The quantity θ_{00} is clearly the total amount of material which has reached a given level z, and this is seen to be subject to the one-dimensional diffusion Equation (5.76) with a certain $K_z(z)$. It is worth noting that θ_{00} is independent of $u(z)$: although the distortion of the cloud in a situation illustrated in Figure 5.1 obviously increases vertical concentration gradients, it does so for both positive and negative gradients and the net effect on total vertical transport at any level is zero.

The first moment θ_{10} defines the position of the centroid of the distribution at given z and t and is subject to the equation (which we obtain by multiplying Equation 5.74 by x and integrating over the x, y plane):

$$\frac{\partial \theta_{10}}{\partial t} - \frac{\partial}{\partial z}\left(K_z \frac{\partial \theta_{10}}{\partial z}\right) = u\theta_{00}. \tag{5.77}$$

This is again a one-dimensional diffusion equation but it also contains a source-term, $u(z)\theta_{00}(z, t)$. In a given problem one first solves Equation (5.76) for θ_{00}, then Equation (5.77) for θ_{10}. The x-coordinate of the centroid of a cloud-slice is

$$c_{\mathrm{x}}(z, t) = \theta_{10}/\theta_{00}. \tag{5.78}$$

Multiplying Equation (5.74) now by x^2 and integrating we arrive at the equation for the second moment θ_{20}:

$$\frac{\partial \theta_{20}}{\partial t} - \frac{\partial}{\partial z}\left(K_z \frac{\partial \theta_{20}}{\partial z}\right) = 2K_{\mathrm{x}}\theta_{00} + 2u\theta_{10} \tag{5.79}$$

which is again a one-dimensional diffusion equation with two source-terms. The x-wise spread of the cloud is conveniently characterized by the standard deviation $\sigma_{\mathrm{x}}(z, t)$ where

$$\sigma_{\mathrm{x}}^2 = \theta_{20}/\theta_{00} - c_{\mathrm{x}}^2. \tag{5.80}$$

When the release takes place from a *point* source, the y-moments θ_{01}, θ_{02}, etc., are also of interest, as may be the cross moments θ_{11}, etc. In parallel, horizontally uniform flow, however, the symmetry of the arrangement requires $\theta_{01} = 0$, $\theta_{11} = 0$, etc., while

θ_{02} is subject to:

$$\frac{\partial\theta_{02}}{\partial t} - \frac{\partial}{\partial z}\left(K_z \frac{\partial\theta_{02}}{\partial z}\right) = 2K_x\theta_{00}\,. \tag{5.81}$$

Thus θ_{02} is determined by a simpler source term than θ_{20}. The lateral spread may be characterized by a standard deviation σ_y:

$$\sigma_y^2 = \theta_{02}/\theta_{00} \tag{5.82}$$

the y-coordinate of the center of gravity c_y being zero. Similar equations may be developed for higher moments θ_{30}, etc., but their practical significance seems to be relatively small.

Equations (5.76), (5.77), (5.79) above embody the concentration moment method. For given $u(z)$, $K_x(z)$ etc., distributions and specified boundary conditions they may be solved in succession and as much information as desired built up on the concentration field. The method's relative simplicity rests on the fact of having a one-dimensional diffusion equation to deal with. By, e.g., Laplace transform methods this may be reduced to an ordinary differential equation which is much simpler to solve than the partial differential Equation (5.74).

5.11. Shear-Augmented Diffusion in a Channel

The simplest and practically most important application of the above equations is to diffusion in a 'channel' or 'sandwich,' such that the vertical flux vanishes at $z=0$ and $z=h$. In the atmosphere this phenomenon occurs in the presence of an inversion lid or ceiling at height h, which stops upward transport of pollutants by means of turbulence suppression. Below the ceiling there is a mixed layer in which clouds of marked fluid may drift and diffuse horizontally. A similar mixed layer often occurs in the sea above the thermocline. At a long enough time t after release a point or line-source cloud distributes itself vertically uniformly over the mixed layer, and its further drift and diffusion proceed along the x-and y-axes. In this *asymptotic* stage therefore the particularly simple solution of Equation (5.76) applies:

$$\theta_{00} = \text{const.} = q/h\,, \tag{5.83}$$

where q is the total amount of material released. The bulk motion of the entire cloud may be characterized by the averaged first moment:

$$\hat{\theta}_{10} = \frac{1}{h}\int_0^h \theta_{10}\,\mathrm{d}z\,. \tag{5.84}$$

From Equation (5.77) one finds for this quantity

$$\frac{\partial\hat{\theta}_{10}}{\partial t} = \frac{1}{h}\int_0^h u\theta_{00}\,\mathrm{d}z = (q/h)\,\hat{u} = \text{const.}\,, \tag{5.85}$$

where

$$\hat{u} = \frac{1}{h}\int_0^h u \, dz$$

is the layer-average velocity. Note that the circumflex generally shall denote such layer averages. From Equation (5.85) we see that the cloud drifts bodily along with constant velocity $\hat{u}$. The centroid at any level z is, however, displaced relative to the average position $\hat{u}t$ by a certain distance which may be calculated from Equation (5.77). Let us write

$$\theta_{10}(z, t) = q/h[\hat{u}t + \phi(z)]. \tag{5.86}$$

We find now from Equation (5.77) that

$$\phi(z) = \int_{z_1}^{z} \frac{dz}{K_z} \int_0^z (\hat{u} - u) \, dz \tag{5.87}$$

with the lower limit z_1 being defined so as to give

$$\int_0^h \phi(z) \, dz = 0. \tag{5.87a}$$

The centroid of a given horizontal cloud slice is therefore at

$$c_x = \theta_{10}/\theta_{00} = \hat{u}t + \phi(z).$$

The individual slices are stacked upon one another displaced as specified by the function $\phi(z)$, but they drift along as a fixed assembly at the velocity $\hat{u}$. The relative displacement of cloud slices is the physical reason for increased longitudinal growth according to our earlier intuitive argument and indeed the function $\phi(z)$ will be seen to enter the expression describing the longitudinal expansion of the cloud.

In order to analyze the spread of the cloud along the x-axis we define an averaged second moment:

$$\hat{\theta}_{20} = \frac{1}{h}\int_0^h \theta_{20} \, dz. \tag{5.88}$$

According to Equation (5.80) this is related to a layer-average variance as follows:

$$\widehat{\sigma_x^2} = \frac{1}{h}\int_0^h \sigma_x^2 \, dz = \frac{h}{q}\hat{\theta}_{20} - \hat{u}^2 t^2. \tag{5.89}$$

On integrating Equation (5.79) across the whole layer $z=0$ to h we find

$$\frac{d\hat{\theta}_{20}}{dt} = 2\frac{q}{h}\left(\hat{K}_z + \hat{u}^2 t + \frac{1}{h}\int_0^h u\phi \, dz\right). \tag{5.90}$$

From the last two relationships now

$$\frac{d\widehat{\sigma_x^2}}{dt} = 2\hat{K}_x + \frac{2}{h}\int_0^h u\phi \, dz \tag{5.91}$$

or integrating with respect to time:

$$\widehat{\sigma_x^2} = 2t(\hat{K}_x + K_e) \tag{5.92}$$

where

$$K_e = \frac{1}{h}\int_0^h u(z)\,\phi(z)\,dz. \tag{5.92a}$$

Equation (5.92) shows that the layer-average variance grows as in molecular diffusion with diffusivity $\hat{K}_x + K_e$. We legitimately identify K_e as 'effective' diffusivity due to shear. It may now also be shown without difficulty that the variance σ_x^2 at a given level z is the sum of $\hat{\sigma}_x^2$ and a z-dependent distribution:

$$\sigma_x^2 = \hat{\sigma}_x^2 - \phi^2(z) - \int_0^z \frac{dz}{K_z}\int_0^z \psi(z)\,dz, \tag{5.93}$$

where

$$\psi(z) = 2(K_x - \hat{K}_x) + 2(u - \hat{u})\,\phi - \frac{2}{h}\int_0^h (u - \hat{u})\,\phi\,dz.$$

In other words, the length of horizontal slices into which a puff may be cut up also changes with z, but the difference between the sizes at two given levels remains the same in the course of diffusion. We emphasize that all these results hold only for the asymptotic stage, once θ_{00} has become vertically uniformly distributed.

The formula for shear diffusivity Equation (5.92a) with $\phi(z)$ as defined in Equation (5.87), was first derived by Taylor on the basis of a more intuitive argument and verified experimentally for diffusion in a pipe. Its content is best illuminated by specific choices of the distributions $u(z)$ and $K_z(z)$. If we assume a power-law velocity distribution (Equation (5.11)) and the Reynolds analogy, $K_z = K_m$, with K_m as given by Equation (5.46) we find

$$K_e = \frac{\alpha^3}{(\alpha+1)^3(\alpha+2)(3\alpha+2)}\frac{U_h^3 h}{u^{*2}} = \frac{B}{c_f}U_h h \tag{5.94}$$

with $B = $ const., $c_f = $ friction coefficient. For $\alpha = 0.15$ the value of the constant B is 4.22×10^{-4}. The friction coefficient c_f is of a similar order of magnitude, which means that K_e is of the order of $U_h h$, i.e., typical *mean* velocity times channel depth. By contrast, the eddy diffusivity K_x is of order *turbulent* velocity times channel depth and is therefore about one order of magnitude smaller. Because σ_x is proportional to

$\sqrt{(K_x + K_e)}$, we conclude that shear-augmented diffusion in a two-dimensional channel results in a cloud some 3 or 4 times longer than it would be due to turbulent diffusion alone. The exact details vary only slightly with the value of the exponent α.

An experimental study of Elder (1959) provided very good confirmation for the formula in Equation (5.92a). Figure 5.10 shows one of Elder's observed dye-patches in a horizontal plane, clearly showing the elongation of the patch in the direction of flow, with an axis ratio of the order of 3. (The diffusion in the cross-flow direction is of course due to turbulence alone and we may assume this to be about the same as it would be in the x-direction, if the mean flow difference were not there to elongate the patch.)

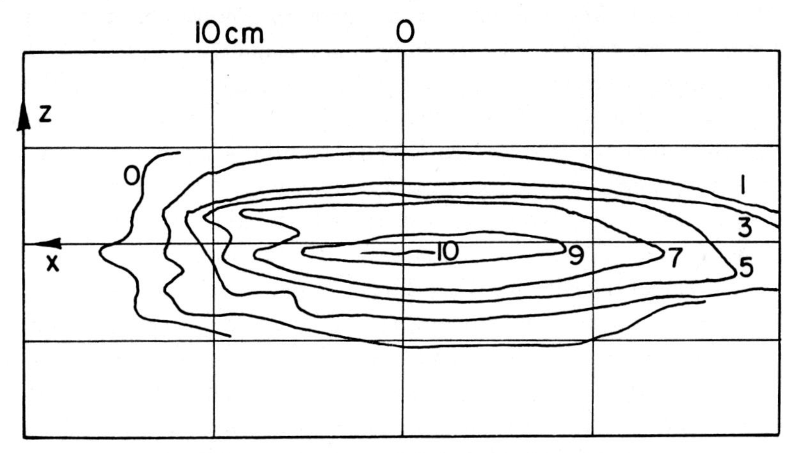

Fig. 5.10. Plan view of a drop of dye diffusing in the turbulent flow in an open channel. Distribution of concentration C, normalized to have a maximum of 10. The flow is to the left. $h = 1.43$ cm, $x/h = 90$ (Elder, 1959).

The same theoretical argument has been applied by Bowden (1965) to the problem of longitudinal mixing in an estuary. In estuaries there is usually a considerable vertical density gradient (a stable one) and this results in particularly low vertical diffusivities K_z. The presence of K_z in the *denominator* in Equation (5.87) leads in such a situation to an even higher K_e/K_x ratio (calculated from Equation (5.92a) than we have deduced from Equation (5.94) for the case of K_x and K_z equal in order of magnitude. Bowden shows that the diffusivity by Equation (5.92a) agrees with observations if the vertical diffusivity K_z is assumed of order 20 times lower than K_x or K_y. Given a high positive Richardson number this is certainly not unreasonable, but it should be pointed out that a similar increase in K_e (as compared to the two-dimensional channel formula with $K_x \approx K_z$) may also be caused by *horizontal* variations of velocity, to the consideration of which we are turning in the next section.

5.12. Dispersion in Natural Streams

In the natural environment the formula Equation (5.92a) finds its probably most important practical application in dispersion predictions in natural streams. As shown by Fisher (1968) and others, longitudinal diffusion in rivers depends mainly on the mean velocity differences which exist *horizontally* across the stream, the physical mechanism being exactly as described above. In a typical laboratory channel (such as that used by Elder in his experiments) the depth is constant and intentionally made much smaller than the width of the channel, so that the side-wall influences are effectively eliminated. Diffusion observations are only carried out on clouds not in contact with the side walls. Mean velocity variations affecting a diffusing cloud are then only those existing along the vertical direction z, velocities being constant along y. In a natural stream, by contrast, the depth and velocity vary continuously and widely horizontally across the stream and we may expect these variations to have much the same physical effects as the vertical velocity changes in a channel.

The combination of velocity differences along both the horizontal and the vertical may be expected to complicate further the shear flow diffusion problem. Fortunately, it is not difficult to see that the *vertical* velocity variations in a natural stream are of subordinate importance, on account of the usually large width to depth ratio. Natural streams are usually well mixed from top to bottom and eddy diffusivities in all directions K_z, K_y and K_z are of the same order of magnitude. Above we have seen (Equation (5.94)) that the effective shear diffusivity due to velocity variations across a channel is proportional to channel depth h. If we apply the same reasoning to horizontal velocity variations alone, we arrive at an effective shear diffusivity proportional to channel width, b. If the ratio b/h is large, the shear effect due to horizontal velocity variations may be expected to be dominant. One would then expect to be able to ignore vertical velocity variations. Indeed it transpires that the formula in Equation (5.92a) suitably applied to horizontal velocity changes in rivers gives estimates of effective diffusivity in reasonable agreement with observation.

Therefore, in order to simplify the analysis of longitudinal diffusion in rivers, it is legitimate to ignore vertical velocity variations and eliminate them from the analysis by introducing a vertically integrated transport as the variable replacing the velocity in formula Equation (5.92a):

$$U(y) = \int_{-h}^{0} u(y, z)\,\mathrm{d}z, \tag{5.95}$$

where the horizontal cross-stream coordinate is y, the surface $z = 0$ is the free surface and $h(y)$ is the depth of the river. Let the width of the river be b, and its cross-sectional area A:

$$A = \int_{0}^{b} h(y)\,\mathrm{d}y. \tag{5.96}$$

The key mixing coefficient is now K_y for which empirical data show that

$$K_y(y) = 0.23\, u^* h, \tag{5.97}$$

where $u^*(y)$ is the friction velocity at a given position y, which is in practice usually replaced by a cross-sectional average value. Let the circumflex now denote width averages, e.g.

$$\hat{U} = \frac{1}{b}\int_0^b U(y)\,dy. \tag{5.98}$$

The definition of a function $\phi(y)$ analogous to Equation (5.87) is now

$$\phi(y) = \int_{y_1}^{y} \frac{dy}{K_y h} \int_0^y (\hat{U} - U)\,dy, \tag{5.99}$$

where y_1 is chosen so as to give a zero integral of $\phi(y)$ across the entire stream.
The effective eddy diffusivity for longitudinal dispersion becomes then

$$K_e = \frac{1}{A}\int_0^b U(y)\,\phi(y)\,dy. \tag{5.100}$$

If typical values of the velocity and depth are U and h, it is not difficult to show from the last result, in analogy with Equation (5.94) that

$$K_e = B\,\frac{b^2}{h^2}\cdot\frac{U}{u^*}\cdot Uh, \tag{5.100a}$$

where $B =$ const., the order of magnitude of which is again about 10^{-3}. Using Equation (5.97) we may also write

$$\frac{K_e}{K_y} = \frac{B}{0.23}\cdot\frac{b^2}{h^2}\cdot\frac{1}{c_f}, \tag{5.100b}$$

where $c_f = u^{*2}/U^2$ is a friction coefficient based on the velocity scale U. We observe that K_e may be several hundred times K_y given that the width to depth ratio of rivers is often 10 or more.

It has been shown by Fisher (1968), Thackston and Krenkel (1967) and others that Equation (5.100) gives good predictions of effective longitudinal diffusivity in natural streams with quite complex velocity distributions. It bears repetition to emphasize, however, that this formula only applies after the marked fluid has distributed itself over the entire available cross section of the river.

5.13. Shear-Augmented Dispersion in Unlimited Parallel Flow

The above relatively simple results were obtained for the asymptotic stage of diffusion, after a cloud has occupied the entire available depth or width of a channel or stream. At much shorter times, or for flow fields large in comparison with the size of a cloud the effect of the distant boundary(-ies) is not yet felt and longitudinal dispersion behaves in a fundamentally different way. The calculations however become more difficult and some specific mean velocity and diffusivity distribution must be adopted to elucidate the behavior of diffusing clouds.

Saffman (1962) has given a solution for the following simple case: instantaneous point or line source release at ground level into a vertically semi-infinite atmosphere in which the velocity varies linearly with height while the diffusivity is constant:

$$\begin{aligned} &u = \lambda z \quad (\lambda = \text{const.}) \\ &K_z = \text{const.} \quad K_x = \text{const.} \\ &\text{boundary conditions: } \partial\chi/\partial z = 0 \quad (z = 0) \\ &\qquad\qquad\qquad\qquad\quad \chi \to 0 \quad (z \to \infty). \end{aligned} \tag{5.101}$$

The solution of the zeroth moment Equation (5.76) simply yields a Gaussian distribution, in accordance with the previously noted independence of the θ_{00} distribution of the velocity profile, and the assumed constancy of K_z:

$$\theta_{00}(z, t) = \sqrt{\frac{2}{\pi}} \frac{q}{\sigma_z} e^{-z^2/2\sigma_z^2}, \tag{5.102}$$

where

$$\sigma_z^2 = 2K_z t. \tag{5.102a}$$

The centroids of thin cloud slices, however, move rather more independently at different levels than in the asymptotic phase described in the previous section, and also their longitudinal radius of inertia σ_x grows in a different way. For the simple model of Equation (5.101) the distribution of $c_x(z, t)$ and $\sigma_x(z, t)$ may be expressed in terms of parabolic cylinder functions. Of greatest interest is the centroid and radius of inertia at ground level, which are given by:

$$\begin{aligned} c_x(0, t) &= \frac{1}{4}\sqrt{\frac{\pi}{2}}\, \lambda t \sigma_z \\ \sigma_x^2(0, t) &= \left(\frac{7}{6} - \frac{\pi}{32}\right) \lambda^2 t^2 \sigma_z^2 + 2K_x t \end{aligned} \tag{5.103}$$

(σ_z as defined by Equation(5.102a)).

Thus the progress of the cloud, described by the movement of $c_x(0, t)$ observable at ground level, is as if it were moving with a convection velocity equal to $u(z)$ at $z = 0.33\,\sigma_z$. Its longitudinal growth shows a more complex behavior: at short times t it is dominated by the $2K_x t$ term, i.e., it is indistinguishable from its growth in the absence of shear. At long diffusion times, by contrast, the term dependent on λ dominates the expression for σ_x, and to a good approximation

$$\sigma_x(0, t) \cong 0.136\, \lambda t \sigma_z \quad (\lambda t \gg 1). \tag{5.103a}$$

This and Equation (5.102a) show that σ_x grows as $t^{3/2}$ or much faster than σ_z. Indeed the ratio σ_x/σ_z (or σ_x/σ_y; for a point source, σ_y behaves essentially as σ_z) tends to infinity linearly with time, so that diffusing clouds become more and more elongated in the direction of the flow as they drift downstream.

The behavior of σ_z according to Equation(5.102a) (and a similar behavior for σ_y, given $K_y = \text{const.}$) is of course unrealistic at short diffusion times. However, we may regard this blemish of the theory removable by an appropriate stretching of the time axis, exactly analogous to the stretching of the x-axis we discussed earlier for continuous sources. Such a transformation of the time variable would not affect our previous conclusion regarding the dominance of the shear-diffusion term, except insofar as from what point on Equation(5.103a) becomes valid. Whatever stretching of the time axis is necessary, cloud growth along x in a field with unlimited variation of the x-component of the velocity remains very much faster than growth due to turbulence alone.

In Chapter IV we have examined experimental evidence on horizontal diffusion in the sea and concluded that patches of marked fluid grow rather faster than they should at large diffusion times in a homogeneous field of turbulence. The conventional explanation in terms of relative diffusion theory (in a homogeneous field) requires that we postulate the presence of an inexhaustible supply of increasingly larger eddies. In other words, we must regard large-scale flow features in the sea as parts of geophysical 'turbulence', which is really stretching a point.

As our results above show, the functional form $\sigma_x(t)$ in a flow field with linear $U(z)$ (or $U(y)$) distribution is asymptotically identical (by a coincidence) with the form of $s_x(t)$ in *relative* diffusion in a homogeneous field, provided that the active diffusing eddies belong to the inertial subrange. Without attributing too great a significance to this coincidence, we may at least conclude that an alternative explanation for cloud growth decisively faster than $t^{1/2}$ is the non-uniformity of the *mean* velocity distribution affecting the cloud. In invoking this explanation we must postulate only that the spatial variations in the *mean* velocity component in question (in the direction of observed faster than $t^{1/2}$ growth and falling within the range of the cloud) increase as the cloud expands. In the sea, both horizontal radii of inertia σ_x and σ_y, of a diffusing cloud appear to grow in this manner. It seems reasonable to attribute these observed facts to spatial variations in both components of the mean horizontal velocity.

For a horizontally infinite field with a linear velocity variation an exact solution of the diffusion equation is available, providing more detail than the moment method. Carter and Okubo (1965) have considered the following simple model of oceanic diffusion: a mean velocity distribution given by

$$u(y) = U_0 + \gamma y \tag{5.104}$$

with $U_0, \gamma = \text{const.}$; horizontal diffusion of a patch assumed to be subject to the equation

$$\frac{\partial \chi}{\partial t} + (U_0 + \gamma y)\frac{\partial \chi}{\partial x} = \frac{\partial}{\partial x}\left(K_x \frac{\partial \chi}{\partial x}\right) + \frac{\partial}{\partial y}\left(K_y \frac{\partial \chi}{\partial y}\right). \tag{5.105}$$

Also assuming $K_x = K_y = K = \text{const.}$, a simple solution of Equation (5.105) is

$$\chi(x, y, t) = \frac{q}{2\pi\sigma_x\sigma_y} \exp\left\{-\frac{(x - U_0 t - \frac{1}{2}\gamma t)^2}{2\sigma_x^2} - \frac{y^2}{2\sigma_y^2}\right\}, \tag{5.106}$$

where

$$\begin{aligned} \sigma_y^2 &= 2Kt \\ \sigma_x^2 &= \sigma_y^2 (1 + \tfrac{1}{12}\gamma^2 t^2). \end{aligned} \tag{5.106a}$$

The behavior of σ_x is therefore in accordance with Equation (5.103), although the constant factor multiplying γt is different. The field of an instantaneously released cloud as given by Equation (5.106) may also be regarded as applying to relative diffusion (because the diffusion equation may be used with the same justification and with the same caveats as in absolute diffusion). This model may therefore be used to represent diffusion of a patch of dye in the sea, if the linear velocity distribution is at all a realistic approximation of the flow field.

A continuous plume model in such a flow field may be built up by integrating many patches released at successive times. Okubo and Karweit (1969) illustrate the resulting plume: two of their illustrations are shown here in Figures 5.11 and 5.12. Although the plume has more complex characteristics than in a homogeneous field in uniform flow, the maximum concentration behaves much as in the absence of velocity shear, i.e. decreases at large distances as $\chi_a = \text{const. } x^{-1}$ because cross-current diffusion is as in the absence of shear. The more striking effects of shear-flow diffusion are reserved for the direction of a spatially nonuniform velocity, but in a continuous plume the x-component diffusion is of little importance.

5.14. Diffusion in Skewed Shear Flow

The distinctive feature of shear-augmented diffusion is that it is produced by the non-

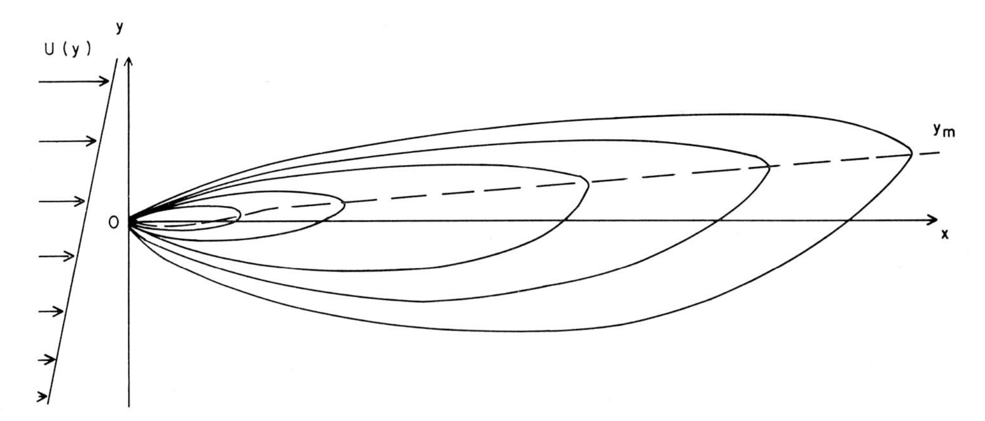

Fig. 5.11. Lines of equal concentration in continuous plume in shear flow (Okubo and Karweit, 1969).

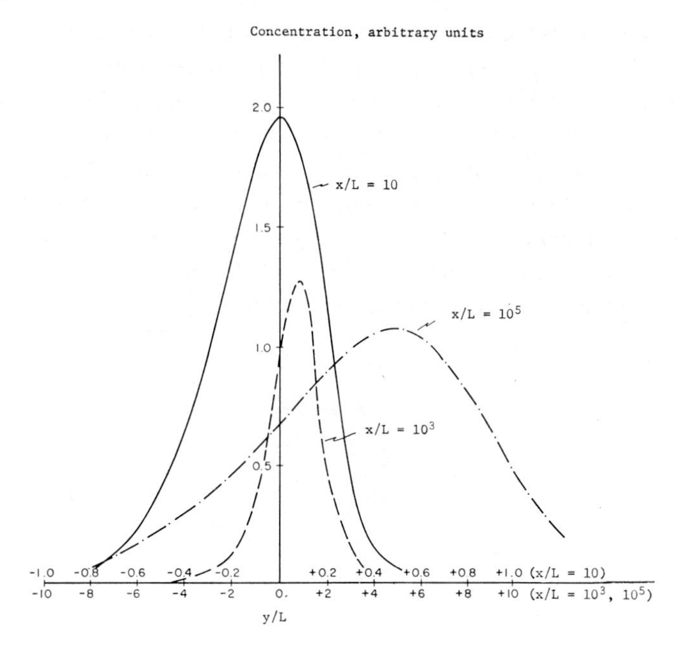

Fig. 5.12. Concentration distribution across plume in shear flow (Okubo and Karweit, 1969).

uniformity in space of a given velocity component and that the growth-enhancement is confined to the direction of that velocity component. In one-directional shear flow thus the increased growth occurs along the x-axis, laid along the mean wind, which is only important for the growth of puffs but rather unimportant for continuous plumes, the width of which remains more or less unaffected by the shear flow.

When both horizontal components of the flow vary in space (for example with the vertical coordinate) it is intuitively evident that diffusion along both horizontal axes will be enhanced by shear. We have already seen that in the outer portion of the planetary boundary layer the direction as well as the magnitude of the horizontal velocity vector changes with height. Boundary layers with this kind of structure are sometimes referred to as 'skewed'. A similar, Ekman-type skewed boundary layer also occurs at the sea surface, where the velocity distribution is as illustrated in Figure 5.13.

When a continuous plume is produced in this kind of skewed shear flow, the convection velocity changes with distance from the source (as the plume expands vertically) in direction as well as magnitude, so that the plume center line appears curved. Diffusion is enhanced by shear both along and across such a plume. The greater effective diffusivity along the plume direction is of little consequence, as always in 'slender' plumes, but the enhanced rate of plume growth across its width means enhanced dilution (a more rapid drop-off of concentration along the axis of the plume) and is therefore practically important.

Experimental evidence on lateral diffusion in skewed shear flow (obtained in the surface layers of Lake Huron) was discussed in Csanady (1966). The information was collected in experiments with continuous plumes of fluorescent dye. The growth rate of the plume and the rate at which the center line concentration dropped could be related to the angular divergence of flow in the top few meters. Figure 5.14 shows an illustration of effective lateral diffusivity in such surface-released plumes vs. angle included by current vectors at the surface and at a depth of 0.5 m. Numerical estimates of effective shear diffusivity in this case must be based on variations of the cross-flow mean velocity component $V(z)$. Using the formula in Equation(5.92a) for this component reasonably realistic estimates could be made, probably because the vertical expansion of a diffusing cloud is effectively limited by the thermocline (a 'diffusion

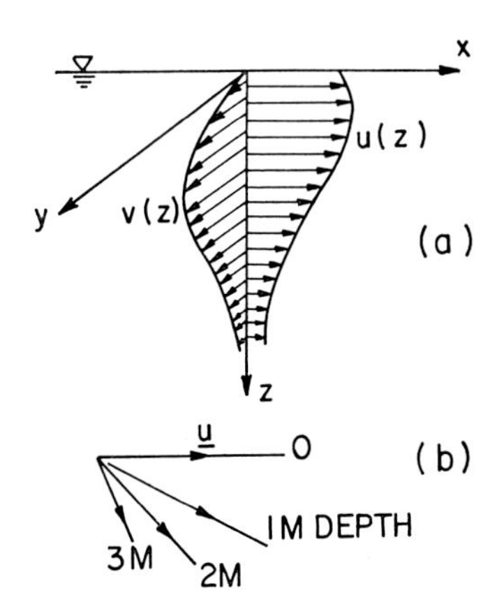

Fig. 5.13. Velocity distribution in typical lake currents: (a) velocity profiles, and (b) hodograph.

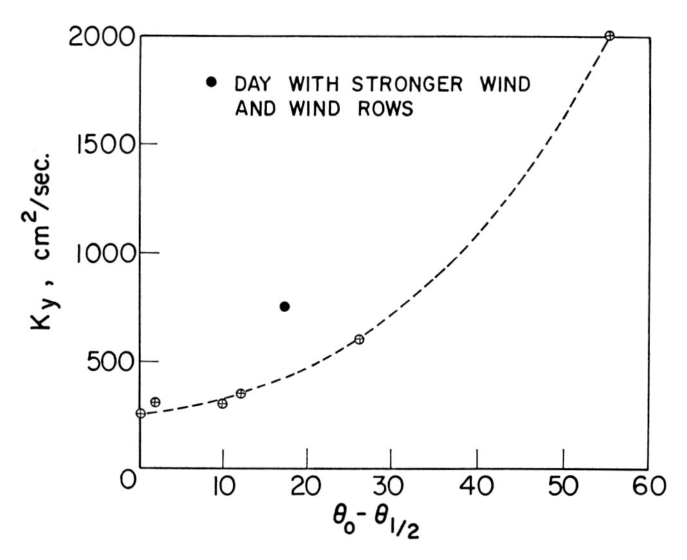

Fig. 5.14. Dependence of apparent horizontal diffusivity on angular divergence of top layers (1964 experiments).

floor'). However, velocity distributions in the surface layers of a large lake or the ocean are quite complex and not well understood and it is difficult to take the analysis much beyond such qualitative arguments.

We are in a somewhat better position to analyze diffusion in the atmospheric Ekman layer (outer part of the planetary boundary layer) for which, under conditions of uniform potential temperature, the 'classical' Ekman theory provides a qualitatively correct model. In this model a constant eddy viscosity $K = K_z$ is assumed and Equation (5.1) solved to give the velocity distributions

$$\begin{aligned} u &= u_g (1 - e^{-z/h}) \cos z/h \\ v &= v_g e^{-z/h} \sin z/h , \end{aligned} \tag{5.107}$$

where the x-axis has been laid along the geostrophic wind and h is the scale depth of the Ekman layer:

$$h = \sqrt{\frac{2K}{f}} . \tag{5.108}$$

As may be seen from Equation(5.107), the velocity tends to the constant u_g at heights large compared to h, so that all velocity variations are effectively confined to a layer of finite thickness. When a cloud is instantaneously released at ground level into this flow field, it is free to grow vertically to a height larger than h, but the total mean velocity differences acting on it remain bounded, unlike in a model with a linear velocity distribution. Without further investigation we cannot say what the asymptotic behavior of such a cloud is because neither Equation(5.92a) nor (5.103a) apply.

The problem may be attacked by the concentration moment method already described, using constant eddy diffusivities, K_z being assumed equal to the eddy viscosity K and applying, for example, a Laplace transform method. The algebra is relatively cumbersome (Csanady, 1969) so that we shall only quote the results as they relate to the ground-level path and spread of a point-cloud released at ground level.

The analysis yields the coordinates of the center of gravity of the cloud, scaled by the length

$$L_e = 2\pi u_g / f . \tag{5.109}$$

A 'typical' magnitude of this length at mid-latitudes is 1000 km. The calculated path of the cloud's center of gravity at ground level is illustrated in Figure 5.15, showing y/L_e vs x/L_e. The surface wind direction is along $y = x$, and the cloud follows this initially, but departs to the right and asymptotically becomes a parabola remaining in the sector between surface wind and geostrophic wind. The constant K-model is of course unrealistic on two counts: (1) near the origin it overestimates cloud growth, and (2) it overestimates the angle of surface wind/geostrophic wind by about a factor of 2. On the first count, the actual cloud behavior may be recovered by an appropriate stretching of the time axis, which means that the cloud follows the surface wind for a longer period. On the second count, the departure from surface wind direction will in

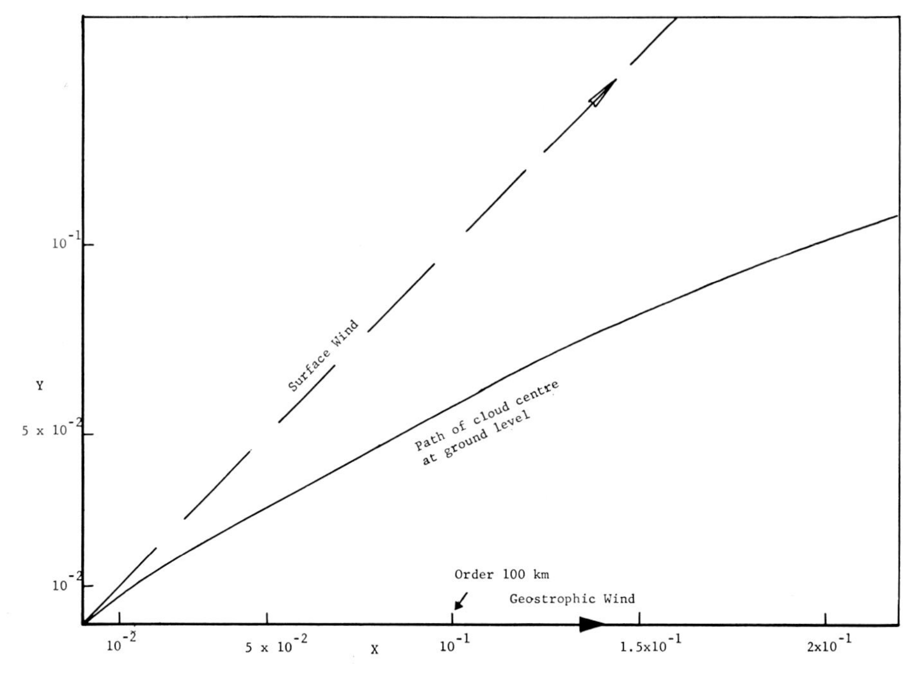

Fig. 5.15. Ground-level track of center of gravity of a diffusing cloud (Csanady, 1969).

fact be less than calculated, but very likely the cloud still remains permanently in the sector between the two wind vectors.

The spread (rms dispersion) of the cloud in the y-direction (perpendicular to the geostrophic wind) may also be calculated by the moment method and the results are illustrated in Figure 5.16, showing σ_y/h vs.

$$\gamma = (2\pi)^{-1/2} \frac{\sigma_z}{h}$$

with (5.110)

$$\sigma_z = \sqrt{2Kt}\,.$$

The parameter γ is thus a ratio of vertical cloud size to Ekman depth. Figure 5.17 shows σ_y/h vs. nondimensional diffusion time, $\tau = Kt/h^2$, also indicating 'typical' corresponding values of actual diffusion time. Generally, these results show an initial phase of cloud behavior in which σ_y depends on turbulent diffusion alone ($\sigma_y{}^2 \cong 2K_y t$). As the cloud grows, the shear effect quickly becomes dominant and produces a behavior akin to that found by Saffman in an unbounded atmosphere with linear velocity variation, (cf., Equation (5.103a)), i.e.,

$$\left(\frac{\sigma_y}{h}\right)^2 = 0.037\,\frac{u_g^2 K t^3}{h^4} = 0.0185\,\frac{u_g^2 \sigma_z^2}{h^4}\,. \tag{5.111}$$

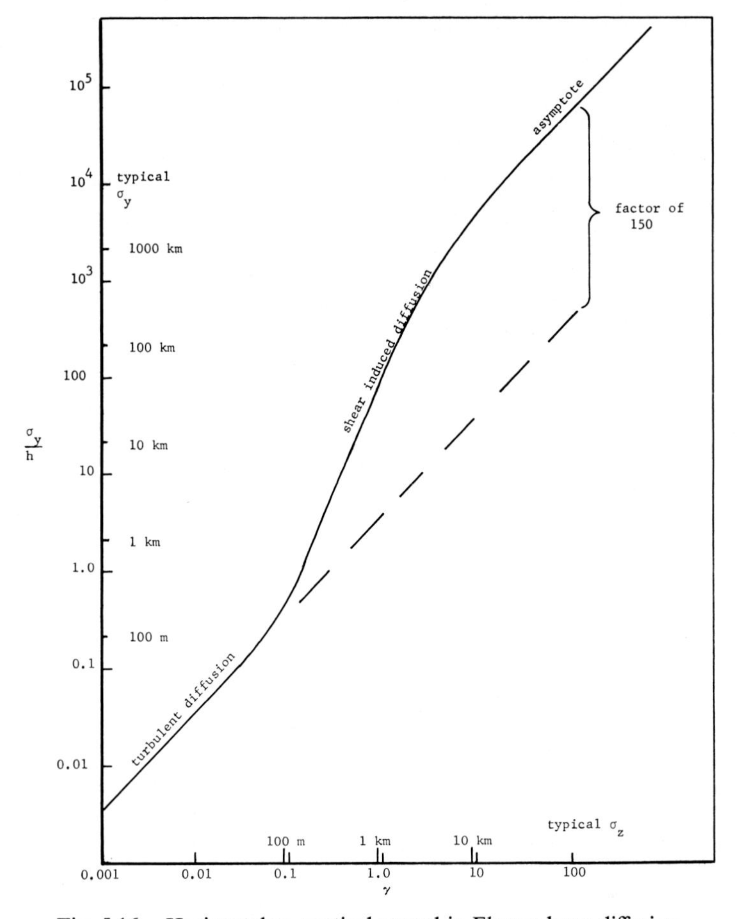

Fig. 5.16. Horizontal vs. vertical spread in Ekman-layer diffusion.

As the cloud becomes large compared to the Ekman depth, however, its further growth becomes very much as in a layer of constant height (cf. Equation (5.94)):

$$\sigma_y^2 = 2K_e t \quad (t \to \infty), \tag{5.112}$$

where the effective shear diffusivity is

$$K_e = 0.2146 \frac{u_g^2}{f}. \tag{5.113}$$

Under 'typical' conditions K_e is about three orders of magnitude greater than K_y. A point to be noted is that the effective diffusivity is *not* dependent on the vertical diffusivity $K_z = K$ (in a layer of constant thickness we recall that $K_e \propto K_z^{-1}$), the dependence being cancelled out by the fact that the Ekman depth is also determined by K.

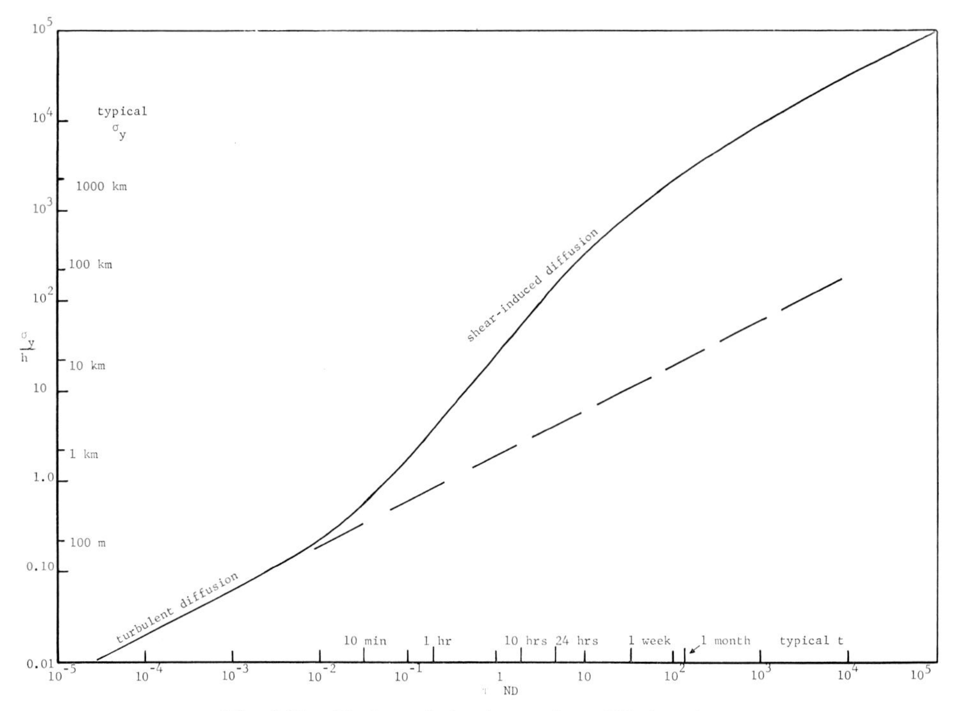

Fig. 5.17. Horizontal cloud spread vs. diffusion time.

A recent analysis of large scale diffusion data by Randerson (1972) shows that the above theoretical conclusions are in good qualitative accord with observation.

References

Aris, R.: 1956, *Proc. Roy. Soc. London* **A235**, 67.
Batchelor, G. K.: 1959, *Adv. Geophys.* **6**, 449.
Bowden, K. F.: 1965, *J. Fluid Mech.* **21**, 83.
Businger, J. A., Wyngaard, J. C., Izumi, Y., and Bradley, E. F.: 1971, *J. Atmospheric Sci.* **28**, 181.
Carter, H. H. and Okubo, A.: 1965, 'A Study of the Physical Processes of Movement and Dispersion in the Cape Kennedy Area', Cheasepeake Bay Inst., Johns Hopkins Univ. Ref. 65-2, 150 pp.
Cermak, J. E.: 1963, *J. Fluid. Mech.* **15**, 49.
Clauser, F. H.: 1956, *Adv. Appl. Mech.* **4**, 1.
Csanady, G. T.: 1966, *J. Geophys. Res.* **71**, 411.
Csanady, G. T.: 1969, *J. Atmospheric Sci.* **26**, 414.
Csanady, G. T., Hilst, G. R., and Bowne, N. E.: 1968, *Atm. Env.* **2**, 273.
Davar, K. S. and Cermak, J. E.: 1964, *Int. J. Air Wat. Poll.* **8**, 339.
Elder, J. W.: 1959, *J. Fluid Mech.* **5**, 544.
Fisher, H. B.: 1968, *J. Sanit. Eg. Div. ASCE* **94**, 927.
Gifford, F. A.: 1962, *J. Geophys. Res.* **67**, 3207.
Lumley, J. L. and Panofsky, H. A.: 1964, *The Structure of Atmospheric Turbulence*, Interscience Publishers, New York, 239 pp.
Malhotra, R. C. and Cermak, J. E.: 1964, *Int. J. Heat Mass Transfer* **7**, 169.
Monin, A. S.: 1970, *Ann. Rev. Fluid Mech.* **2**, 225.
Morkovin, M. V.: 1965, *Int. J. Heat Mass Transfer* **8**, 129.

Okubo, A. and Karweit, M. J.: 1969, *Limnol. Oceanog.* **14**, 514.
Panofsky, H. A.: 1968, in F. V. Hansen and J. D. Shreve (eds.) *Symposium on the Theory and Measurement of Atmospheric Turbulence and Diffusion in the Planetary Boundary Layer*, Sandia Laboratories, Albuquerque, N.M., p. 47.
Pasquill, F.: 1962, *Atmospheric Diffusion*, D. van Nostrand Co., New York, 297 pp.
Plate, E. J.: 1971, *Aerodynamic Characteristics of Atmospheric Boundary Layers*, U.S. Atomic Energy Comm., 190 pp.
Poreh, M. and Cermak, J. E.: 1964, *Int. J. Heat Mass Transfer* **7**, 1083.
Randerson, D.: 1972, *J. Appl. Meteorol*, to be published.
Saffman, P. G.: 1962, *Quart. J. Roy. Meteorol. Soc.* **88**, 382.
Schlichting, H.: 1960, *Boundary Layer Theory*, McGraw-Hill Book Co. New York, 647 pp.
Smith, F. B.: 1957, *J. Fluid Mech.* **2**, 49.
Sutton, O. G.: 1932, *Proc. Roy. Soc.* **A135**, 143.
Sutton, O. G.: 1953, *Micrometeorology*, McGraw-Hill Book Co., New York, 333 pp.
Taylor, G. I.: 1953, *Proc. Roy. Soc. London* **A219**, 186.
Thackston, E. L. and Krenkel, P. A.: 1967, *J. Sanit. Eng. Div. ASCE* **93**, 67.
Townsend, A. A.: 1956, *The Structure of Turbulent Shear Flow*, Cambridge University Press, 315 pp.
Wieghardt, K.: 1948, *Z. Angew. Math. Mech.* **28**, 346.

CHAPTER VI

EFFECTS OF DENSITY DIFFERENCES ON ENVIRONMENTAL DIFFUSION

6.1. Introduction

Heating or cooling of fluid usually causes density changes which in turn affect body forces acting on fluid elements, notably the force of gravity. A particle warmer than its environment usually becomes lighter and experiences an Archimedean buoyancy force, while a cooler and therefore heavier parcel becomes negatively buoyant. Such phenomena affect environmental diffusion in three different ways:

(1) By influencing the energy balance of turbulent motions. Buoyancy forces may produce or absorb turbulent energy and are therefore a factor in determining the intensity of atmospheric and oceanic turbulence.

(2) By influencing the velocity history of wandering fluid elements. Once a particle is lighter or heavier than its environment, it becomes subject to systematic vertical accelerations (upward or downward) and its velocity-autocorrelation is affected, with appropriate consequences for the diffusion of clusters of particles.

(3) By influencing the bodily motion of larger hot or cold clouds, such as those discharged from factory chimneys. Buoyant plumes are known to rise and remove pollutants from ground level.

The last one is an obvious effect; that the first two interfere with the dispersal of pollutants should also be clear from our earlier discussions. In the present chapter we shall systematically discuss these effects, relying mostly on the equations of motion with a buoyancy term included. As a prelude to our later discussions we shall at first review the fundamental equations and their simplifications appropriate to atmospheric and oceanic problems.

6.2. Fundamental Equations

The equations of fluid motion are derived from a consideration of the balance of mass, momentum and energy for an arbitrary 'control volume' of the fluid. They are conveniently stated in suffix notation using the summation convention (see e.g., Batchelor, 1967). The density ϱ, hydrostatic pressure P, absolute temperature T and velocity components u_i $(i=1, 2, 3)$ are considered to be functions of the position coordinates x_i $(i=1, 2, 3)$ and of time t. The principle of the conservation of mass yields the continuity equation:

$$\frac{\partial \varrho}{\partial t} + \frac{\partial}{\partial x_j}(\varrho u_j) = 0, \tag{6.1}$$

where the repeated suffix j implies summation over the three terms, $j=1, 2, 3$.

The application of the law of motion leads to three component momentum equations ($i=1, 2, 3$):

$$\frac{\partial}{\partial t}(\varrho u_i) + \frac{\partial}{\partial x_j}(\varrho u_i u_j) =$$

$$= \varrho X_i - \frac{\partial P}{\partial x_i} + \frac{\partial}{\partial x_j}\left\{\mu\left(\frac{\partial u_i}{\partial x_j} + \frac{\partial u_j}{\partial x_i}\right) - \tfrac{2}{3}\mu\frac{\partial u_k}{\partial x_k}\right\}. \tag{6.2}$$

In these equations X_i are the components of any body forces per unit mass (e.g., gravitational acceleration) and μ is dynamic viscosity which may be also a function of position.

The law of the conservation of energy is most concisely stated as an equation for the specific entropy S:

$$\varrho T\left(\frac{\partial S}{\partial t} + u_j\frac{\partial S}{\partial x_j}\right) = \varrho\Phi + \frac{\partial}{\partial x_j}\left(k\frac{\partial T}{\partial x_j}\right), \tag{6.3}$$

where Φ is the rate at which mechanical energy is converted into heat within the fluid by viscosity (per unit mass) and k is thermal conductivity. The relationship of entropy to pressure and temperature is

$$\mathrm{d}S = \frac{c_p}{T}\,\mathrm{d}T - \frac{\beta}{\varrho}\,\mathrm{d}P, \tag{6.4}$$

where c_p is specific heat at constant pressure and

$$\beta = -\frac{1}{\varrho}\left(\frac{\partial\varrho}{\partial T}\right)_P$$

is the thermal expansion coefficient of the fluid.

The density appears in the above equations in a number of places so that its variations may be expected to affect the flow in quite complex ways. In atmospheric and oceanic applications such variations are caused by changes in pressure, temperature and in the composition of the fluid. For a fluid of constant composition, an 'equation of state' describes the dependence of density on pressure and temperature. For atmospheric air the 'perfect' gas equation may be used:

$$P/\varrho = RT, \tag{6.5}$$

where $R=c_p-c_v$ is the gas constant. For gases obeying the perfect gas law the value of the volume expansion coefficient is $\beta=1/T$. For liquids the relationship is more complex, but for small changes in temperature (and moderate changes in pressure) a linear relationship of density to temperature is often assumed, while the effects of pressure are ignored:

$$\varrho - \varrho_0 = -\varrho_0\beta(T - T_0) \tag{6.6}$$

with $\beta \cong$ const., appropriately chosen for the range of temperature in question. Here ϱ_0 and T_0 are reference density and temperature. The last relationship is quite poor for fresh water near its temperature of maximum density (4 °C), where a quadratic relationship gives much better results (i.e., β varying linearly with temperature). Changes in composition may be accommodated by appropriate changes in R and β in Equations (6.5) and (6.6). Significant changes of composition occur in estuaries and other brackish bodies of water where the salinity varies with location and in the lowest layers of the atmosphere where specific humidity may be fairly high and variable with altitude or horizontal coordinates.

6.3. Approximate Forms of the Equations

For air and water at ordinary temperatures the numerical value of the expansion coefficient β is quite small, of the order of 10^{-3} to 10^{-4} per °C. For modest temperature variations therefore (say those less than 10 °C) the changes of density are 1% or less and can be neglected in most places in the above equations. The changes in viscosity, conductivity and specific heat with temperature are of a similar order so that these coefficients may then also be treated as constants. The resulting simplification of the equations of continuity, motion and energy is known as the 'Boussinesq approximation' (see e.g., Chandrasekhar, 1961).

An important qualification to the neglect of small density changes is that buoyant accelerations produced by them must be taken into account because they are comparable to other important acceleration terms. The acceleration of gravity is large compared to most other accelerations and is balanced largely by a strong vertical static pressure gradient in both the atmosphere and the ocean. When even a small fraction of the force of gravity is unbalanced, it can initiate fluid motions of considerable intensity. Accordingly the Boussinesq approximation consists of ignoring density changes in the above fundamental equations (and also changes of viscosity, conductivity, etc), *except* in the external force term.

To develop the approximate forms of the basic equations we assume that gravity is the only body force of importance and lay the x_3 axis along the vertical, positive upward. In suffix notation the body force X_i is then

$$X_i = -g\delta_{i3}, \tag{6.7}$$

where δ_{ij} is the Kronecker delta. Most of this force is balanced by a hydrostatic pressure gradient which has nothing to do with any buoyant motions. Let the mean density of the environment at a given level x_3, away from any specific buoyant parcel we may be analyzing, be $\varrho_a(x_3)$. Then a pressure distribution is established in accordance with the hydrostatic equation (Equation (6.2) with $u_i = 0$):

$$\mathrm{d}P_s/\mathrm{d}x_3 = -g\varrho_a. \tag{6.8}$$

In consequence of small density differences and the ensuing motions the actual pressure P in and near a buoyant parcel differs from P_s by an amount known as the 'reduced' pressure p:

$$p = P - P_S. \tag{6.9}$$

The gradient of the reduced pressure is comparable to the other acceleration terms in Equation (6.2). Subtracting Equation (6.8) from (6.2), setting $\varrho \cong \varrho_0$ in all terms except the body force term, and dividing through by ϱ_0 we find that perturbations on the basic state of hydrostatic equilibrium are governed by:

$$\frac{\partial u_i}{\partial t} + u_j \frac{\partial u_i}{\partial x_i} = -\frac{1}{\varrho_0}\frac{\partial p}{\partial x_i} + \frac{\varrho - \varrho_a}{\varrho_0} X_i + \nu \nabla^2 u_i, \tag{6.10}$$

where ϱ_0 is a convenient reference density and $\nu = \mu/\varrho_0$ is kinematic viscosity. The reference density may, for example, be the equilibrium density of the environment at the origin, $\varrho_0 = \varrho_a(0)$, where absolute static pressure is $P_0 = P_S(0)$, the absolute temperature of the environment at the same level being $T_0 = T_a(0)$.

Setting $\varrho = \varrho_0 = \text{const.}$ in the continuity equation (Equation (6.1)) we obtain the well known form

$$\partial u_i / \partial x_i = 0. \tag{6.11}$$

The viscous dissipation term in the energy equation only becomes significant at high velocities and may be neglected in the applications we are considering. If the motions take place at essentially constant static pressure (i.e. if they are confined to a sufficiently thin layer of atmosphere or ocean), changes of entropy with pressure may be neglected and Equation (6.4) replaced by

$$\mathrm{d}S \cong \frac{c_p}{T}\,\mathrm{d}T. \tag{6.4a}$$

With these simplifications the energy equation becomes

$$\frac{\partial T}{\partial t} + u_j \frac{\partial T}{\partial x_j} = \kappa \nabla^2 T, \tag{6.12}$$

where $\kappa = k/\varrho_0 c_p$ is 'thermal diffusivity.'

When large elevation differences may be involved in the buoyant motions under study, the $\mathrm{d}P$ term in Equation (6.4) may not be neglected. By far the largest portion of pressure changes is then caused by vertical displacements and in Equation (6.4) we may replace $\mathrm{d}P$ by $\mathrm{d}P_S$, the change in equilibrium static pressure.

In the absence of any external heat supply ($\mathrm{d}S = 0$) a pressure change $\mathrm{d}P_S$ would produce a temperature change, from Equation (6.4):

$$\mathrm{d}T = \frac{\beta T}{\varrho c_p}\,\mathrm{d}P_S. \tag{6.13}$$

Let this temperature change due to isentropic compression or expansion alone be designated $\mathrm{d}T_i$. The difference between the total change in temperature in a given flow problem and the isentropic one is then

$$\mathrm{d}(T - T_\mathrm{i}) = \mathrm{d}T - \frac{\beta T}{\varrho c_\mathrm{p}} \mathrm{d}P_\mathrm{S} = \frac{T}{c_\mathrm{p}} \mathrm{d}S. \tag{6.14}$$

Therefore the left hand side of the energy Equation (6.3) may be regarded as describing the total rate of change (following the motion of a fluid element) of the quantity $\varrho c_\mathrm{p}(T - T_\mathrm{i})$. It is then convenient to regard

$$\theta = T - T_\mathrm{i} \tag{6.15}$$

as the relevant 'excess' temperature in the presence of significant pressure variations. This quantity is known as the 'potential' temperature. A convenient choice of origin for θ is $\theta = 0$ at reference density, pressure and temperature, i.e., $T_\mathrm{i} = T_0$ and $\theta = 0$ at $x_3 = 0$.

The vertical distribution of T_i may be found by an integration of Equation (6.13). Substituting Equation (6.8) we find

$$\mathrm{d}T_\mathrm{i} = -\frac{\beta T g}{c_\mathrm{p}} \mathrm{d}x_3. \tag{6.16}$$

For moderate elevation differences the factor $\beta Tg/c_\mathrm{p}$ may often be regarded as constant, resulting in a linear distribution of T_i with height:

$$T_i \cong T_0 - \frac{\beta T_0 g}{c_\mathrm{p}} x_3. \tag{6.17}$$

To the same approximation therefore

$$\nabla^2 (T - T_\mathrm{i}) = \nabla^2 \theta \cong \nabla^2 T. \tag{6.18}$$

Substituting this on the right hand side of the energy Equation (6.3) we have the simple result

$$\frac{\partial \theta}{\partial t} + u_\mathrm{j} \frac{\partial \theta}{\partial x_\mathrm{j}} = \kappa \nabla^2 \theta \tag{6.19}$$

or an equation identical in form with Equation (6.12), only replacing absolute temperature by potential temperature.

In order to close the above system of equations it is necessary to relate the excess density $(\varrho - \varrho_\mathrm{a})$ appearing in Equation (6.10) to the field of temperature T, or rather potential temperature θ. For small differences from the reference temperature T_0 Equation (6.6) is sufficiently accurate (for gases as well as liquids) with $\beta = \text{const}$. The buoyant acceleration term in Equation (6.10) then becomes

$$\frac{\varrho - \varrho_\mathrm{a}}{\varrho_0} X_\mathrm{i} = \beta g (\theta - \theta_\mathrm{a}) \delta_{\mathrm{i}3}. \tag{6.20}$$

In deriving the approximate form of the energy equation (Equation (6.12)) we have left open the question, just how thin a layer of fluid had to be for the approximation $dP \cong 0$ to be valid. We now see that the simpler form holds as long as variations in T_i across the layer are negligible compared to temperature differences caused by external heating or cooling. When this is the case, also $(\theta - \theta_a)$ in Equation (6.20) may be replaced by $(T - T_a)$.

For 'perfect' gases (a sufficiently good approximation for atmospheric air) $\beta T = 1$, and the isentropic temperature gradient becomes

$$\frac{dT_a}{dx_3} = -\frac{g}{c_p}. \tag{6.21}$$

In water the value of βT varies from zero at 4 °C to rather higher values. At 20 °C, for example, for fresh water $\beta = 2.1 \times 10^{-4}\ \mathrm{C}^{-1}$, $\beta T = 0.0615$ and the isentropic gradient is about 0.25 °C 100 m^{-1}. In the atmosphere the isentropic gradient is near 0.01 °C m^{-1}. Thus in layers of the atmosphere or ocean of a depth of 10 m or less the distinction between absolute and potential temperatures becomes usually unimportant, but not so for layers of the order of 100 m in depth.

6.4. Equations for Turbulent Flow

Although the above equations apply also to the details of atmospheric and oceanic turbulent flow, they are not directly useful on account of the complexity of these details. Exactly as in dealing with turbulent diffusion, we have to define appropriate ensemble mean quantities if we want to develop a theory which can be compared with experiment. Particularly in geophysical applications a certain amount of arbitrariness is unavoidable in the definition of ensemble means (what precise ensemble one is to use) and certain plausible choices may lead to unexpected consequences (see e.g., Starr, 1968). We shall ignore these difficulties here, however, and assume that suitable stable mean quantities may be defined. Thus let the local velocity, potential temperature, and reduced pressure within the turbulent field of motion of a buoyant parcel be resolved into mean and fluctuating components:

$$\begin{aligned} u_i &= \bar{u}_i + u_i' \\ \theta &= \bar{\theta} + \theta' \\ p &= \bar{p} + p'. \end{aligned} \tag{6.22}$$

These quantities may be substituted into the equations of motion and an averaging process carried out on each term. The result is Reynolds' form of the equations of motion (e.g., Lumley and Panofsky, 1964):

$$\frac{\partial \bar{u}_i}{\partial t} + \bar{u}_j \frac{\partial \bar{u}_i}{\partial x_j} = -\frac{1}{\varrho_0}\frac{\partial \bar{p}}{\partial x_i} + \beta g(\bar{\theta} - \theta_a)\,\delta_{i3} + \nu \nabla^2 \bar{u}_i - \frac{\partial}{\partial x_j}\overline{(u_i' u_j')}. \tag{6.23}$$

Here we have already used the Boussinesq approximation to avoid unnecessary complexity. The only new term compared to Equation (6.10) is the Reynolds stress divergence:

$$\frac{\partial}{\partial x_j} \overline{(u_i' u_j')}.$$

This term more or less replaces the viscous one, $\nu \nabla^2 \bar{u}_i$, because the latter is negligible in turbulent flow except in the immediate neighborhood of solid boundaries. Unfortunately, there is no accurate relationship between the Reynolds stresses and the mean velocity field. In 'free' turbulent flows, such as jets, wakes and buoyant plumes, however, fairly good results are obtained by assuming that the effects of turbulent movements are similar to those of molecular agitation in the sense that the stresses are proportional to velocity gradients:

$$\overline{u_i' u_j'} = - \nu_T (\partial \bar{u}_i / \partial x_j), \tag{6.24}$$

where ν_T is a turbulent kinematic viscosity. The value of the latter is proportional to the length and velocity scales of the motion, L^* and u^* (Townsend, 1956):

$$\nu_T = \gamma_m L^* u^*. \tag{6.25}$$

The value of the constant γ_m must be determined empirically. It is different for wakes than for jets or for buoyant plumes, although this has no very fundamental significance on account of the arbitrariness inherent in the definition of length and velocity scales.

A similar averaging process may be carried out on the energy Equation (6.19) and results in:

$$\frac{\partial \bar{\theta}}{\partial t} + \bar{u}_j \frac{\partial \bar{\theta}}{\partial x_j} = \kappa \nabla^2 \bar{\theta} - \frac{\partial}{\partial x_j} \overline{(u_j' \theta')}. \tag{6.26}$$

The new term is the divergence of the turbulent heat flux:

$$\frac{\partial}{\partial x_j} \overline{(u_j' \theta')}$$

which again replaces $\kappa \nabla^2 \bar{\theta}$, the latter being negligible except very close to solid boundaries. An eddy thermal diffusivity κ_T may be introduced in analogy with Equation (6.24):

$$\overline{u_j' \theta'} = - \kappa_T \frac{\partial \bar{\theta}}{\partial x_j}, \tag{6.27}$$

where

$$\kappa_T = \gamma_h L^* u^*. \tag{6.28}$$

The constant γ_h referring to the turbulent transport of heat is usually similar in magnitude to γ_m, which relates to momentum transport (Equation (6.25)). Indeed one often sets $\gamma = \gamma_m = \gamma_h$ as a sufficient approximation (turbulent Prandtl number unity) in view of the crudeness of the eddy viscosity and thermal diffusivity model. This model actually describes a physical process somewhat simpler than turbulent transfer, which is yet sufficiently similar to the real phenomenon to provide valuable insight.

The structure of the continuity equation remains unchanged on averaging. Thus the approximate equations of motion for the mean component of turbulent flow become (Boussinesq approximation plus eddy coefficients):

$$\frac{\partial \bar{u}_j}{\partial x_j} = 0$$

$$\frac{\partial \bar{u}_i}{\partial t} + \bar{u}_j \frac{\partial \bar{u}_i}{\partial x_j} = -\frac{1}{\varrho_0} \frac{\partial \bar{p}}{\partial x_i} + \beta g (\bar{\theta} - \theta_a) \delta_{i3} + \frac{\partial}{\partial x_j}\left(\nu_T \frac{\partial \bar{u}_i}{\partial x_j}\right) \tag{6.29}$$

$$\frac{\partial \bar{\theta}}{\partial t} + \bar{u}_j \frac{\partial \bar{\theta}}{\partial x_j} = \frac{\partial}{\partial x_j}\left(\kappa_T \frac{\partial \bar{\theta}}{\partial x_j}\right).$$

In spite of the drastic simplifications carried out above these equations remain a complicated nonlinear set, containing coupling terms between the mean velocity and the temperature fields.

6.5. Turbulent Energy Equation

When the averaged equation of motion (Equation (6.23)) is subtracted from the original form (Equation (6.10)), an equation for the fluctuating turbulent velocity components u_i' results. By a standard manipulation of these equations it is possible to derive what is in effect a balance equation for the kinetic energy present in turbulent movements (Townsend, 1956). The details of this manipulation and some of the arguments justifying the accepted physical interpretation of the terms in the resulting turbulent energy equation are relatively cumbersome and seem to be clearly beyond the scope of the present treatise. We shall merely quote and discuss the results, referring the reader to Townsend (1956) and Lumley and Panofsky (1964) for the mathematical details.

The kinetic energy of turbulent movements is, per unit mass of the fluid

$$e_k = \tfrac{1}{2} u_i' u_i'. \tag{6.30}$$

The mean value of this quantity, $\bar{e}_k$, changes according to the turbulent energy equation:

$$\frac{\partial \bar{e}_k}{\partial t} + \bar{u}_j \frac{\partial \bar{e}_k}{\partial x_j} = -\overline{u_i' u_j'} \frac{\partial \bar{u}_i}{\partial x_j} + \beta g \overline{u_3' \theta'} - \varepsilon - \frac{\partial F_i}{\partial x_i}, \tag{6.31}$$

where ε is the rate of viscous dissipation of turbulent energy and $F_i (i=1, 2, 3)$ are the components of an energy transport vector which depends on velocity and pressure fluctuations (or rather various mean products involving these) in a relatively complex way. The Boussinesq approximation has been used in the derivation of this equation.

The left hand side of Equation (6.31) contains the rate of change of kinetic energy, following the mean motion of the fluid. On the right, the last two terms are dissipation and divergence of transport. Our interest centers on the first two terms on the right which are potential production terms of energy.

The first term on the right represents an interaction between mean flow and turbulence. For small to medium scale problems in the lower atmosphere or the ocean (including most diffusion problems of interest) the 'mean flow' is usually defined so as to make the typical value of $\bar{u}_i$ large compared to that of u_i'. For example, the 2 m level 'mean wind' (hourly mean or so) usually has a typical velocity large compared to the turbulent fluctuations in it, which are mostly of mechanical origin, caused by 'roughness elements' (including trees and houses) at ground level. The interaction between this kind of mean flow and what might be described as 'mechanical' turbulence is always such as to cause a flow of energy *from* mean flow *to* turbulence. In other words the first term on the right of Equation (6.31) is under such conditions positive. If we substitute the Reynolds stress from Equation (6.24) into this term we find

$$- \overline{u_i' u_j'} \frac{\partial \bar{u}_i}{\partial x_j} = \nu_T \left(\frac{\partial \bar{u}_i}{\partial x_j} \right)^2 \tag{6.32}$$

which is positive definite if ν_T is positive. 'Negative viscosity phenomena' exist at much larger scales, where energy flow may be from small scale motions (defined as a form of 'turbulence') to 'mean flow' (Starr, 1968).

The second term on the right of Equation (6.31) is the averaged product of buoyant acceleration and vertical displacement and is therefore the rate of work done by buoyant forces. The sign of this may be positive or negative, representing either production or absorption of energy by buoyant forces. For positive β (true of air and water, except fresh water between 0 °C and 4 °C) a positive value of $\overline{u_3' \theta'}$ implies energy production, a negative one energy absorption. We have seen in connection with Equation (6.26) that $\overline{u_3' \theta'}$ is proportional to the vertical component of the turbulent heat flux. Thus an upward heat flux is associated with turbulent energy generation by buoyant forces (the result being at least partially 'thermal' rather than mechanical turbulence), whereas a downward heat flux implies energy absorption by the same forces. For fresh water between 0 °C and 4 °C the roles of upward and downward heat flux are reversed.

From the point of view of our central topic, diffusion in the environment, the importance of the above facts lies in that supply and demand of turbulent energy determine the intensity of turbulence present in any flow field. When there is an extra 'thermal' supply of turbulent energy, the intensity of turbulence increases. When the turbulent energy is continuously depleted by work against the force of gravity, the intensity of turbulence diminishes and it might disappear altogether. We have seen in previous chapters that the spreading of a cloud is directly proportional to the intensity of turbulence (to the rms turbulent velocities, to be precise), which we have just seen to be directly affected by buoyant forces. With turbulence completely suppressed, any diffusion is due to molecular movements and is a very much slower process than its turbulent counterpart.

A convenient parameter measuring the relative importance of buoyancy effects is the ratio of the two potential production terms in Equation (6.31). In atmospheric and oceanic applications the mean flow is usually horizontal while its maximum rate of change occurs along the vertical. If we lay the x_1-axis along the mean wind, it turns out that the term

$$\overline{u'_1 u'_3}\, \frac{\partial \bar{u}_1}{\partial x_3}$$

is much larger than any of the others in the double sum expressing mechanical energy production (Equation (6.32)). The ratio of thermal to mechanical energy production is then

$$\mathrm{Ri} = \frac{\beta g \overline{u'_3 \theta'}}{\overline{u'_1 u'_3}\, \dfrac{\partial \bar{u}_1}{\partial x_3}} \tag{6.33}$$

which is known as the 'Richardson number.' The denominator of this expression is always negative and therefore Ri is negative for upward heat flux when buoyant forces *produce* energy and positive when they *absorb* it. At a certain large enough positive value of Ri the turbulence intensity drops to zero. This limiting or critical value of the Richardson number is currently believed to be about 0.21.

6.6. Diffusion Floors and Ceilings

In the atmosphere and ocean regions of turbulent flow are often bounded by thin layers in which there is no measurable turbulence, on account of the stabilizing effect of the force of gravity discussed in the previous section. For such 'stable sheets' the value of the Richardson number is locally above critical and diffusion proceeds by molecular agitation alone. (Note that in the definition of Ri (Equation (6.33)) *molecular* flux of heat and momentum must then be substituted in place of the turbulent flux components.) Because of low molecular diffusivities, very sharp gradients develop in such stable sheets and any admixture to the air or water appears to be confined for all practical purposes to one side. A well known example is the layer of haze due to particulate pollution over cities: this usually extends upwards to a stable sheet in the atmosphere, above which the air abruptly becomes clear, as every air traveler knows. The stable sheet in this case acts as a diffusion 'lid' or 'ceiling'. Its counterpart is a diffusion 'floor' in the ocean (or large lakes, such as those reported in Lake Huron (Csanady, 1970)) which stops the downward diffusion of impurities. Over a larger depth range in either the atmosphere or the ocean several stable sheets usually exist, between which diffusion of any admixture is confined to a turbulent layer of moderate depth.

It is of some practical importance to consider the effect of such ceilings or floors on the concentration field of concentrated sources. As a practical example, we may keep in mind a factory chimney, emitting pollutants into a turbulent layer which is no more than 2 or 3 times as deep as the chimney is high. As such a chimney plume grows in cross section, it eventually encounters the diffusion ceiling which may be assumed to 'reflect' the cloud much as the solid ground does (to a certain degree of approximation, of course, there being always some slight loss by molecular diffusion, breaking of internal waves on the stable interface, etc.). The practical question is, how does the presence of the ceiling affect ground level concentrations.

The main features of the problem are brought out by a simple model, in which wind and turbulence are assumed uniform between ground and inversion lid, i.e., floor and ceiling. The 'transport lines' (effective streamlines for the pollutants (Csanady, 1957)) are reflected both by floor and ceiling, as illustrated in Figure 6.1. When either reflected

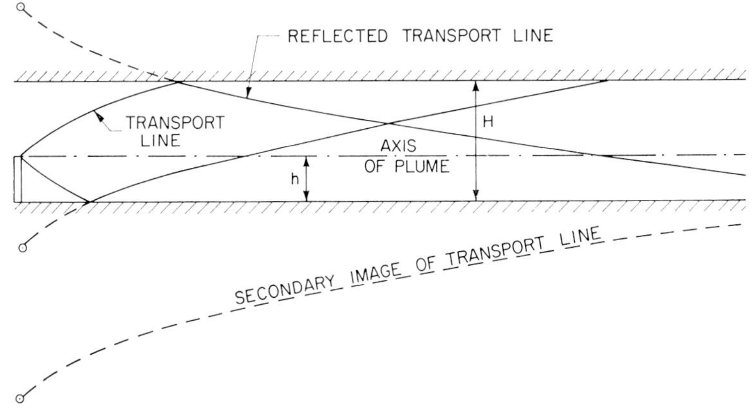

Fig. 6.1. 'Reflection' of diffusing cloud by ground level and by inversion lid.

streamline reaches the opposite boundary it is reflected again and so on. Mathematically, each reflection may be represented by an image source, the primary images being on the far side of either boundary at the same distance as the physical source. Each image source gives rise to a secondary image on reflection with respect to the opposite boundary, then to a tertiary one by a further reflection of the secondary image, and so on. The result is an infinite series of exponential terms. For a point source at a height h above ground, if the total depth of the turbulent layer is H, the concentration field in a uniform wind is, by a simple extension of Equation (3.31):

$$\chi = \frac{Q}{2\pi\sigma_y\sigma_z U} \exp\left\{-\frac{y^2}{2\sigma_y^2}\right\} \cdot \sum_{n=-\infty}^{\infty} \left[\exp\left\{-\frac{(z-h+2nH)^2}{2\sigma_z^2}\right\} + \right.$$

$$\left. + \exp\left\{-\frac{(z+h+2nH)^2}{2\sigma_z^2}\right\}\right]. \qquad (6.34)$$

At ground level, $z=0$, and along the axis of the plume, $y=0$, this gives the following variation of concentration with distance:

$$\chi_a = \frac{Q}{\pi\sigma_y\sigma_z U} \sum_{n=-\infty}^{\infty} \exp\left\{-\frac{(h-2nH)^2}{2\sigma_z^2}\right\}. \tag{6.35}$$

At large distances from the source the vertical distribution of concentration across the turbulent layer must become uniform, so that the infinite sum in Equation (6.34) must tend to an asymptotic value independent of z as $\sigma_z \to \infty$. A simple mass balance shows the corresponding asymptotic concentration χ_∞ to be

$$\chi_\infty = \frac{Q}{\sqrt{2\pi}\,\sigma_y UH} \exp\left\{-\frac{y^2}{2\sigma_y^2}\right\}. \tag{6.36}$$

Therefore the infinite sum in Equation (6.34) must tend to $\sqrt{2\pi}\,\sigma_z/H$. This is easily shown directly by noting that for $\sigma_z \to \infty$ the sum must approach the integral with respect to the variable n. The integral may be evaluated and duly gives the answer arrived at above from a physical argument.

Figure 6.2 shows the resulting variation with cloud size of the factor

$$\phi\left(\frac{\sigma_z}{h}, \frac{H}{h}\right) = \frac{h}{\sigma_z} \sum_{n=-\infty}^{\infty} \exp\left\{-\frac{(h-2nH)^2}{2\sigma_z^2}\right\} \tag{6.37}$$

to which the axial ground level concentration is related by, according to Equation (6.35):

$$\chi_a = \frac{Q}{\pi\sigma_y hU}\,\phi. \tag{6.37a}$$

For constant source height h, the curves thus give the variation of ground level concentration from a cross wind line source with different ceiling heights and for varying distances from the source (actually for varying cloud size σ_z, related to distance x by an appropriate empirical relationship, which we discussed in Chapter III.)

One may note from this figure that maximum ground level concentrations are only affected if the ceiling is lower than $H = 2h$. However, the decay of concentration with distance at large distances from the source is rather slower than without a ceiling: according to Equation (6.36) χ_∞ decreases only as σ_y^{-1}, for the obvious physical reason that diffusion in the vertical space dimension is eliminated.

6.7. Diffusion in a Continuously Stratified Fluid

The rather extreme case of complete turbulence suppression by buoyancy forces discussed in the previous section is quite important in environmental applications, but it is an effect localized in stable sheets. In between any such stable sheets, or between ground level and the lowest ceiling, the fluid usually remains turbulent, but in the presence of upward or downward heat flux the density distribution becomes non-

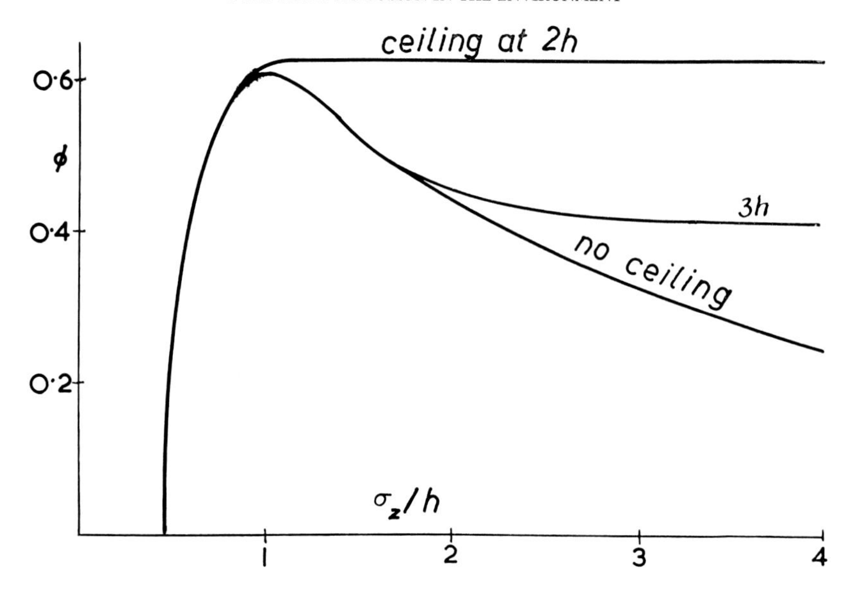

Fig. 6.2. Variation of ground-level concentration downwind of a crosswind line source at height *h*, for various mixed layer depths *H*.

neutral, unstable or stable. In such 'continuously' stratified layers (in contrast to the relatively abrupt jumps at stable sheets) the value of the Richardson number is below critical but different from zero and thus the intensity of turbulence is affected. This has some immediate consequences for turbulent diffusion (on account of the proportionality of σ_y to i_y, etc.) which may be deduced from our results in earlier chapters. There is, however, also a more subtle effect arising from the influence of buoyancy forces on the velocity history of individual diffusing elements, which appears as a modification of the Lagrangian correlation coefficient.

The simple statistical theory developed in Chapter III (and its extensions in Chapters IV and V) is valid strictly only for a homogeneous fluid, or in the case of larger elevation differences, when the density distribution is neutral, so that no systematic effects are exerted by buoyant forces. However, we have also seen in Chapter III that any direct or indirect effects of atmospheric stratification could be taken into account by an empirical adaptation of the theory. Steps to this effect included choosing the right 'stability parameter' *n* in Sutton's model or introducing data on horizontal and vertical standard deviations as purely empirical functions. We have pointed out at the same time that some of the empirical functions $\sigma_z(t)$ – especially those characterizing diffusion under unstable conditions – were inconsistent with the Lagrangian velocity correlation functions characterizing laboratory diffusion, or indeed with any reasonably chosen correlation function for a stationary process. It is possible to elucidate the physical reasons for this at first sight quite unsatisfactory state of affairs by analyzing the displacement of diffusing particles subject to buoyancy forces.

As the simplest theoretical model of diffusion in a stratified fluid, we consider a layer of homogeneous turbulence of constant mean velocity U, across which there is a uniform *vertical* heat flux, maintaining a constant mean potential temperature gradient $\partial\bar{\theta}/\partial z$. Note that such steady state conditions are compatible with the energy Equation (6.29). Associated with the flow of heat there are small temperature fluctuations θ', correlated with vertical velocity fluctuations w' ($\equiv u_3'$; note that we are reverting to the simpler x, y, z, and u, v, w notation), which carry the heat flux:

$$H = \varrho c_{\mathrm{p}} \overline{w'\theta'} . \tag{6.38}$$

The departures from the local mean temperature θ' also give rise to buoyancy forces per unit mass of $\beta g\theta'$, which tend to accelerate or decelerate fluid particles. The vertical acceleration is (time derivative of velocity *following* the motion of particles), from Equations (6.10) and (6.20), identifying θ_{a} with $\bar{\theta}$:

$$\mathrm{d}w'/\mathrm{d}t = \beta g\theta' + Z(t), \tag{6.39}$$

where

$$Z(t) = -\frac{1}{\varrho_0}\frac{\partial p'}{\partial z} + \nu\nabla^2 w'$$

is a random force arising from the turbulent pressure and velocity fluctuation field. It is reasonable to suppose that this random force will have statistical properties very much as in the neutral case, for a given intensity of turbulence. The viscous term in $Z(t)$ is usually rather smaller than the pressure gradient term. In regard to the latter, it is important to note that p' is the net product of the entire irregular velocity field of the turbulent flow. In a homogeneous fluid it may be calculated from (Batchelor, 1953):

$$p'(\mathbf{x}) = \frac{\varrho_0}{4\pi}\int \frac{\partial^2 (u_i' u_j')}{\partial x_i \partial x_j} \frac{\mathrm{d}\mathbf{x}^*}{|\mathbf{x}^* - \mathbf{x}|}, \tag{6.40}$$

where the double divergence in the integrand is to be evaluated at radius vector $\mathbf{x}^*$ and integration is over all space. The influence of momentum flux variations at points $\mathbf{x}^*$ distant from the point $\mathbf{x}$ where the pressure is evaluated only drops slowly (inverse distance $|\mathbf{x}^* - \mathbf{x}|$ in integrand). Thus the irregular accelerations $Z(t)$ are *not directly* related to the local excess velocity w'.

The structure of Equation (6.39) above is quite similar to Langevin's Equation (2.13) in that there is an 'organized' and a 'random' component of vertical force acting on a small fluid parcel, much as such forces were supposed to act on Brownian particles. The equations of motion for horizontal velocity fluctuations do not contain the relatively 'organized' buoyant force.

The rate at which θ' changes following a moving fluid element may be deduced by subtracting the third of Equations (6.29) from (6.19):

$$\frac{\mathrm{d}\theta'}{\mathrm{d}t} = -w'\frac{\partial\bar{\theta}}{\partial z} + \kappa\nabla^2\theta' . \tag{6.41}$$

As we have seen before, the Laplacian expresses the difference in temperature between the element and its immediate neighborhood. The molecular heat flux term in the last equation may therefore be written

$$\kappa \nabla^2 \theta' = - k\theta' , \tag{6.42}$$

where k is a quantity of dimension time^{-1}, the exact value of which depends on the sharpness of the instantaneous temperature gradients in the vicinity of the fluid element. Thus in a turbulent field k is a random variable. However, k is unlikely to be significantly correlated with either w' or θ' and it is therefore not likely to lead to appreciable error if we replace the actual random value of k by its constant mean value. With this assumption k becomes a kind of 'decay constant' for temperature fluctuations θ'. A similar approach has been successfully used by Priestley (1959) in dealing with 'buoyant parcels' in the atmosphere.

Writing now

$$G = \mathrm{d}\bar{\theta}/\mathrm{d}z = \text{const.}$$

for our supposedly uniform potential temperature gradient, the energy equation takes on the simple form:

$$\mathrm{d}\theta'/\mathrm{d}t = - Gw' - k\theta' . \tag{6.43}$$

This equation, together with Equation (6.39) describes fluctuations of temperature and velocity for a wandering fluid element, provided that the random turbulent accelerations $Z(t)$ are prescribed in some way. As in the case of Brownian motion, the details of the velocity and temperature history of a wandering particle are without interest. What we require from the theoretical analysis is a velocity covariance $\overline{w'(t)\, w'(t+\tau)}$ for an ensemble of diffusing particles, because we have seen that this can be related to the dispersion of the cluster by Taylor's theorem.

6.8. Velocity Autocorrelation and Particle Spread in Stratified Fluid Model

In order to arrive at a velocity correlation on the basis of Equations (6.39) and (6.43) it is necessary to have some information on the statistical properties of $Z(t)$. It is reasonable to postulate that the pressure and stress pulses due to the eddying motion retain the same basic character in the presence of stratification (at least if the latter is of moderate intensity) which they had under neutral conditions. In a homogeneous and stationary field without stratification (laboratory turbulence) the Lagrangian correlation coefficient was approximately of the exponential-decay type. Therefore the random 'mechanical' turbulence process:

$$w'_{\mathrm{m}}(t) = \int_{t_0}^{t} Z(t')\,\mathrm{d}t' \tag{6.44}$$

may be assumed to have an exponential-decay type correlation function, i.e.,

$$\overline{w'_{\mathrm{m}}(t)\, w'_{\mathrm{m}}(t+\tau)} = \overline{w'^2_{\mathrm{m}}} \exp\left(-\,|\tau|/t_{\mathrm{L}}\right), \tag{6.45}$$

where t_L is the Lagrangian time scale. It is possible to deduce from this assumption and Equations (6.39) and (6.43) the form of the velocity correlation function in the presence of buoyant accelerations (Csanady, 1964).

One may easily eliminate $w'(t)$ from Equations (6.39) and (6.43) and find:

$$\frac{d^2\theta'}{dt^2} + k\frac{d\theta'}{dt} + G\beta g\theta' = -GZ(t). \tag{6.46}$$

This is readily integrated between some initial instant t_0 when a particle is considered 'released' and t. The result contains exponential functions $\exp[r_1(t-t_0)]$ and $\exp[r_2(t-t_0)]$ where $r_{1,2}$ are roots of the characteristic Equation of (6.46):

$$r_{1,2} = -\frac{k}{2} \pm \sqrt{\frac{k^2}{4} - G\beta g}. \tag{6.47}$$

When the solution of Equation (6.46) is introduced into Equation (6.39) one finds:

$$w'(t) = \beta g \int_{t_0}^{t} \theta'(\tau)\, d\tau + w'_0 + w'_m, \tag{6.48}$$

where w'_0 is the value of w' at $t = t_0$ and w'_m is as defined in Equation (6.44). Clearly, $w'(t)$ again contains exponential terms $\exp[r_1(t-t_0)]$ and $\exp[r_2(t-t_0)]$.

If the temperature distribution is stable, $G>0$, by Equation (6.47) both roots r_1 and r_2 are negative. Under such conditions the terms depending on the initial parameters w'_0, θ'_0 tend to zero for large $(t-t_0)$. We may regard these 'initial' values to be a result of a long prior process of turbulent motion and allow $t_0 \to -\infty$ in the integrated expressions. The remaining process $w'(t)$ is then a linear transformation of the 'mechanical' turbulence process, $w'_m(t)$ viz.:

$$w'(t) = w'_m(t) - \frac{G\beta g}{r_1 - r_2} \int_{-\infty}^{t} [e^{r_1(t-\tau)} - e^{r_2(t-\tau)}] \cdot w'_m(\tau)\, d\tau. \tag{6.49}$$

Thus under *stable* stratification the velocity history $w'(t)$ is again a stationary process. Further results become particularly instructive if we introduce the well-known Brunt-Vaisala frequency N:

$$N^2 = G\beta g. \quad (G > 0) \tag{6.50}$$

The mean square velocity, $\overline{w'^2}$ may now be calculated from Equation (6.49), also using Equation (6.45):

$$\overline{w'^2} = \overline{w'^2_m}\left[1 + \frac{(Nt_L)^2}{kt_L}\, \frac{1 - kt_L}{1 + kt_L + (Nt_L)^2}\right]. \tag{6.51}$$

This result shows that the mean square velocity of a wandering fluid element in a stably stratified fluid may be smaller or greater than that due to 'mechanical' impulses alone. The impulses are, however, themselves products of the turbulent velocity field. By Equation (6.40) – assuming that pressure pulses supply the dominant contributions to $Z(t)$ – their statistical properties may be related to Eulerian mean quantities, such as mean square velocity $\overline{w_E'^2}$ at a fixed point, spatial correlation lengths, and the like. In a neutral field these impulses produce a mean square Lagrangian velocity identical with the Eulerian one. It is therefore reasonable to identify $\overline{w_m'^2}$ in the non-neutral case also with the Eulerian velocity-square. Equation (6.51) may then be looked upon as a relationship between Eulerian and Lagrangian mean square velocities. The two are different, unless it so happens that

$$kt_L = 1. \tag{6.52}$$

Because k is a decay constant for temperature fluctuations and t_L^{-1} an essentially similar physical quantity for velocity fluctuations, one would be tempted to conclude that Equation (6.52) should be at least approximately true. This would amount to assuming the Reynolds analogy for the transfer of heat and momentum between a fluid element and its neighborhood. In view of the different physical processes governing the two phenomena, however, this is a dubious assumption in the present context. Some crude order of magnitude estimates (Csanady, 1964) have given a ratio $k:t_L^{-1}$ as low as 0.13. Such a low value of kt_L would produce a large difference between Eulerian and Lagrangian velocities even for moderate stability. The Lagrangian velocities would be larger on account of harmonic oscillations of fluid elements. However, no experimental evidence is available to encourage belief that such a curious state of affairs in fact occurs in nature and it is more reasonable at present to assume that Equation (6.52) is true to a sufficient approximation to ensure that Lagrangian and Eulerian mean square velocities are equal.

On the basis of this last assumption a fairly simple expression for the Lagrangian autocorrelation may be derived for stable conditions. Even for moderate stability the square root in Equation (6.47) becomes imaginary and it is preferable to use

$$\omega = \sqrt{N^2 - \frac{k^2}{4}}. \tag{6.53}$$

The Lagrangian autocorrelation function calculated from Equation (6.49), with Equation (6.52) assumed, becomes then:

$$R(\tau) = e^{-|t/2t_L|}\left(\cos\omega\tau - \frac{1}{2\omega t_L}\sin\omega\tau\right). \tag{6.54}$$

From Taylor's theorem we find now the mean square dispersion in the vertical:

$$\sigma_z^2 = 2\overline{w'^2}\int_0^t\int_0^{t'} R(\tau)\,d\tau\,dt' =$$

$$= \frac{2\overline{w'^2}}{N^2}\left\{1 - e^{-|t/2t_L|}\left(\cos\omega t - \frac{1}{2\omega t_L}\sin\omega t\right)\right\}. \tag{6.55}$$

It may be seen that the dispersion tends asymptotically to the limit

$$\sigma_z^2 = 2\overline{w'^2}/N^2 \quad (t \to \infty) \tag{6.56}$$

which is an indication that the total area under the $R(\tau)$ curve is zero (from $\tau=0$ to ∞). Physically, the sinusoidal oscillations in $R(\tau)$ arise because the force of gravity acts as a restoring force for displaced fluid elements. The result that the area under the $R(\tau)$ curve is exactly zero only holds for $kt_L=1$; if $kt_L<1$ (which was suggested by some numerical estimates), σ_z^2 tends asymptotically to

$$\sigma_z^2 = \frac{1-(kt_L)^2}{(1+N^2t_L^2)^2-(kt_L)^2}\, 2\overline{w_E'^2}\, t_L(t-t_L) \quad (t \to \infty) \tag{6.57}$$

which is equivalent to saying that the asymptotic eddy diffusivity becomes what it is without stratification, times the fraction multiplying the right hand side:

$$K_z = \frac{1-(kt_L)^2}{(1+N^2t_L^2)^2-(kt_L)^2}\, K_z^*, \tag{6.58}$$

where $K_z^* = \overline{w_E'^2}\, t_L$ is the asymptotic eddy diffusivity due to 'mechanical' impulses alone.

With a kt_L less than one, and even a moderate degree of stratification K_z may become considerably less than K_z^* because Nt_L can easily be of the order of 3 to 10. Thus the asymptotic slope of $\sigma_z^2(t)$ may be quite low and it may vanish altogether. These findings are consistent with some data already summarized in Chapter III. To quote some further evidence, it has been reported by Stewart *et al.* (1958) that under stable conditions they were unable to find any trace of material emitted at an 'effective' height of 135 m above ground, up to distances of 10 km from the source. Hilst and Simpson (1958) have found considerable negative loops in $R(\tau)$ under stable conditions and in one case of strong stability observed σ_z to tend to a constant asymptote, at a value of some 6.5 m. Using Equation (6.56), the observed temperature gradient, and assuming 3% vertical gustiness one finds for their case $\sigma_z=5$ m. The gustiness was not measured, but under stable conditions it is known to have approximately the above value. It is reasonable to conclude that the theoretical model provides considerable insight into the pecularities of vertical diffusion in a stably stratified fluid. It would, however, be desirable to have laboratory verification of the results (especially of Equation (6.58)) under the simple conditions assumed in the theory.

When the stratification is *unstable* $G<0$, one of the roots of the characteristic equation is positive (say $r_1>0$) and the solution of Equation (6.48) contains terms proportional to

$$e^{r_1(t-t_0)}.$$

These terms, and with them the mean square velocity, grow beyond all bounds, so that the velocity-history of a diffusing particle is not a stationary process, but an *evolutionary* one. When a particle is given an initial velocity, either up or down, after a certain time it will, on the average, have a greater velocity in the same direction, in direct consequence of the unstable arrangement of the fluid. This is not incompatible with our earlier assumption that the Eulerian mean square velocity is constant: individual particles may cross the unstable layer in upward or downward movements which are accelerated in the mean square, while steady state conditions prevail at any fixed point. As in the stable case, Eulerian and Lagrangian velocities need not have the same statistical properties when buoyancy forces act on particles.

On applying the argument which produces Taylor's theorem, but without assuming a stationary process one has

$$\sigma_z^2 = \int_0^t \int_0^{t'} \overline{w(t')\, w(t'')}\, dt'\, dt''. \tag{6.59}$$

If the velocities grow exponentially this equation shows that σ_z exhibits essentially the same behavior, the leading terms in the solution of Equations (6.48) and (6.59) being then of the form:

$$\sigma_z = \text{const. } e^{r_1(t-t_0)}. \tag{6.60}$$

The time-scale of the exponential growth is r_1^{-1}, which is from Equation (6.47):

$$r_1 = \sqrt{\frac{k^2}{4} - G\beta g} - \frac{k}{2} \tag{6.61}$$

with $G<0$, as noted above. With even moderate gradients k is probably small compared to $-G\beta g$ and one may write approximately

$$r_1 \cong \sqrt{-G\beta g}. \tag{6.62}$$

This quantity is the analogue of the Brunt-Vaisala frequency under unstable conditions, its order of magnitude with moderate instability being $10^{-2}-10^{-3}\ \text{s}^{-1}$. Thus in the vertical expansion of a cloud the exponential growth stage should become evident at 10^2 to 10^3 s after release, i.e. if wind speed is 5 m s^{-1}, at downwind distances of a few hundred meters to a few kilometers. This is precisely the character of the observed σ_z curves in unstable air already presented in Chapter III. The theory therefore also provides a qualitative explanation for the peculiar character of vertical diffusion in an unstable atmosphere.

We should note here that our discussion in this section clearly related to *absolute* diffusion, i.e., to the spread of the concentration field observable by fixed samplers. In practice this is produced by irregular vertical movements ('meandering') of entire cloud parcels. Our results show that such meandering becomes particularly intense under unstable conditions, beginning at distances from a few hundred meters to a few kilometers from a source. Chimney plumes exhibiting this precise behavior are known as 'looping' plumes, observed to be associated with unstable conditions (*Meteorology and Atomic Energy*, 1955).

6.9. Bodily Motion of Buoyant and Heavy Plumes

Next we turn to the practically very important problem of the mean motion of larger buoyant and heavy parcels. When such a parcel is released, e.g., from a chimney or a sewage outfall, a systematic vertical displacement is superimposed on its irregular meanderings. Stating this in terms of the probability distribution $P(\mathbf{x}, t)$ of particle displacements (which we introduced in Chapter III), the mean position of the cloud

$$\bar{\mathbf{c}}(t) = \int \mathbf{x}P(\mathbf{x}, t)\,\mathrm{d}\mathbf{x} \tag{6.63}$$

varies in time in such cases, the changes in $\bar{\mathbf{c}}$ being governed by the dynamics of the buoyant parcel. In applying a Gaussian model (say) to predict the mean concentration field, due account has then to be taken of the displacement of the parcel's center of gravity. Ground level pollution intensity in the atmosphere depends, for example, very markedly on the buoyant rise of smoke plumes.

In an individual realization, the rate at which any upward motion of a larger buoyant cloud (or the corresponding downward motion of a heavy cloud; this equivalence will be understood in the discussion below) changes depends mainly on the total buoyancy and the total 'effective' mass of the cloud. As turbulent motions spread out the excess temperature over a larger region of space, the total effective mass increases with the third power of the cloud's linear dimensions. Thus the vertical motion of a cloud slows down in time, even though its total vertical momentum steadily increases under the continued action of the buoyancy force.

An important point is that the dynamic behavior of a buoyant parcel (i.e., its effective instantaneous excess temperature) depends on its *realized* size and not on the length-scale of the mean $P(\mathbf{x}, t)$ distribution, because the latter also includes a contribution from the bodily meandering of parcels. When using the averaged equation of motion (Equation (6.29)), it is therefore necessary to describe the fluxes of heat and momentum in a frame of reference moving with the center of gravity: we have already seen that averaging in such a frame conserves the important physical quantity of cloud size. This may be accomplished approximately by substituting in Equation (6.29) eddy diffusivities of heat and momentum appropriate to various stages of *relative diffusion*. We discussed the properties of this process at some length in Chapter IV, on the assumption that a field of stationary and homogeneous field of turbulence acts on the diffusing cloud.

In buoyant parcels the additional complication arises that the motions due to buoyancy are themselves turbulent and generate a field of turbulence which is neither stationary nor homogeneous. In atmospheric or oceanic buoyant motions both this 'self-generated' turbulence of the parcel and the environmental turbulence are present and affect both the bodily movement of parcels and their growth. A relatively simple analysis is only possible if either kind of turbulence dominates. In the case of chimney plumes, for example, the motion close to the chimney is often dominated by the buoyancy of the plume, almost as if the atmosphere were quiescent. Further downwind, however, the buoyant motions decay in intensity and with them the turbulence generated by these motions. Subsequently the mechanical or thermal turbulence naturally present in the wind may become the dominant dispersing mechanism. We shall see below that these two regimes are characterized by different rates of growth and different upward bulk velocities.

In an individual realization the center of gravity of a diffusing cloud is at

$$\mathbf{c} = (1/Q) \int \mathbf{x} N(\mathbf{x}, t)\, \mathrm{d}\mathbf{x}, \tag{6.64}$$

where $N(\mathbf{x}, t)$ is, as in Chapter IV, the instantaneously realized concentration field and Q is total mass of marked fluid released. The vector $\mathbf{c}$ changes at the rate (noting that the space integral of $\mathrm{d}N/\mathrm{d}t$ is zero on account of the continuity equation):

$$\mathrm{d}\mathbf{c}/\mathrm{d}t = (1/Q) \int \mathbf{u}(\mathbf{x}, t)\, N(\mathbf{x}, t)\, \mathrm{d}\mathbf{x}, \tag{6.65}$$

where $\mathbf{u}(\mathbf{x}, t)$ is (absolute) fluid velocity at $(\mathbf{x}, t)$. Introducing coordinates referred to the center of gravity $\mathbf{y} = \mathbf{x} - \mathbf{c}$ we may rewrite the last result simply as

$$\mathrm{d}\mathbf{c}/\mathrm{d}t = (1/Q) \int \mathbf{u}(\mathbf{y} + \mathbf{c}, t)\, N(\mathbf{y}, t)\, \mathrm{d}\mathbf{y}. \tag{6.66}$$

Taking ensemble averages of both sides we find in the integrand the product of the means, $\bar{\mathbf{u}} \cdot \bar{N}$ and the correlation $\overline{\mathbf{u}'N'}$ which is a hybrid turbulent flux containing the *absolute* velocity and concentration fluctuations at fixed $\mathbf{y}$. Writing $\mathbf{v} = \mathbf{u} - \mathrm{d}\mathbf{c}/\mathrm{d}t$ for the velocity relative to the center of gravity the space integral of this flux term is seen to be

$$\int \overline{\mathbf{u}'N'}\, \mathrm{d}\mathbf{y} = \overline{\mathrm{d}\mathbf{c}'/\mathrm{d}t \int N'\, \mathrm{d}\mathbf{y}} + \int \overline{\mathbf{v}'N'}\, \mathrm{d}\mathbf{y}. \tag{6.67}$$

We observe that the first integral on the right is zero by definition. The flux $\overline{\mathbf{v}'N'}$ in the moving frame may be assumed to be proportional to the gradient $\nabla \bar{N}$†, the space integral of which is also zero. The mean velocity of the center of gravity is then

$$\mathbf{W} = \mathrm{d}\bar{\mathbf{c}}/\mathrm{d}t = (1/Q) \int \bar{\mathbf{u}}(\mathbf{y} + \mathbf{c})\, \chi(\mathbf{y}, t)\, \mathrm{d}\mathbf{y} \tag{6.68}$$

where we have written $\chi(\mathbf{y}, t)$ for the ensemble-mean concentration field in a frame of reference moving with the center of gravity of the buoyant parcel. Note, however, that $\bar{\mathbf{u}}(\mathbf{y} + \mathbf{c})$ is mean *absolute* velocity at a fixed distance y from the center of gravity, because we are interested in the absolute displacements of a diffusing cloud. Of course, the velocities in the equations of motion (Equation (6.29)) are also absolute velocities. The *mean* velocity $\bar{\mathbf{u}}$ at a fixed $(\mathbf{y}, t)$ is what is produced by the systematic action of buoyant forces, while fluctuations $\mathbf{u}'$ are at least partly due to the bodily meandering of buoyant clouds. It is legitimate to determine this 'organized' part $\bar{\mathbf{u}}$ of the total velocity field $\mathbf{u} = \bar{\mathbf{u}} + \mathbf{u}'$ by a solution of Equation (6.29). The bulk upward velocity of a buoyant parcel is then given by Equation (6.68).

From the point of view of environmental pollution the most important case of

† Or at least to be antisymmetrical about a central vertical plane.

buoyant motions occurs in a continuous plume released from a chimney or sewage outfall into a wind or current which may in a first approximation be regarded as steady and uniform. The instantaneous appearance of such a plume and its time-mean path are illustrated in Figure 6.3. In experimental work on chimney plumes the time-mean path is usually determined by taking many photographs of the plume, projecting them on a screen, tracing each apparent center line, then averaging a sufficient number of exposures to determine a mean curve $Z(x)$. (Here Z is vertical displacement above the source, x distance along the mean wind.)

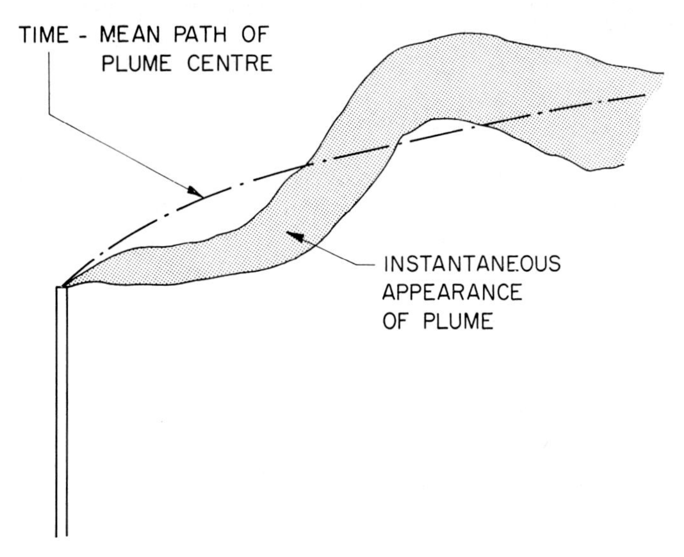

Fig. 6.3. Schematic picture of buoyant plume in cross flow, showing instantaneous appearance and mean path.

If the wind speed at plume level is U, the experimentally determined upward velocity of sections of the plume is therefore:

$$W = \mathrm{d}Z/\mathrm{d}t = U(\mathrm{d}Z/\mathrm{d}x), \tag{6.69}$$

where $\mathrm{d}Z/\mathrm{d}x$ is a directly observed quantity.

The two definitions in Equations (6.68) and (6.69) are clearly equivalent for all practical purposes. The concentration distribution $\chi(\mathbf{y}, t)$ gives the average density of 'marked' fluid and it is the average vertical velocity of this which is described by either Equation (6.68) or (6.69), with $Z(x)$ determined empirically as described.

Except in cases of very light winds, most buoyant chimney plumes bend over into the wind soon after their release and their slope $\mathrm{d}Z/\mathrm{d}x$ becomes rather less than unity. Closer to the source the slope is greater, but here usually also the excess temperature is considerable and the Boussinesq approximation not valid. Thus where a reasonably simple dynamical analysis is possible, i.e., where the temperature excess is suitably small, the upward motion of most plumes is section by section very similar to that of a 'line thermal' generated by the instantaneous release of heat along a straight horizontal line-source in a uniform fluid. The flow pattern in a cross section of the plume may then be analyzed with sufficient accuracy as if a line thermal were involved. The heat content per unit length of the 'equivalent' line thermal is easily determined by calculating the heat content of an elementary slice of the plume of length $\mathrm{d}x$:

$$H\,\mathrm{d}t/\mathrm{d}x = H/U\,, \tag{6.70}$$

where H is heat release rate per unit time at the source. A study of such line thermals provides considerable insight into the behavior of buoyant chimney or sewage effluent plumes: this is the topic we shall turn to in the next section.

6.10. Dynamics of a Line Thermal

Consider the instantaneous release of a quantity of heat along the x-axis in a fluid at rest at time $t=0$. The volume expansion of the fluid leads to buoyant motions, which, after a sufficiently long time t past release, may be adequately described by equations based on the Boussinesq approximation. We assume also that the quantity of heat released is sufficient to cause the buoyant motions to be turbulent, the Reynolds number (based on appropriate length and velocity scales of the flow) high by the time these equations are valid. Solid boundaries are supposed to be far enough not to influence the motion. The heat release per unit length of the x-axis is assumed uniform so that there is no mean motion along x nor any heat or momentum transport, the mean flow being two-dimensional.

In similar problems of 'free' turbulent flow (particularly in two-dimensional ones), the eddy viscosity and conductivity assumption is known to supply quite adequate answers. We shall use this approximation in the equations below. It should be noted, however, that the similarity arguments of the following section apply even if the eddy coefficients are not introduced at all, i.e., if the turbulent flux terms are left in their original form $\overline{v'w'}$, $\overline{v'\theta'}$, etc., assuming only that they are governed by the same velocity, length, and temperature scales as the mean flow.

The equations governing motion in a line thermal may be written down from Equation (6.29), assuming two-dimensional flow in a y–z plane and deleting overbars on the mean quantities for simplicity:

$$\begin{aligned}
&\frac{\partial v}{\partial y} + \frac{\partial w}{\partial z} = 0\\
&\frac{\partial v}{\partial t} + v\frac{\partial v}{\partial y} + w\frac{\partial v}{\partial z} = -\frac{1}{\varrho_0}\frac{\partial p}{\partial y} + \nu_T\nabla_1^2 v\\
&\frac{\partial w}{\partial t} + v\frac{\partial w}{\partial y} + w\frac{\partial w}{\partial z} = -\frac{1}{\varrho_0}\frac{\partial p}{\partial z} + \nu_T\nabla_1^2 w + \beta g(\theta - \theta_a)\\
&\frac{\partial \theta}{\partial t} + v\frac{\partial \theta}{\partial y} + w\frac{\partial \theta}{\partial z} = \kappa_T\nabla_1^2\theta \qquad \left(\nabla_1^2 = \frac{\partial^2}{\partial y^2} + \frac{\partial^2}{\partial z^2}\right).
\end{aligned} \tag{6.71}$$

We have reverted to (xyz) notation again on account of its familiarity. The equation of continuity may be satisfied by introducing a stream function:

$$v = -\partial\psi/\partial z \qquad w = \partial\psi/\partial y \tag{6.72}$$

in terms of which the vorticity is:

$$\xi = \partial w/\partial y - \partial v/\partial z = \nabla_1^2\psi . \tag{6.73}$$

The pressure may be eliminated by taking the curl on the second and third of Equation (6.71):

$$\left(\frac{\partial}{\partial t} - \frac{\partial\psi}{\partial z}\frac{\partial}{\partial y} + \frac{\partial\psi}{\partial y}\frac{\partial}{\partial z}\right)\nabla_1^2\psi = \nu_T\nabla_1^4\psi + \beta g\frac{\partial\theta}{\partial y}. \tag{6.74}$$

This equation and the last of Equations (6.71) (with v, w expressed in terms of ψ) constitute two equations for the two unknowns ψ and θ. In the absence of boundaries thc velocities must vanish at sufficiently large distances from the thermal, i.e.,

$$\nabla\psi = 0 \quad \text{as} \quad y, z \to \infty . \tag{6.75}$$

The distribution of potential temperature far from the thermal need not be as simple: in general, $\theta = \theta_a(z)$ as $y, z \to \infty$, where θ_a is the atmospheric potential temperature. Such variations in θ would certainly influence the upward motion of a thermal. In the first instance we shall consider the simplest case of a *neutral atmosphere*, $\theta_a = 0$, i.e.,

$$\theta = 0 \quad \text{as} \quad y, z \to \infty . \tag{6.76}$$

With this last assumption an integration of the energy equation over all (two-dimensional) space and the application of the divergence theorem yields:

$$\partial/\partial t \iint_{-\infty}^{\infty} \theta \, \mathrm{d}y \, \mathrm{d}z = 0 . \tag{6.77}$$

The physical meaning of this result is that total excess heat is conserved in a neutral atmosphere:

$$H/U = \iint_{-\infty}^{\infty} \varrho c_p \theta \, \mathrm{d}y \, \mathrm{d}z = \text{const.}, \tag{6.78}$$

where we have used our result from the previous section that the heat content per unit depth of the line thermal modeling a chimney plume is H/U, with H = heat released at the chimney per unit time into a uniform wind of velocity U.

The basic equations contain the buoyant acceleration $\beta g\theta$. The space integral of this is also conserved (to the accuracy of the Boussinesq approximation) and is

$$\iint_{-\infty} \beta g\theta \, \mathrm{d}y \, \mathrm{d}z = \frac{\beta g}{\varrho c_p}\frac{H}{U}. \tag{6.79}$$

length scales, V and L, the distribution of u/V, etc., in terms of y/L etc., remains identical in the course of the development of the flow. If such moving equilibrium were to apply in the course of the development of a line-thermal, it would also imply a self-similar excess temperature distribution, i.e., $\theta/\Theta = \text{func}(y/L, \text{etc.})$, where Θ is a suitable temperature scale. There is no compelling logical reason why the flow pattern in a line thermal should be self-similar for all stages of the motion (past an initial development phase), but experience with other simple kinds of free turbulent flows suggests this as a likely possibility, provided only that self-similarity is compatible with the equations of motion. As we shall see below, experiment does confirm the predictions of the similarity theory, so that the notion of a 'moving equilibrium' in a line thermal in a neutral atmosphere may be accepted as a verified fact.

If the length, velocity and temperature scales of the flow in a line thermal are $L(t)$, $V(t)$ and $\Theta(t)$, nondimensional (starred) velocities, etc., may be introduced by

$$\begin{aligned} &v^* = v/V \qquad w^* = w/V \\ &y^* = y/L \qquad z^* = z/L \\ &\psi^* = \psi/LV = \text{func}(y^*, z^*) \qquad \theta^* = \theta/\Theta = \text{func}(y^*, z^*) \end{aligned} \tag{6.86}$$

noting that ψ^* and θ^* are not functions of time. Substituting into Equations (6.71) and (6.74) two nondimensional equations for ψ^* and θ^* may be obtained:

$$\begin{aligned} &\frac{1}{V^2}\frac{\mathrm{d}(LV)}{\mathrm{d}t}\nabla_1^{*2}\psi^* - \frac{1}{V}\frac{\mathrm{d}L}{\mathrm{d}t}\left(y^*\frac{\partial}{\partial y^*} + z^*\frac{\partial}{\partial z^*}\right)\nabla_1^{*2}\psi^* - \\ &\quad -\left(\frac{\partial\psi^*}{\partial z^*}\frac{\partial}{\partial y^*} - \frac{\partial\psi^*}{\partial y^*}\frac{\partial}{\partial z^*}\right)\nabla_1^{*2}\psi^* = \frac{\nu_\mathrm{T}}{VL}\nabla_1^{*4}\psi^* + \frac{\beta g\Theta L}{V^2}\frac{\partial\theta^*}{\partial y^*} \\ &\frac{L}{V\Theta}\frac{\mathrm{d}\Theta}{\mathrm{d}t}\theta^* - \frac{1}{V}\frac{\mathrm{d}L}{\mathrm{d}t}\left(y^*\frac{\partial}{\partial y^*} + z^*\frac{\partial}{\partial z^*}\right)\theta^* - \\ &\quad -\left(\frac{\partial\psi^*}{\partial z^*}\frac{\partial}{\partial y^*} - \frac{\partial\psi^*}{\partial y^*}\frac{\partial}{\partial z^*}\right)\theta^* = \frac{\kappa_\mathrm{T}}{VL}\nabla_1^{*2}\theta^*. \end{aligned} \tag{6.87}$$

These equations contain several nondimensional combinations of the space, velocity, and temperature scales in addition to the starred variables. If the flow is to be self-similar, the solutions ψ^* and θ^* must be independent of time. This is only possible if the various nondimensional products in Equations (6.87) are also all independent of time.

The nondimensional combinations in Equation (6.87) containing eddy transport coefficients are reciprocal eddy Reynolds and Péclet numbers, while the one containing $\beta g\Theta$ is the ratio of buoyant forces to inertia forces, and may also be regarded as a Rayleigh number divided by Reynolds number and Péclet number

$$\begin{aligned} \nu_\mathrm{T}/VL &= \mathrm{Re}_\mathrm{T}^{-1} \\ \kappa_\mathrm{T}/VL &= \mathrm{Pé}_\mathrm{T}^{-1} \\ \beta g\Theta L/V^2 &= A = \mathrm{Ra}\ \mathrm{Pé}_\mathrm{T}^{-1}\ \mathrm{Re}_\mathrm{T}^{-1}. \end{aligned} \tag{6.88}$$

Given only that these nondimensional combinations remain constant during the growth of a line thermal some quite specific deductions may be made in regard to the behavior of the length, velocity, and temperature scales. The temperature scale Θ may be chosen to be the 'average' excess temperature in the following sense:

$$\Theta = \frac{1}{L^2} \iint_{-\infty}^{\infty} \theta \, \mathrm{d}y \, \mathrm{d}z = \frac{F}{\beta g U L^2} \tag{6.89}$$

where we have made use of Equations (6.79) and (6.80).

A convenient velocity scale is the bulk vertical velocity W of the marked fluid defined by Equation (6.68). In view of the conservation of total heat content (valid in a neutral atmosphere) it is legitimate to assume that excess temperature θ 'marks' the plume as well as any other conservative property. Therefore the definition, Equation (6.68), may be modified to read

$$W = \frac{\iint w\theta \, \mathrm{d}y \, \mathrm{d}z}{\iint \theta \, \mathrm{d}y \, \mathrm{d}z} = V \frac{\iint w^*\theta^* \, \mathrm{d}y^* \, \mathrm{d}z^*}{\iint \theta^* \, \mathrm{d}y^* \, \mathrm{d}z^*} . \tag{6.90}$$

Setting $W = V$ we see that both nondimensional integrals in Equation (6.90) have the value unity (the one in the denominator on account of Equation (6.89). A convenient length scale L is the rms lateral dispersion of marked fluid, already defined by Equation (6.84):

$$L = \sigma_y . \tag{6.91}$$

Writing $A = \mathrm{Ra}\mathrm{Pé}_\mathrm{T}^{-1}\mathrm{Re}_\mathrm{T}^{-1} = \text{const.}$ we find then from Equations (6.88) and (6.89)

$$V^2 L = F/UA = \text{const.} \tag{6.92}$$

From Equation (6.85) we obtain now

$$\mathrm{d}L^2/\mathrm{d}t = 2(\kappa_\mathrm{T} + \kappa_\mathrm{m}) = 2\mathrm{Pé}_\mathrm{T}^{-1} VL + 2BVL , \tag{6.93}$$

where

$$B = \text{const.} = \iint_{-\infty}^{\infty} y^* v^* \theta^* \, \mathrm{d}y^* \, \mathrm{d}z^*$$

the exact value of B depending on the details of the flow pattern. Defining the new constant

$$\alpha = \mathrm{Pé}_\mathrm{T}^{-1} + B \tag{6.93a}$$

we may write Equation (6.93) as

$$\mathrm{d}L/\mathrm{d}t = \alpha V \tag{6.94}$$

which may be regarded as a 'law of growth' for the line thermal. Similar laws of growth apply to other simple free turbulent flows, e.g., jets, mixing layers, and wakes. They follow directly by dimensional argument, if we reason that the rate of expansion of the flow region, dL/dt, is itself a property of the turbulent field and hence must be determined by length and velocity scales L, V. The proportionality factor α may be expected to remain constant if the turbulent flow conserves its structure in the course of its development. However, different patterns of turbulent flow may be expected to be characterized by different growth constants α.

The momentum integral relationship of Equation (6.82) may be stated as

$$\gamma \frac{\mathrm{d}(VL^2)}{\mathrm{d}t} = \frac{F}{U}, \tag{6.95}$$

where

$$\gamma = \int\int_{-\infty}^{\infty} w^* \, \mathrm{d}y^* \, \mathrm{d}z^*$$

is a 'momentum constant'. The result in Equation (6.95) is consistent with Equations (6.92) and (6.94) if

$$\gamma = 2A/3\alpha . \tag{6.96}$$

From Equations (6.89), (6.92), and (6.94) it is now possible to deduce the time-dependence of the three scales, L, V and Θ using the initial condition:

$$L \to 0 \quad \text{as} \quad t \to 0 . \tag{6.97}$$

The results are, after replacing A by γ on the basis of Equation (6.96):

$$\begin{aligned} L &= \left(\frac{3\alpha}{2\gamma} \frac{F}{U} t^2\right)^{1/3} \\ V &= \left(\frac{4}{9\alpha^2\gamma} \frac{F}{U} t^{-1}\right)^{1/3} \\ \Theta &= \left(\frac{1}{\beta g} \frac{4\gamma^2}{9\alpha^2} \frac{F}{U} t^{-4}\right)^{1/3} . \end{aligned} \tag{6.98}$$

The vertical distance covered by the line thermal since its release is further

$$Z = \int_0^t W(t')\,\mathrm{d}t' = \left(\frac{3}{2\alpha^2\gamma} \frac{F}{U} t^2\right)^{1/3} = \frac{L}{\alpha} . \tag{6.99}$$

The consequences of the similarity theory of line thermals are sometimes developed from this last relationship, $L = \alpha Z$, postulated *à priori* on the grounds that Z is the only 'natural' length scale available (e.g., Scorer, 1958). If such an argument were valid, it would have to apply equally well to the width of two-dimensional wakes, for example, with Z replaced by distance behind the body causing the wake (at large distances, where the size of the body itself no longer matters). This is known not to be the case

(Townsend, 1956). A law of growth of the form of Equation (6.94) does, on the other hand, apply to such wakes, but this, together with the conservation of momentum yields a relationship of the form $L = \text{const.} \sqrt{x}$, i.e., a square root relationship rather than a linear one appearing in Equation (6.99).

From the above results it is now a simple matter to show that the other three non-dimensional products in Equation (6.87), viz.,

$$\frac{1}{V^2}\frac{\mathrm{d}(LV)}{\mathrm{d}t}, \quad \frac{1}{V}\frac{\mathrm{d}L}{\mathrm{d}t} \quad \text{and} \quad \frac{L}{V\Theta}\frac{\mathrm{d}\Theta}{\mathrm{d}t}$$

are also constant, being respectively equal to $(3\alpha/2)$, α and (-2α). We conclude that self-similarity in a line thermal is (a) consistent with the equations of continuity, motion, and energy and, (b) implies the relationships given by Equations (6.98) and (6.99). We note that these relationships contain two constants α and γ, the values of which depend on the exact flow pattern within the thermal and must, in our present incomplete state of knowledge of turbulent flow, be determined by experiment.

6.12. Bent-Over Chimney Plumes

We have developed the above similarity theory with a view to applying it to buoyant plumes in the atmosphere or ocean. To put our results in perspective, we shall discuss here in some detail the physical factors involved in the movement and dispersion of buoyant gases discharged from an industrial chimney into a turbulent atmosphere. Somewhat similar (but not quite identical) considerations apply to continuous heavy or buoyant plumes in the ocean, but we shall not consider this second application in detail.

Hot gases usually leave the chimney tops with a vertical velocity w_0 of order 10 m s^{-1}, entering a cross-wind of similar speed U. If the wind is very strong, or chimney exit velocity low (i.e., if the ratio w_0/U is less than approximately 1.0) or if the airflow is very irregular, eddies in the wake of the chimney entrain some of the effluent. Rapid vertical movement occurs in the separated flow region behind the chimney stack and this usually communicates with the larger similar region behind the buildings where the boiler plants, etc., are housed. Effluents are thus brought to ground level very close to the chimney and in undesirably high concentrations, a phenomenon known as 'aerodynamic downwash' (Sherlock and Stalker, 1940). The avoidance of downwash is an aerodynamic design problem and is best carried out with the aid of wind tunnel model studies of a plant to be built (or to be doctored to eliminate downwash). We shall not discuss this phenomenon further, assuming that the gases on leaving the chimney clear the wake of both chimney stack and any buildings.

The effluent gases constitute a buoyant jet entering a turbulent cross-wind. Rapid mixing with atmospheric air takes place and the jet axis bends over into the wind as the effluent gases acquire the horizontal momentum of the ambient air. Observations show that the transfer of *horizontal* momentum is essentially complete within a few chimney diameters from the exit, portions of the plume thereafter traveling horizontally at a mean speed indistinguishable from wind speed. Their *vertical* velocity relative to the ambient fluid does not, however, disappear as quickly, mainly because of the continued action of the buoyancy force.

Some order of magnitude figures may illuminate the above remarks. From careful observation of the size of chimney plumes near the chimney top it may be inferred that the effective mass of the effluent gases increases through vigorous mixing with ambient air by something like a factor of 30 within a distance of 3 to 5 chimney diameters or so. By this time the deficiency of horizontal momentum compared to ambient air is therefore a negligible 3%. The vertical velocity due to initial momentum is also 3% of the original w_0, say of order 30 cm s^{-1}. However, if the chimney diameter is not much less than say 3 m, the buoyancy force has had several seconds to act in the time the gases have moved some 3 to 5 times 3 m. The initial buoyant acceleration is usually of the order of 10 m s^{-2}, and although this also reduces through mixing in the same proportion as initial momentum, it generates an appreciable vertical velocity within the first few seconds. Indeed if the average acceleration between leaving the chimney top and 30-fold increase in mass is only 1 m s^{-2}, and if this initial adjustment phase lasts at least 1 s, the buoyant contribution to vertical velocity becomes 1 m s^{-1} or 3 times larger than that due to initial momentum. We should note that this conclusion holds only for chimneys discharging substantial quantities of heat. An initial buoyancy of order 10 m s^{-2} implies an initial excess temperature of order 300 °C. Given a 3 m diameter chimney and an exit velocity of 10 m s^{-1} this corresponds to a considerable rate of heat release (order 6000 kcal s^{-1}). The same conclusions do not apply to much smaller chimneys (order 1 m in diameter or less) nor to those which discharge their gases with a small buoyant acceleration. In such non-buoyant cases any plume rise is due to initial momentum and is essentially complete within 10 chimney diameters or so. We shall confine our further remarks to large heat sources which produce 'buoyancy dominated' plumes (see, however, the appendix to this chapter).

The above considerations suggest that in buoyancy dominated plumes neither the radius R_0 of the chimney, nor the initial vertical velocity w_0 are likely to be of much consequence in determining the path of the plume beyond the earliest mixing phase. The total excess heat, however, is likely to be of dominant importance. Another point is that beyond the first 3 to 5 chimney diameters or so the excess temperature is small enough even in plumes generated by large heat sources for the Boussinesq approximation to be valid.

Choosing the coordinate axes in the usual way (Figure 6.4) we have seen that for a heat release rate H at the chimney, the heat content of the plume per unit distance along the x-axis is H/U. This affects plume dynamics by exerting a total buoyant force on a given plume slice of

$$\frac{F}{U} = \frac{\beta g}{\varrho c_p} \frac{H}{U}$$

as we have already seen in Section 6.9.

In a neutral atmosphere, or for suitably small vertical plume displacements, and

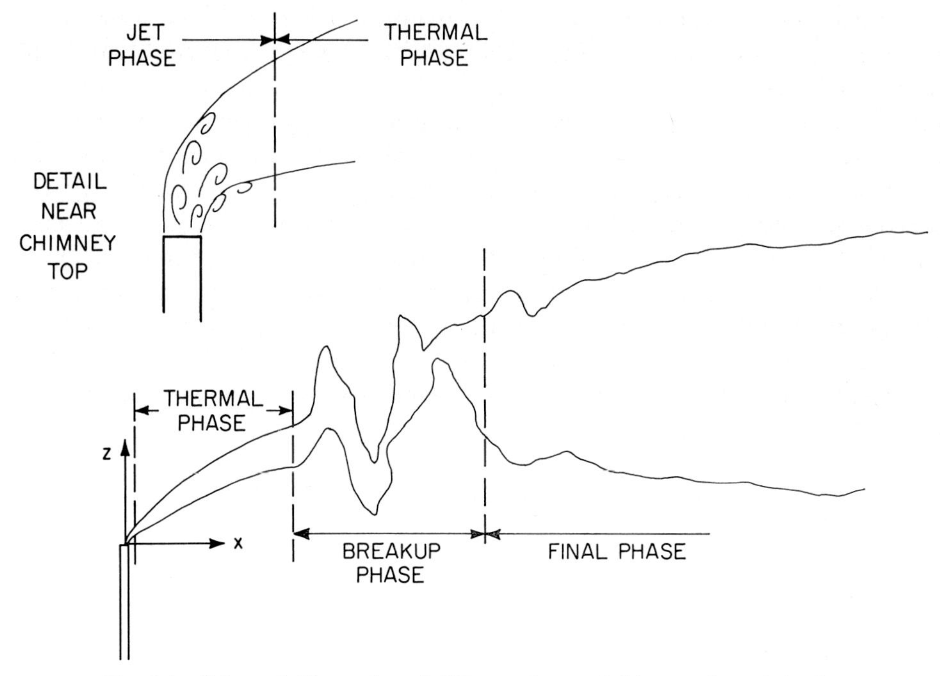

Fig. 6.4. Schematic illustration of different phases of chimney plume behavior.

beyond the above discussed initial adjustment or 'jet' phase, the flux of buoyancy F may be regarded a constant. However, in a stable atmosphere the potential temperature of the environment increases as the plume rises, so that the plume's excess temperature and hence its total buoyancy decreases. The converse is the case in an unstable atmosphere.

Observations show that most buoyancy dominated plumes retain smooth outlines and a moderate slope (of the order of 0.2 and less) against the horizontal for some distance after leaving the chimney. In this phase it turns out to be reasonable to regard segments of the plume as if they were segments of a line thermal moving upward through quiescent surroundings. Mixing in this phase must therefore be mainly due to the self-generated turbulence of buoyant movements, with atmospheric turbulence exerting no significant effects.

A key experimental fact revealed by careful observations on buoyant plumes is that this relatively regular 'phase 1' of plume rise comes to an end at some predictable distance from the source and rather more vigorous mixing sets in, where the plume often breaks up into several distinct parcels. This 'breakup' phase of the plume is more pronounced in strong atmospheric turbulence and also occurs closer to the chimney. It is clear that the large eddies which lead to the breakup are those naturally present in the wind (they are too large to be produced by the plume's own motion) and also that their mixing action dwarfs the effects of the self-generated turbulence. When 'breakup' is pronounced, it leads to an almost stepwise increase in average plume diameter. A little further downwind the distinct parcels merge again into a larger, more diffuse plume, the subsequent growth of which is relatively slow. The short breakup phase may be called 'phase 2', the final diffuse phase 'phase 3' of plume behavior. In both these phases atmospheric turbulence clearly dominates mixing.

The mean distance Z by which the plume rises above the chimney top is a function of the downwind distance x and a number of other physical parameters which differ from phase to phase of plume behavior, and which are listed in Table VI.1, summarizing our discussion above. We assume here that the plume is sufficiently far above ground so that the flow pattern within it is not disturbed by the ground. Otherwise also the distance $(h_s + Z)$ $(h_s = \text{stack height})$ may influence further plume rise (on this point see some later remarks in Section 6.15).

Dimensional analysis now leads to the following functional relationship in the four phases:

phase 0 ('jet phase')

$$\frac{Z}{R_0} = \text{func}\left(\frac{x}{R_0}, \text{Fr}, \frac{w_0}{U}, \frac{\varrho_g}{\varrho_c}\right), \tag{6.100}$$

where

$$\text{Fr} = w_0 \left(R_0 g \frac{\varrho_a - \varrho_g}{\varrho_g}\right)^{-1/2}$$

is a densimetric Froude number.

phase 1 ('thermal phase')

$$\frac{Z}{l} = \text{func}\left(\frac{x}{l}, \frac{g}{T_a} \frac{d\theta_a}{dz} \frac{l^2}{U^2}\right), \tag{6.101}$$

where

$$l = F/U^3$$

is a convenient length scale of the buoyancy dominated phase.

phase 2 ('breakup phase')

$$\frac{Z}{l} = \text{func}\left(\frac{x}{l}, i_y, \frac{l}{L_t}\right) \tag{6.102}$$

phase 3 ('diffuse phase')

$$\frac{Z}{l} = \text{func}\left(\frac{x}{l}; i_y; \frac{}{L_t}; \frac{g}{T_a} \frac{d\theta_a}{dz} \frac{l^2}{U^2}\right). \tag{6.103}$$

In the above we have represented the atmospheric potential temperature distribution by its gradient, but this of course may itself be a function of height. Also, as we remarked before, the effect of a solid boundary on the buoyant movements in a large plume (possibly extending down to ground level) has been ignored. Another simplification was that we have assumed distinct 'phases,' which must in reality continuously merge into one another. Thus even the above complex scheme is highly idealized.

TABLE VI.1

Physical factors involved in plume behavior (buoyancy dominated plumes)

Phase 0	Phase 1	Phase 2	Phase 3
Description Bent-over buoyant jet phase	Buoyancy dominated or 'thermal' phase (self-generated turbulence causes mixing)	Breakup phase (atmospheric eddies cause rapid growth)	Diffuse phase (slow further growth due to atmospheric eddies)
Downwind extension (order of magnitude) 5 diameters	300 m	100 m	Indefinite
Important variables Wind speed U Exit speed w_0 Chimney radius R_0 Density of effluent ϱ_g Density of air ϱ_a Buoyant acceleration $g(\varrho_a - \varrho_g)/\varrho_g$	Wind speed U Flux of buoyancy F Atmospheric potential temperature gradient $d\theta_a/dz$ (this enters in the combination $g/T_a \, d\theta_a/dz$, cf. later discussion)	Wind speed U Flux of buoyancy F Atmospheric turbulence intensity i_y Scale of atmospheric turbulence L_t	Wind speed U Flux of buoyancy F Atmospheric turbulence intensity i_y Scale of atmospheric turbulence L_t Atmospheric potential temperature gradient $d\theta_a/dz$

The line-thermal analogy applies to one of the four phases of this idealized scheme. This is the only phase for which we have an adequate theory.

6.13. Theory of Buoyancy Dominated Plumes in a Neutral Atmosphere

When the atmospheric potential temperature gradient is negligible, Equation (6.101) takes on the simple form:

$$Z/l = \text{func}(x/l). \tag{6.104}$$

This relatively simple relationship should therefore characterize phase 1 of a buoyancy dominated plume in a neutral atmosphere. To deduce the functional form of the relationship we note that in this phase the line thermal analogy is reasonable, atmospheric turbulence being unimportant. An element of the plume at distance x from the source may be regarded as a slice of a line thermal which has been developing for a time $t = x/U$. Applying Equation (6.99) we find then at once

$$\frac{Z}{l} = C\left(\frac{x}{l}\right)^{2/3}, \tag{6.105}$$

where $C = (3/2\alpha^2\gamma)^{1/3}$ is a constant depending on the exact flow pattern in the self-preserving plume-section.

It is also of interest to observe that Equations (6.98) and (6.99), characterizing the behavior of a line thermal have in effect been derived from the following three relationships:

$$\beta g \Theta L^2 = \frac{F}{U} = \text{const.} \tag{6.89a}$$

$$\gamma \frac{\mathrm{d}}{\mathrm{d}t}(WL^2) = \frac{F}{U} \tag{6.95a}$$

$$\mathrm{d}L^2/\mathrm{d}t = 2\alpha WL. \tag{6.93a}$$

All three follow from the self-similarity of the profiles *and* integral relationships for heat and momentum.

The above three equations may also be established by an elementary approximate argument, due to Morton *et al.* (1956). Consider a slice of the plume (Figure 6.5) and assume that this slice moves upward with a constant velocity W and a temperature excess Θ constant 'within' the plume, zero without (this is sometimes stated in the form that both velocity and temperature have 'top hat' profiles across a plume slice). The main idealization involved in this model is that, although in reality the flow and temperature patterns are continuous, an artificial distinction is introduced between an 'identifiable plume' and the ambient fluid. The identifiable plume is then assumed to grow by 'entrainment' of the ambient fluid, the rate of entrainment being governed by an entrainment velocity v_e at the perimeter of the plume. If the radius of the supposedly circular plume is R, its growth rate is, by inflow of ambient fluid

$$\frac{\mathrm{d}}{\mathrm{d}t}(\pi R^2) = 2\pi R v_e. \tag{6.106}$$

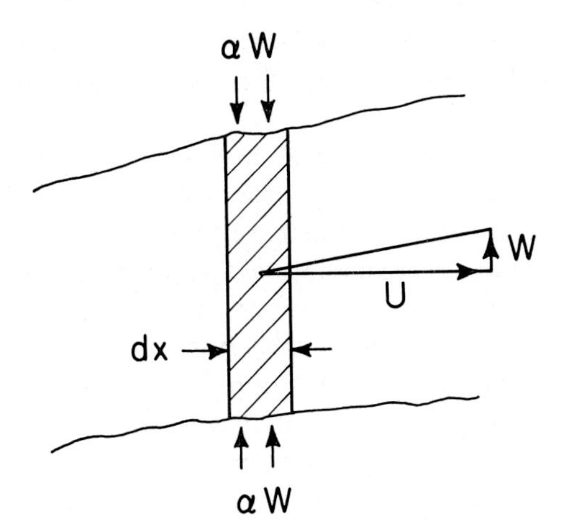

Fig. 6.5. Entrainment into element of nearly horizontal plume.

Because the entrainment is due to the plume's self-generated motion, it is reasonable to assume $v_e = \alpha W$ with $\alpha = \text{const}$. It is then seen that Equation (6.106) becomes identical with Equation (6.93a), on replacing L by R. From the balance of heat and momentum also Equations (6.89) and (6.95) follow, with $\gamma = 1$, and the factor π multiplying the left hand side of Equation (6.89) (Slawson and Csanady, 1967).

In spite of the rather crude assumptions involved, the 'entrainment' model of a plume therefore possesses essentially the same physical content as the similarity theory. One advantage of the elementary approach is the ease with which it may be applied to other situations: for example it is easy to write down equations similar to Equations (6.89), (6.95) and (6.93) for a plume element *inclined* at an arbitrary angle against the vertical. Another point is that the same approach may be used as an approximation in cases where self-similarity does not hold.

We turn next to phases 2 and 3 of a buoyancy dominated plume. The view that a plume segment is almost like a line thermal remains valid, but the mixing with ambient air must now be attributed to atmospheric turbulence, rather than to the self-generated turbulence of the plume itself. Equations (6.89a) and (6.95a) do not involve any turbulence parameters and remain valid in a neutral atmosphere, with the proviso that γ is a momentum constant in these latter phases, where the flow pattern may well be different from phase 1, so that γ is not necessarily the same from phase to phase.

To assess the rate of growth of the plume in phases 2 and 3 we assume that the effects of the environmental turbulence now dwarf those due to an organized flow pattern as well as to the self-generated turbulence, i.e., that κ_T and κ_m in Equation (6.93) are in these phases negligible compared to an eddy diffusivity K_T due to atmospheric turbulence. It is reasonable to assume that atmospheric turbulence is stationary and homogeneous in the neighborhood of a plume element. The corresponding eddy diffusivity is then that appropriate to *relative* diffusion in a homogeneous field, which we have discussed at length in Chapter IV. We have found that while the diffusing cloud is small compared to the eddy length scale ($L \ll L_t$) the diffusivity K_T increases rapidly in time. For the inertial subrange a growth rate proportional to t^2 has been determined, which also implied $L = \text{const.}\ t^{3/2}$, a faster than linear growth of cloud size. The duration of such a phase, however, is found to be very limited. When the cloud becomes large compared to the atmospheric eddies ($L \gg L_t$), K_T becomes constant, proportional to the product of rms turbulent velocity and eddy length scale.

It is clear from the account of experimental facts on plume rise given earlier that we may identify phase 2 of plume growth with the phase of accelerated relative diffusion, when K_T grows in time. A qualitative theoretical model of this phase is provided by the inertial subrange formulas for K_T. In phase 3 of plume growth, on the other hand, we may assume that K_T has reached its asymptotic value.

It turns out to be most informative to focus on the plume *slope* dZ/dx in phases 2 and 3. On integrating the momentum Equation (6.95a) in time we find

$$\frac{dZ}{dx} = \frac{W}{U} = \frac{lx}{\gamma L^2}, \tag{6.107}$$

where again $l = F/U^3$ is the length scale based on buoyancy flux. Consider now the changes in plume slope during phase 2. If at the onset of this phase the plume size L is rather smaller than the eddy length-scale L_t, rapid growth occurs in a relatively short distance along the x-axis. For example, between $x = 300$ and 400 m the plume radius might increase by a factor of 3. From Equation (6.107) it follows then that plume slope decreases in the ratio 27:4 or approximately by a factor of 7. It is thus possible for the slope to become negligible (10^{-2} or less), by the end of phase 2.

On the other hand, such behavior is certainly not necessary: if the plume is large enough at the time when atmospheric turbulence becomes the dominant cause of mixing further growth is relatively slow. 'Phase 2' may under such circumstances be missing and a third phase set in at once, where the growth is described approximately by the asymptotic relationship:

$$L^2 = 2K_T t = 2i_y L_t x, \tag{6.108}$$

where i_y is lateral turbulence intensity.

The corresponding final slope is

$$\left.\frac{dZ}{dx}\right|_{x\to\infty} = \frac{l}{2i_y \gamma L_t} = \frac{F}{2i_y \gamma L_t U^3}. \tag{6.109}$$

We observe that the asymptotic slope is highly sensitive to wind speed.

It should also be noted here that in phases 2 and 3 strict self-similarity cannot be expected to hold, because the eddy Reynolds and Péclet numbers K_T/VL vary in time or along plume axis. The entire argument then rests on approximate equations involving the assumption of an 'identifiable plume,' on the basis of the approach of Morton *et al.* (1956) which we described briefly above.

6.14. Comparison with Observation

Laboratory studies of line thermals have been carried out by Richards (1963) and Tsang (1971). Their results generally supply an excellent confirmation of the predictions of the similarity theory, Equations (6.98) and (6.99). Both studies made use of a water tank with transparent sides, the marked fluid having been sometimes negatively, sometimes positively buoyant. Bulk flow parameters as well as detailed flow patterns were obtained. The main observed bulk parameters were the displacement of the leading edge of the thermal, x_e and the apparent plume radius R_a.

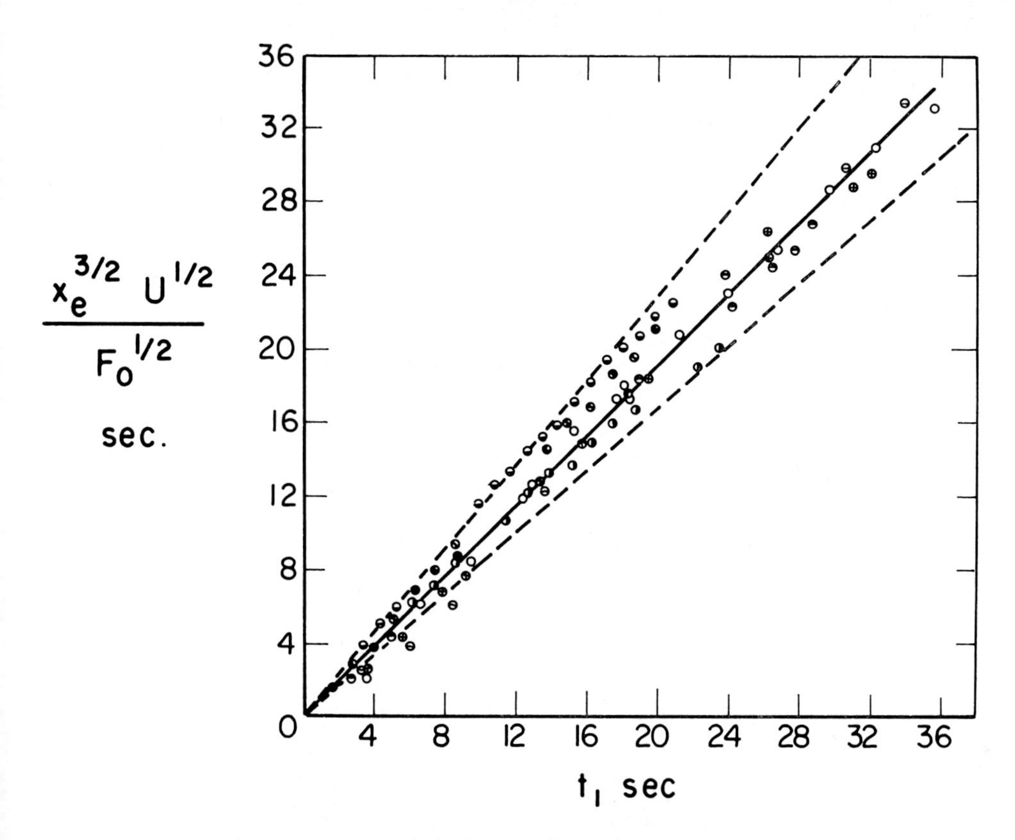

Fig. 6.6. Observed behavior of a line thermal's leading edge. Laboratory experiments of Tsang (1971).

Figures 6.6 and 6.7 show sample illustrations of observed $x_e(t)$ and $R_a(x_e)$ relationships. For the purpose of comparing these with our theoretical results the following approximate identification may be made (Figure 6.8):

$$\begin{aligned} R_a &= 2L \\ x_e &= Z + R_a = Z + 2L\,. \end{aligned} \tag{6.110}$$

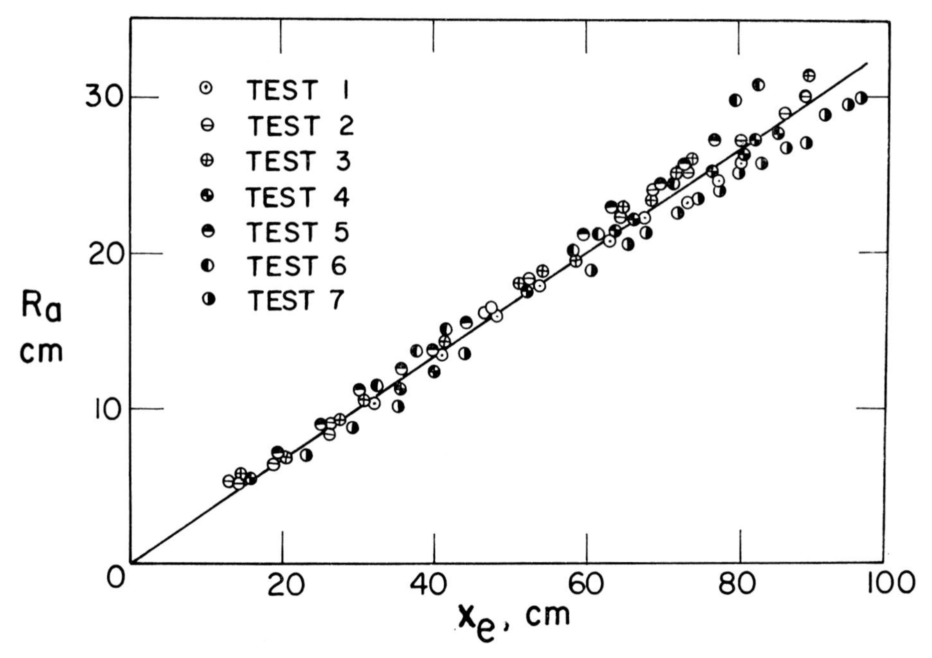

Fig. 6.7. Observed relationship of a line thermal's apparent radius R_a to its leading edge displacement x_e. Laboratory experiments of Tsang (1971).

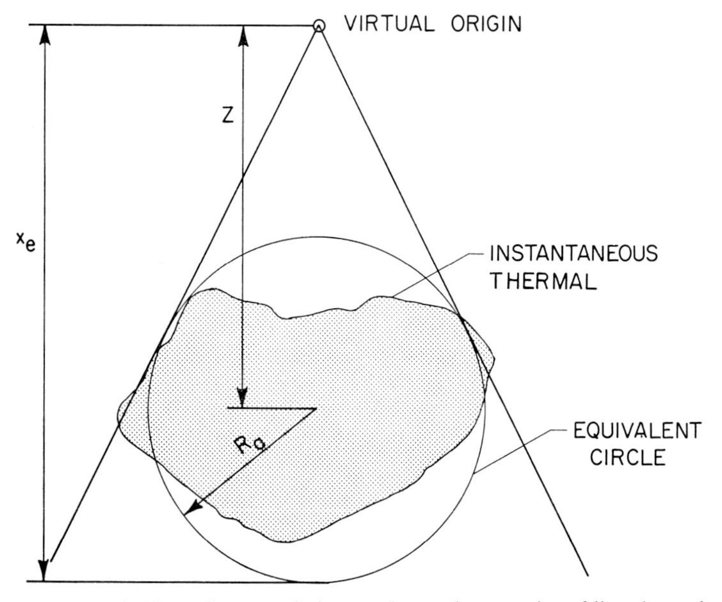

Fig. 6.8. Definition of terms relating to observed properties of line thermals.

The observed results are in accordance with Equation (6.99) which we repeat here:

$$L = \alpha Z$$

$$Z = C\left(\frac{F}{U}\right)^{1/3} t^{2/3} \quad \text{where } C = (3/2\alpha^2\gamma)^{1/3}. \tag{6.99a}$$

Using the approximations of Equation (6.110) we may deduce from the experimental data the values of the constants α and C applying to each individual thermal. It turns out that these values scatter considerably, apparently in response to the exact manner of generating the thermal. In Richards' experiments α varied between 0.25 and 1.0, C between 0.33 and 1.5. Tsang has shown that by a careful control of release conditions the range of variability may be greatly reduced: his mean values, are $\alpha=0.25$, $C=1.03$, with a spread of no more than 10% either side of these means.

With the aid of the definition of $C=(3/2\alpha^2\gamma)^{1/3}$ it is possible to calculate from a given value-pair α, C the momentum constant γ. The range in the Richards' experiments turns out to be $\gamma=7$ to 40, while Tsang's mean values yield $\gamma=22$.

We emphasize here that the spread in the values of the constants observed by Richards was *not* due to experimental inaccuracies: each individual thermal behaved quite accurately as predicted by the similarity theory. The variation in α and γ is a genuine physical fact and may be attributed to the existence of different flow structures in response to slightly changed initial conditions. One of the challenging current problems of research is to elucidate the physical reasons for these differences.

Turning now to chimney plumes in the atmosphere, the theoretical prediction is that in a neutral atmosphere (or in weakly stable or unstable cases, for small total rise Z) buoyancy-dominated plumes should rise in their relatively regular 'phase 1' according to the $\frac{2}{3}$ power law (Equation (6.105)), the constant C being the same as applies to laboratory line thermals. The available evidence has recently been reviewed in some detail by Briggs (1969) who shows that this is indeed the case. After allowing for a factor $\pi^{1/3}$ (on account of a different definition of the flux of buoyancy, F/π in our notation) Briggs' recommendation is a mean value of the constant $C=1.15$, varying, however between about 0.7 and 1.7. These values are very much as those characterizing laboratory thermals: the small differences may be partly attributed to the inaccuracy of the assumptions in Equation (6.110).

A practically important datum is the extension of 'phase 1' in chimney plumes. According to our previous considerations, phase 1 ends when the velocities of the buoyant movements decay to a level comparable to velocity fluctuations induced by atmospheric turbulence. The transition point should therefore be characterized by:

$$W = bi_z U, \tag{6.111}$$

where $b=$const. of order unity, and i_z is vertical turbulence intensity. Noting that W/U is also the slope of the plume, Equation (6.111) may be stated in the form that, at the 'breakup' point:

$$\left.\frac{dZ}{dx}\right|_b = bi_z. \tag{6.112}$$

The experimental evidence on the 'breakup constant' b is meager, but from the study of Bringfelt (1969) and the data of Slawson and Csanady (1967) we may tentatively infer that

$$b = \alpha^{-1}. \tag{6.113}$$

Applying the $\frac{2}{3}$ power law to determine the slope and coordinates of this breakup point we find further

$$\left.\frac{x}{l}\right|_{b} = \left(\frac{2C}{3bi_z}\right)^3$$

$$\left.\frac{Z}{l}\right|_{b} = \frac{2}{3\,(\alpha b)^2\, i_z^2}.$$

These results show that the transition point coordinates are sensitive to atmospheric turbulence. In the case of moderate turbulence, $i_z = 0.05$, we find $x/l|_b = 1200$, $Z/l|_b = 260$ or so. These values of the transition coordinates were indicated by early observations and confirmed by more recent studies (Csanady, 1961; Bringfelt, 1969). In the presence of more intense environmental turbulence a much shorter phase 1 is found (Slawson and Csanady, 1971).

In regard to phase 2 of plume behavior our earlier discussion indicated that the ratio of plume size to eddy length scale $L/L_t|_b$ determines whether or not a sudden plume growth will in fact take place. Observations have shown (Slawson and Csanady, 1967, 1971) that in phase 2:

(1) The plume slope dZ/dx changes only imperceptibly in some cases, drastically in others. At the transition point $dZ/dx|_b$ is usually of the order of 0.1 in accordance with Equation (6.112). In some cases this slope is retained for long distances past the transition point, in others the slope reduces abruptly to a negligible value.

(2) The cases with an abrupt change of slope are associated with high atmospheric turbulence, usually of thermal origin.

The same observations have also shown the occasional presence of a phase 3 with constant slope, qualitatively in accordance with Equation (6.109). The evidence is insufficient and incomplete to attempt a quantitative verification. We should also note that at large distances the plume usually becomes large enough to be influenced by the presence of the ground, so that Equation (6.109) is unlikely to be of much practical use.

6.15. Flow Pattern within a Plume

The line thermal experiments of Richards and Tsang referred to above have also yielded detailed information on the flow pattern. A typical illustration of streamlines is shown in Figure 6.9 after Tsang (1971). Theoretically, such a pattern could be obtained by solving the non-linear Equations (6.87), after substituting appropriate numerical values for the nondimensional constants. Using Equations (6.98) and the experimentally determined average values of α and γ (0.25 and 22) we find the following constants:

$$\frac{1}{V^2}\frac{\mathrm{d}(LV)}{\mathrm{d}t} = \frac{3\alpha}{2} = 0.38$$

$$\frac{1}{V}\frac{\mathrm{d}L}{dt} = \alpha = 0.25$$

$$\frac{\beta g \Theta L}{V^2} = \frac{3\alpha\gamma}{2} = 8.25\,.$$

The reciprocal Reynolds number may be assumed to be approximately equal to the reciprocal Péclet number. The constant B in Equation (6.93) is likely to be positive, so that

$$\nu_{\mathrm{T}}/VL \cong \kappa_{\mathrm{T}}/VL \leqslant \alpha = 0.25\,.$$

The coefficient of the nonlinear terms in Equation (6.87) is unity.

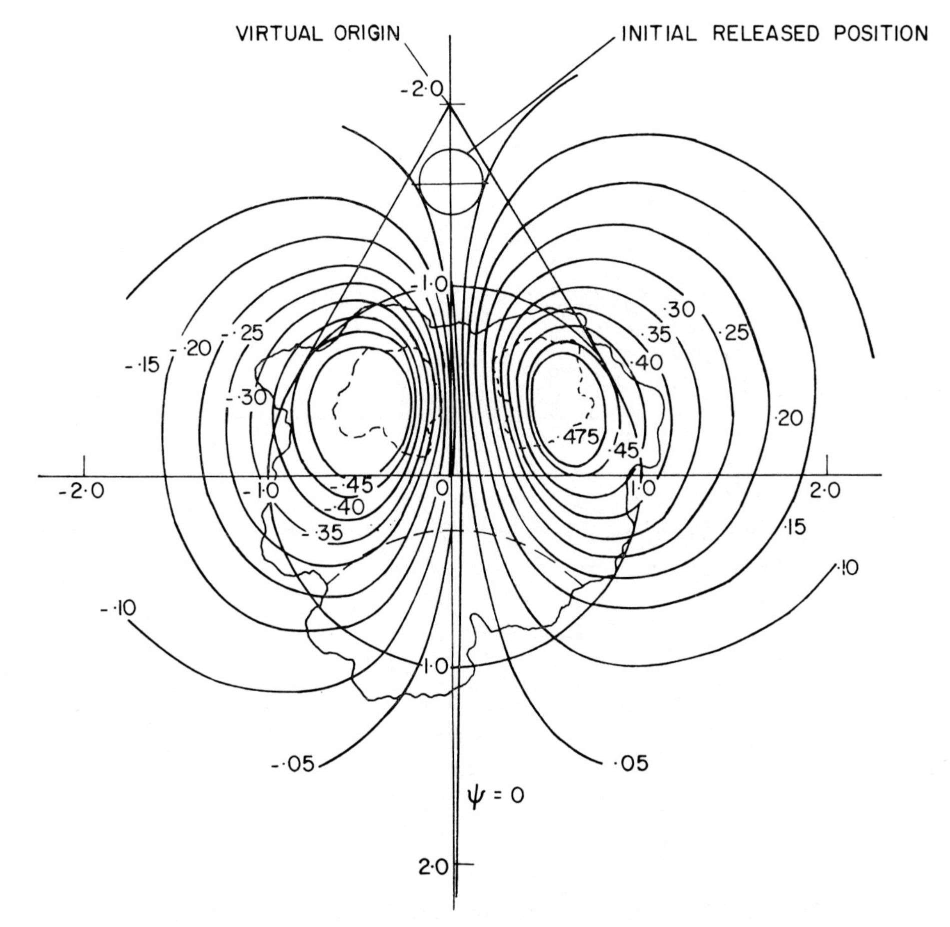

Fig. 6.9. Streamline pattern in line thermal (Tsang, 1971).

The largest term in that equation is thus the buoyancy term, which is presumably balanced mainly by the next largest terms, i.e., the nonlinear terms representing convective change of vorticity. The eddy stress terms are likely to be significant mainly in regions of high velocity gradients.

An analytical solution of the nonlinear Equation (6.87) appears to be difficult. A solution of the linearized problem (nonlinear terms neglected) has been presented (Csanady, 1965), but in view of the foregoing, this is likely to be a poor model. The calculated flow pattern (Figure 6.10, drawn by Tsang, 1971) is nevertheless qualitatively in reasonable accord with the experimental evidence, reproducing the 'double vortex' structure of a line thermal quite well, and having the correct asymptotic behavior at large distances from the core of the thermal.

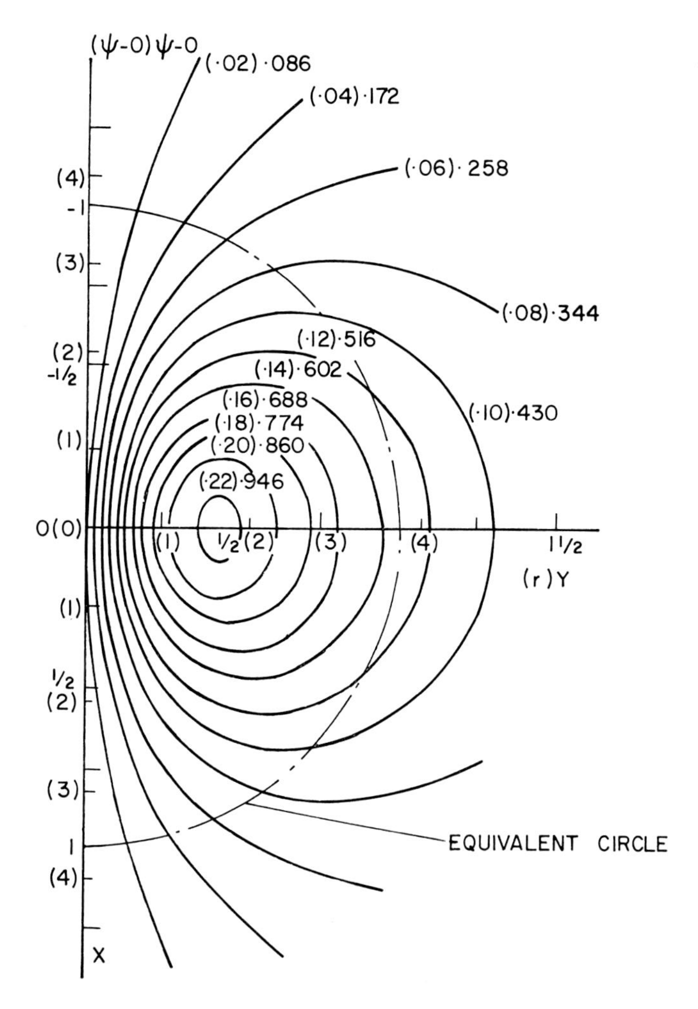

Fig. 6.10. Flow pattern calculated from linearized theory (Csanady, 1965; redrawn to better scale by Tsang, 1971).

A numerical solution of Equation (6.87) has been found by Lilly (1964) by the use of an ingenious method for the case when, in effect, $\alpha=0.45$, $\gamma=10$. For this case Lilly finds good agreement with the experimental pattern if the eddy coefficients are chosen so that

$$\kappa_T/VL = \nu_T/VL = 0.32 .$$

On comparison with Equation (6.93) we find $B=0.13$, justifying to some extent our earlier surmise in connection with this constant. Lilly's flow pattern found with these constants is reproduced in Figure 6.11. This shows rather good agreement with the observed patterns, although it is more symmetrical about the vertical than the latter. The discrepancy may be attributed to the observed fact that entrainment of ambient fluid is rather more vigorous at the front of the visible thermal, where the density stratification is unstable, that at its rear, where the density distribution is stable, a state of affairs not represented by a constant eddy viscosity.

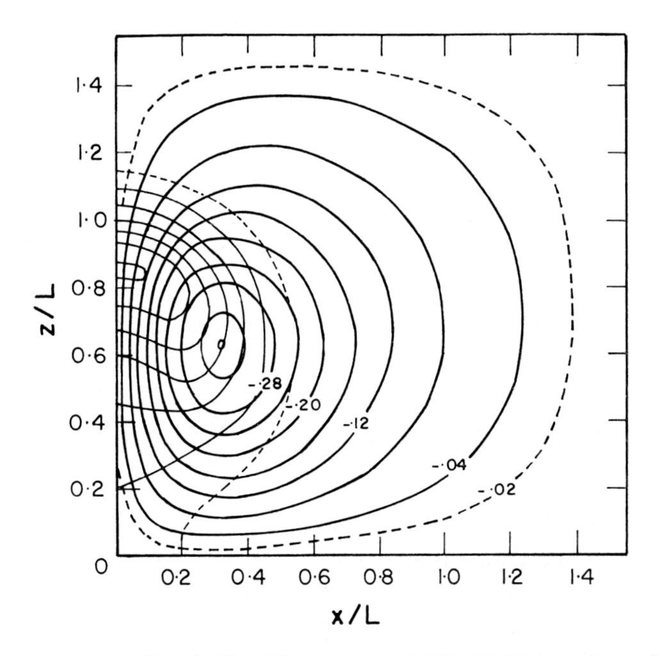

Fig. 6.11. Numerically calculated flow pattern of Lilly (1964) (non-linear theory).

The dominance of the inertial forces suggests an inviscid approximation to the observed flow pattern. This has been noted by Tsang (1971) who also shows that the approximation is certainly good outside the observable boundaries of the thermal. Tsang uses a doublet model, but a vortex pair model is more in accordance with the observed flow pattern (and is of course equivalent to a doublet model at larger distances).

Figure 6.12 illustrates the vortex pair model of a line thermal. Each vortex is characterized by its circulation $\pm\Gamma$ (the two having opposite senses of rotation) and velocities induced by either vortex may be calculated from

$$v = \Gamma/2\pi r\,, \tag{6.115}$$

where r is distance from the vortex core. Each vortex thus induces an upward velocity at the other vortex of a magnitude

$$v_{\mathrm{i}} = \Gamma/4\pi Y\,, \tag{6.115a}$$

where Y is the distance of one vortex from the z-axis. In an inviscid flow model of the buoyant thermal it is consistent to identify the bulk upward velocity with the induced velocity, i.e., $v_{\mathrm{i}} = W$. This step yields the circulation

$$\Gamma = 4\pi Y W\,. \tag{6.115b}$$

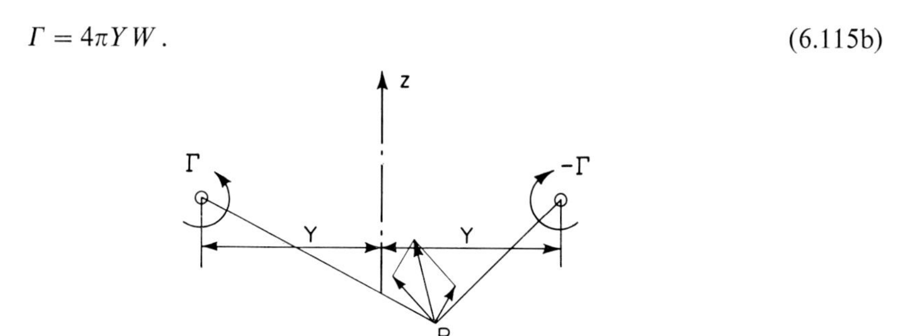

Fig. 6.12. Schematic illustration of vortex pair model of line thermal, also showing total induced velocity at point *P*.

Using the definition of vorticity, it is easily verified that the moment about the vertical of the vorticity distribution is equal to the total vertical momentum:

$$\iint\limits_{-\infty}^{\infty} y\xi \,\mathrm{d}y\,\mathrm{d}z = \iint\limits_{-\infty}^{\infty} w\,\mathrm{d}y\,\mathrm{d}z = \gamma W L^2\,, \tag{6.116}$$

where the last equality follows from the similarity theory, with γ the momentum constant. For the pair of concentrated vortices illustrated the moment of the vorticity distribution is by an application of Stokes' theorem:

$$\iint\limits_{-\infty}^{\infty} y\xi \,\mathrm{d}y\,\mathrm{d}z = 2Y\Gamma = 8\pi Y^2 W\,. \tag{6.116a}$$

From the last two relationships we find now, using $\gamma = 22$:

$$Y/L = 0.94\,.$$

The observed pattern of the vorticity distribution shows the vortex core to lie slightly closer to the z-axis than $\frac{1}{2}R_{\mathrm{a}}$. If we identify L as approximately equal to $\frac{1}{2}R_{\mathrm{a}}$ (as we did in Equation (6.110) above) excellent agreement with our above model results. The center velocity induced by the two vortices together is then $4W$, compared with an observed maximum velocity of about $2.8W$. So close to the vortex cores a discrepancy of this magnitude is to be expected in an inviscid model. At distances of $2L$ or greater, however, the velocity estimates based on this model appear to be quite accurate. Outside the observed boundaries of the thermal (as already remarked) Tsang's results confirm this conclusion in some detail.

The vortex pair model also allows some approximate predictions to be made on the effect of the ground on plume rise. Consider the case that the plume is at a height exactly equal to L above ground level (Figure 6.13). Representing our flow pattern by two concentrated vortices on either side of plume center, at $y=\pm 0.94L$, $z=0$, (the ground being at $z=-L$) we may satisfy the inviscid boundary condition of zero normal velocity at the ground by adding two image-vortices at $z=-2L$, $y=\pm 0.94\,L$, each having the same magnitude of the circulation Γ as the actual vortex above, but with the opposite sense. The image vortices induce a velocity equal to $0.47W$ at the cores of the 'real' vortices ($y=\pm 0.94L$, $z=0$) in the *downward* diection, so that the net upward drift velocity of these vortices reduces to 53% of what it would be in the absence of the ground. While these results are approximate only, they suggest that once a plume has grown large enough to touch ground level, its further rise may be inhibited by the ground effect.

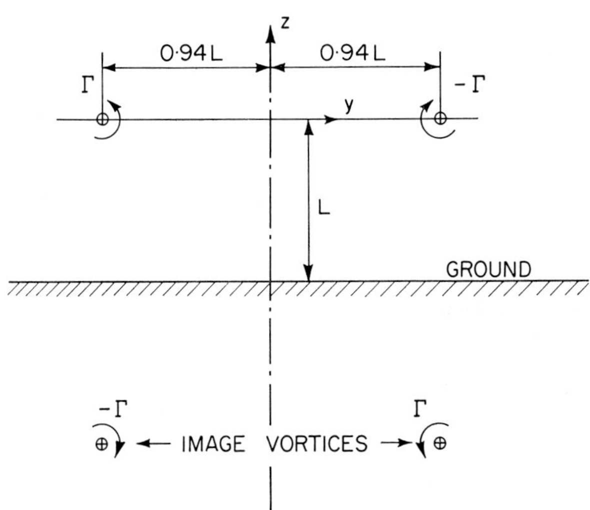

Fig. 6.13. Vortex pair near solid boundary.

6.16. Effect of Atmospheric Stratification

The actual distribution of the atmospheric potential temperature with height is often complex, but, as we have seen in Section 6.7 a simple theoretical model may be constructed by assuming that the potential temperature *gradient* $d\theta_a/dz$ is constant. A theory based on this hypothesis should give valuable insight into the effects of atmospheric conditions on plume rise, but we should again bear in mind that the theoretical model is highly idealized.

Consider then a line thermal in a linearly stratified atmosphere. The boundary condition on temperature at infinity becomes, replacing Equation (6.76):

$$\theta = \theta_a = Gz \quad \text{as} \quad y, z \to \infty, \tag{6.117}$$

where $G = \mathrm{d}\theta_a/\mathrm{d}z$ is the constant potential temperature gradient. It is convenient to introduce the local excess temperature

$$\theta_e = \theta - \theta_a \tag{6.118}$$

in terms of which the energy equation (last of Equation (6.71)) may be written

$$\frac{\partial \theta_e}{\partial t} + \frac{\partial (v\theta_e)}{\partial y} + \frac{\partial (w\theta_e)}{\partial z} = \kappa_T \nabla_1^2 \theta_e - Gw. \tag{6.119}$$

The buoyancy force is proportional to θ_e, cf., Equation (6.71).

On integrating Equation (6.119) over the entire cross-section of a line thermal we find, using the boundary condition Equation (6.117):

$$\mathrm{d}/\mathrm{d}t \iint_{-\infty}^{\infty} \theta_e \,\mathrm{d}y\,\mathrm{d}z = -G \iint_{-\infty}^{\infty} w \,\mathrm{d}y\,\mathrm{d}z. \tag{6.120}$$

Thus the total buoyancy of the thermal is now not constant but changes at a rate proportional to its total momentum. It is convenient to introduce symbols for the flux of total buoyancy and for total momentum (in kinematic units, i.e., divided by density):

$$F/U = \iint_{-\infty}^{\infty} \beta g \theta_e \,\mathrm{d}y\,\mathrm{d}z$$
$$M = \iint_{-\infty}^{\infty} w \,\mathrm{d}y\,\mathrm{d}z. \tag{6.121}$$

Then Equation (6.120) may be written

$$\mathrm{d}/\mathrm{d}t\,(F/U) = -\beta g G M. \tag{6.120a}$$

Close to the source the flux of buoyancy F tends to the initial value F_0:

$$F_0 = \frac{\beta g}{\varrho c_p} H, \tag{6.122}$$

where H is the heat release rate from the chimney. Integrating the vertical momentum equation over the (y, z) plane we find in exact analogy with Equation (6.82):

$$\mathrm{d}M/\mathrm{d}t = F/U, \tag{6.123}$$

where M and F are as defined in Equation (6.121). Combining Equations (6.120) and (6.123) we arrive at

$$\mathrm{d}^2F/\mathrm{d}t^2 + g\beta G F = 0. \tag{6.124}$$

Suppose now that the stratification is *stable*, $G>0$. Then we use the notation $N^2 = g\beta G$ (N = Brunt-Vaisala frequency) as in Section 6.8 and write the solution of Equation (6.124) as

$$F = A\cos Nt + B\sin Nt. \tag{6.125}$$

Given that the buoyancy at $t=0$ is $F=F_0$ and the momentum $M=M_0$ this may be expressed as

$$\begin{aligned} F &= F_0\cos Nt - UNM_0\sin Nt \\ &= F_0\left[1 + \left(U^2N^2M_0^2/F_0^2\right)\right]^{1/2}\cos\left[Nt + \tan^{-1}\left(UNM_0/F_0\right)\right]. \end{aligned} \tag{6.125a}$$

The corresponding time-variation of the momentum is

$$\begin{aligned} M &= \frac{F_0}{UN}\sin Nt + M_0\cos Nt \\ &= \frac{F_0}{UN}\left[1 + \left(U^2N^2M_0^2/F_0^2\right)\right]^{1/2}\sin\left[Nt + \tan^{-1}\left(UNM_0/F_0\right)\right]. \end{aligned} \tag{6.126}$$

The effects of a non-zero initial momentum consist of a phase shift and of an increase in the apparent value of the buoyancy. Both these effects are small if UNM_0/F_0 is small, which is almost invariably the case in buoyancy dominated plumes. Thus to a satisfactory approximation we may use Equations (6.125) and (6.126) in the form:

$$F = F_0\cos Nt \tag{6.125b}$$

$$M = \frac{F_0}{UN}\sin Nt \tag{6.126a}$$

with the proviso that F_0 may have to be adjusted slightly and the origin of t shifted in the event that UNM_0/F_0 becomes appreciable.

Thus there are sinusoidal fluctuations of both buoyancy and momentum at the Brunt-Vaisala frequency, their phases being displaced by 90°.

If the stratification is *unstable*, $G<0$, we write $m^2 = -g\beta G$, and the solutions become

$$\begin{aligned} F &= F_0\cosh mt + mUM_0\sinh mt \\ M &= \frac{F_0}{mU}\sinh mt + M_0\cosh mt. \end{aligned} \tag{6.127}$$

In this case both total buoyancy and momentum grow exponentially, although for periods t so short that $mt \ll 1$ the change from the values characterizing a neutral atmosphere

$$\left.\begin{aligned} F &= F_0 \\ M &= (F_0/U)\,t + M_0 \end{aligned}\right\} \quad (mt \ll 1) \tag{6.127a}$$

is relatively slight. A similar reduction to neutral conditions holds in a stable atmosphere for $Nt \ll 1$.

The above results are instructive by themselves, but in order to assess the influence of stratification on the actual upward motion it is necessary to determine how the size of the thermal changes in time. Some complications are encountered in this attempt. In a neutral atmosphere we used the second moment of the temperature distribution to define a suitable length scale $L = \sigma_y$. If we multiply the energy Equation (6.119) by y^2 and integrate over all y and z we come upon some divergent integrals. Another problem arises with self-similarity. If we assume that velocity, length, and excess-temperature scales are V, L and Θ, and reduce the energy Equation (6.119) to a non-dimensional form similar to Equation (6.87) we find an extra term on the right:

$$(GL/\Theta)\, w^* .$$

The combination GL/Θ has to be independent of time for self-similarity. Another nondimensional combination involving Θ is $\varrho g \Theta L/V^2$ which must also be constant. The two together imply that

$$\beta g G L^2/V^2 = \text{const.} \tag{6.128}$$

or that NL/V or mL/V are constants. However, as we have seen before, L/V has to be proportional to t, and the similarity criterion of Equation (6.128) cannot be satisfied, except approximately if

$$\beta g G L^2/V^2 \ll 1 \tag{6.129}$$

which is true if Nt or mt is suitably small, i.e., in the early part of a thermal's motion when also Equations (6.127a) are valid. In this early phase the thermal does not yet 'feel' the effects of stratification. Self-similarity is not possible beyond this phase. Such a result is not very surprising if one observes that, according to the above considerations, in a stable atmosphere the thermal's momentum fluctuates between positive and negative values, while it grows beyond limit in an unstable environment. Such behavior is clearly incompatible with the notion of 'moving equilibrium' on which the assumption of self-similarity is based.

In the absence of self-similarity we are forced to fall back on approximate arguments of the type indicated by the discussion preceding Equation (6.106).

6.17. Approximate Arguments for Plumes in Stratified Surroundings

It is always possible to define a length scale L for an identifiable plume by some rational prescription (e.g., that distance where the excess temperature drops to 10% of its value at the center) and of course the upward bulk velocity provides a velocity scale V. Thus we may write for our total momentum as in the case of neutral surroundings

$$M = \gamma V L^2 , \tag{6.130}$$

where the momentum 'constant' γ is now not a constant on account of possible variations in the velocity and temperature distributions. Assuming that γ is nearly constant piece by piece of a plume however, we may expect to obtain reasonable approximate results.

For short development times t we have seen that the effects of stratification were unimportant. For the corresponding phase of plume behavior and perhaps slightly beyond, it is reasonable to adopt the 'law of growth' we used for neutral conditions:

$$\mathrm{d}L/\mathrm{d}t = \alpha V \tag{6.94a}$$

and identify the velocity-scale V with the vertical bulk velocity $W = \mathrm{d}Z/\mathrm{d}t$. Equation (6.130) can now be expressed as:

$$M = \frac{\gamma}{\alpha} L^2 \frac{\mathrm{d}L}{\mathrm{d}t} = \frac{\gamma}{3\alpha} \frac{\mathrm{d}L^3}{\mathrm{d}t}\,. \tag{6.131}$$

For *stable* environmental conditions we may integrate Equation (6.126b) now to yield

$$L^3 = \frac{3\alpha}{\gamma} \frac{F_0}{UN^2} (1 - \cos Nt)\,. \tag{6.132}$$

Equation (6.94a) also implies $L = \alpha Z$, so that a formula for plume elevation may at once be written down. From Equation (6.132) it may be seen that Z reaches a maximum where $Nt = \pi$, but here the upward velocity W becomes zero and the above similarity assumptions certainly break down before such a point is reached. Expanding the cosine function in Equation (6.132) we recover the $\frac{2}{3}$ power law for small arguments Nt. At somewhat greater values of Nt the plume path $Z(x)$ $(x = Ut)$ remains *below* the $\frac{2}{3}$ power law curve. At some point in this region the atmospheric turbulence presumably comes to dominate mixing, as the buoyancy of the plume diminishes to zero and, overshooting the equilibrium level, even becomes negative. Note that this happens according to our results in the previous section regardless of the exact manner in which the plume radius grows.

Under stable conditions the intensity and scale of atmospheric turbulence is likely to be moderate and it is realistic to assume that further plume growth by atmospheric eddies may be described by a constant eddy diffusivity. In terms of our earlier description of plume behavior, phase 2 is likely to be absent in a stable environment and phase 3 set in as the intensity of buoyant movements becomes sufficiently low. Certainly in view of the crudeness of our assumptions in this section in the absence of self-similarity an adequate replacement for Equation (6.94a) is, for atmospheric turbulence dominated mixing:

$$\mathrm{d}L^2/\mathrm{d}t = 2K_\mathrm{T}\,, \tag{6.133}$$

where K_T is a constant eddy diffusivity. Substituting into Equation (6.130), identifying V with $W = \mathrm{d}Z/\mathrm{d}t$, substituting Equation (6.126a) for M and integrating we find,

$$Z = \frac{F_0}{2\gamma U K_\mathrm{T} N} (\mathrm{Si}\,(Nt) - \mathrm{Si}\,(Nt_1))\,, \tag{6.134}$$

where $\mathrm{Si}(\lambda)$ is the sine integral and t_1 is the effective beginning of the atmospheric turbulence dominated phase. The shape of the $Z(Nt)$ curve is illustrated in Figure

6.14; it is seen to tend to an asymptotic value (with Si(∞) being equal to $\pi/2$) after a few oscillations of gradually reducing amplitude.

From a practical point of view greatest importance attaches to the estimation of the asymptotic value Z_a reached by the plume. Most of this rise takes place *before* atmospheric turbulence comes to dominate mixing, i.e., in the phase more or less well described by Equation (6.132), with $Z = L/\alpha$. An approximation to Z_a is thus the peak height predicted by Equation (6.132), or at least a large fraction of it, which may be written as:

$$Z_a = C_s (F_0/UN^2)^{1/3}, \tag{6.135}$$

where C_s is a 'stable plume' constant, $C_s \cong 6\alpha^{-2}\gamma^{-1}$.

Empirical data summarized by Briggs (1969) confirm the relationship in Equation (6.135) quite well, with $C_s \cong 2.0$ (after allowing for our slightly different definition of F_0), which is only about half of the value of the theoretical maximum for C_s, given $\alpha = 0.25$, $\gamma = 22$.

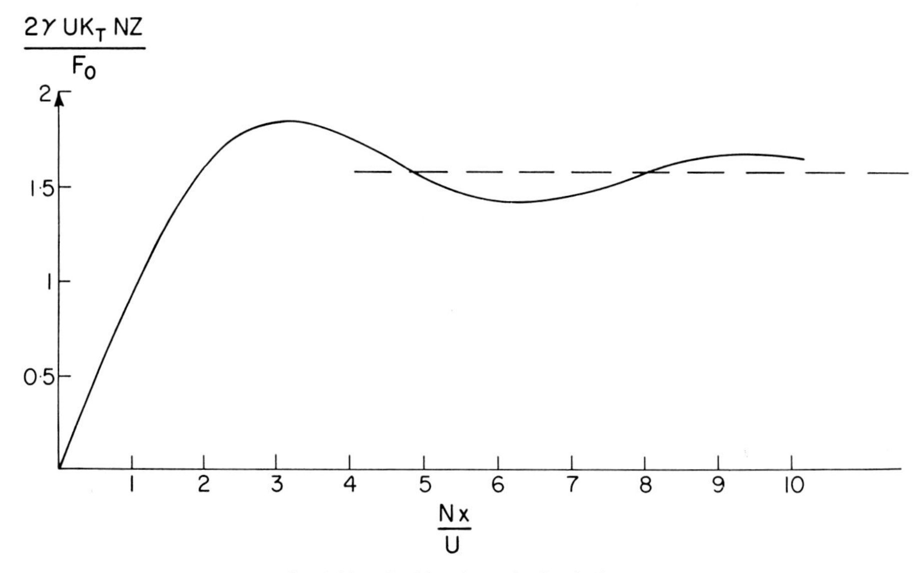

Fig. 6.14. Stable plume in final phase.

Under unstable conditions rather more complex conditions prevail (Slawson and Csanady, 1971). If we assume a 'law of growth' for phase 1 of plume growth (Equation (6.94a)) we may integrate Equation (6.126) again as in the stable case and arrive at

$$Z = (3/\alpha^2\gamma)^{1/3} (F_0/UN^2)^{1/3} \quad (1 - \cosh mt). \tag{6.136}$$

For small mt this is identical with our $\frac{2}{3}$ power law applying to the neutral case, but departs from it on the *high* Z-side as soon as mt is appreciable. One would be tempted to conclude that in an unstable atmosphere chimney plumes rise to a height greater than predicted by the $\frac{2}{3}$ power law. On the other hand, atmospheric turbulence is usually high in unstable conditions for reasons discussed in Section 6.5 and 'phase 2' of plume growth may set in quite early, so early in fact that mt is still small, i.e., the departure from the $\frac{2}{3}$ power law negligible. Plume breakup under unstable conditions is often so vigorous that buoyancy and vertical momentum are for all practical purposes dissipated and no further buoyant rise takes place. In such cases the direct effects of an unstable stratification on plume rise are negligible, the intense atmospheric turbulence simple aborting plume rise. The plume then departs from the $\frac{2}{3}$ power law on the *low* Z-side after breakup and levels out rapidly.

A third possibility is that although atmospheric turbulence is strong enough to break up the plume *before* it rises above the $\frac{2}{3}$ power law curve, so that its center line at first dips *below* that curve, the plume remains coherent enough for our earlier relations to hold approximately, i.e., for Equation (6.127) to remain more or less valid into a more regular phase 3. A growth law according to Equation (6.133) for such a final phase (or indeed any other reasonable growth law) together with Equation (6.127) implies again an exponentially rising plume as the effects of buoyancy and of the unstable environment reassert themselves. The behavior of the plume under such circumstances is illustrated in Figure 6.15 together with the other two possibilities mentioned earlier. Clearly, once a plume embarks on an exponential vertical rise, it causes no further concern in regard to ground level pollution. On the other hand, a plume which levels out a short distance above the chimney top is likely to give rise to serious practical pollution problems, as we shall discuss in greater detail below.

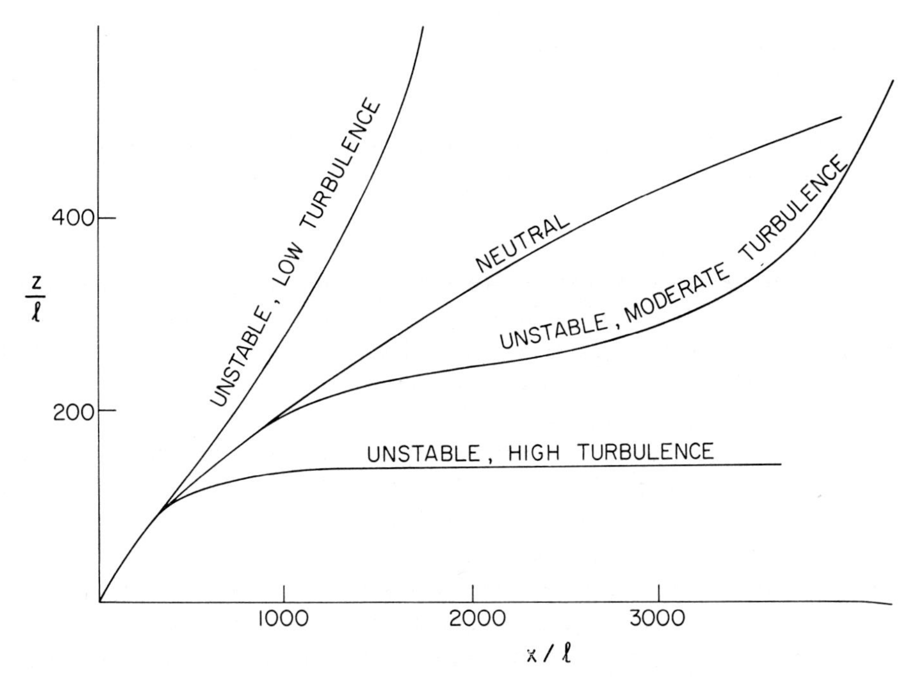

Fig. 6.15. Qualitative illustration of unstable plume shapes.

6.18. Engineering Assessment of Ground Level Pollution from Buoyancy Dominated Plumes

The above few sections have revealed a surprising wealth of complex physical phenomena involved in the seemingly simple problem of the buoyant rise of chimney plumes. This was the case even under highly idealized circumstances such as in uniform wind and turbulence, over uniform and flat terrain, and with constant atmospheric temperature gradient. In actual cases the atmospheric potential temperature is usually a complex function of height with the consequence that layers with widely different turbulence intensities overlie each other. Some illustrations of the resulting further complexities in plume behavior may be found in the paper by Slawson and Csanady (1971), quoted earlier.

If all these complexities of plume rise had to be taken into account in assessing possible effects of ground level pollution, the task would be difficult indeed. As in most engineering calculations, however, it should be sufficient to focus on certain 'critical' conditions which may occur with significant frequency and decide whether or not the expected nuisance or hazard is acceptable. Several classes of 'critical' conditions may be identified on the basis of present evidence and future research may uncover others. Presently identifiable critical categories are:

(1) *Plume 'reflected' by inversion lid.* The conditions discussed in Section 6.6 may apply also if the plume is buoyant, unless the buoyancy of the plume is great enough to penetrate the inversion. By the time a plume reaches the level of an inversion lid, its excess temperature is usually small enough for it to be stopped and reflected by the lid. The net effect downwind is very much as if plume height were coincident with inversion height, in a vertically limited layer as discussed in Section 6.6. Given the normal height of inversion lids (of the order of 500 m and more), it is clear that maximum ground level concentrations due to an *individual source* under a lid are unlikely to be greater than under 'regular dilution' conditions (cf., below). However, should an inversion lid occur over the atmosphere of a large city or conurbanation, and should it be accompanied by light winds, the combined effects of *many pollution sources* could build up to a dangerous level. Most of the well publicized air pollution 'episodes' belong to this category (e.g., the Los Angeles smog). In many places air pollution ordinances are in effect, allowing some arm of local government to order a shutdown of pollution-producing industries, when the health of the public is endangered by such an episode.

The engineer designing a chimney for, say, a power station or smelter plant, can do little at present about this particular critical category. He can recommend that a plant not be built in a meteorologically unfavorable area where air mass stagnation is frequent and where other pollution sources already abound. Nevertheless, alternative locations may be unacceptable for economic reasons. A remarkable fact of life in a.d. 1971 is that many communities actually desire the establishment of pollution sources in view of other benefits, mainly employment. Particularly severe local air pollution problems may be encountered in valleys, because an inversion lid often forms below the top of the mountains on either side, creating an essentially one-dimensional dilution problem, winds also being channeled along the valley.

One technological way to deal with this problem at least in part is to build a chimney so tall that its top rises above the average level of the inversion lid. This is often a very expensive proposition and is also undesirable from the point of view of air traffic safety.

Some suggestions have been advanced recently to the effect that 'pulsing' of chimney effluents would generate large buoyant puffs better able to penetrate inversions than continuous plumes from the same sources. Research on the penetration of inversions is proceeding in several centers and is a fascinating topic in itself, but the results are not so far sufficiently conclusive to establish positively any benefits that may result from pulsing. We have seen, however, that initial conditions have a great deal to do with the exact rate of entrainment of ambient air into a line thermal. By minimizing this rate of entrainment a better chance of penetrating the inversion results, because the excess temperature reduces then at a slower rate. Quite possibly, a method of release may be found in the future which carries buoyant plumes above inversions of normal intensity even when the level difference between chimney top and inversion lid is considerable.

(2) *Fumigation.* When the air is stable around the chimney top and above, we have seen that a buoyant plume reaches an asymptotic level which may be calculated. We have also seen that diffusion under stable conditions is relatively slow and that the vertical standard deviation in particular may not grow beyond a certain limit. Plumes diffusing under such circumstances may remain visible for long distances as the maximum concentration of pollutants carried in them remains relatively high. The actual maximum concentration levels may be calculated on the basis of empirical data relevant to 'very stable' conditions, as outlined in Chapter III. However, with the plume remaining aloft and the vertical diameter small, no ground level pollution is caused.

As was first described by Hewson (1945) in his study of a smelter chimney near Trail, B. C., Canada, the above favorable conditions can suddenly change into a particularly intense form of pollution known as 'fumigation.' Pronounced and persistent stability in the atmosphere usually occurs at night, and is broken up in the morning as solar heating of the ground produces an unstable or at least nearly neutral well-mixed layer first at low levels, but then gradually increasing in depth. As the top of the turbulent layer reaches plume level, pollutants are brought to ground level all along the length of the previously existing stable plume, out to some very large distances from the source and in concentrations comparable to the maxima observed under near-neutral conditions. Within a half hour or so the stable plume has been dissipated and normal dilution prevails again following the fumigation episode.

Although such fumigation episodes may be a considerable nuisance, they do not appear to constitute as critical conditions as more persistent exposure to similar concentrations in a fixed location. The fumigation episode is short and the area affected by it a relatively narrow cone, so that in reasonably open country around the source of the fumigating pollutant the occurrence of such episodes is distributed more or less evenly. An exception is provided by, say, a smelter in a narrow valley, the plume of which is usually channeled along the valley, so that the same spots are fumigated regularly. Such was the case studied by Hewson in Trail, B.C. and damage to vegetation occurred at distances of the order of 50 km from the chimney. It should be added, however, that the intense SO_2 emission of this chimney would in all probability have caused damage to vegetation also in open country, only there it would have been confined to the neighborhood of the source.

We may then tentatively adopt the attitude that if a chimney does not give rise to undesirable ground level effects under 'regular' and 'convective' dilution conditions (cf., below), it will not do so during fumigation episodes. It would of course be more desirable to make separate quantitative estimates of maximum concentrations to be expected in such episodes, but the accuracy of these is likely to be low in view mainly of the many physical factors entering the problem. One could, for example, calculate the stable plume existing just prior to morning breakup, using Equation (6.135) to determine the equilibrium level of plume center line, and the data of Chapter III to estimate the standard deviations σ_y and σ_z, and through those the maximum concentration at plume center line. This preexisting plume could be regarded as an instantaneous distributed source, beginning to diffuse downward at the time of sudden breakup (with an inversion lid remaining at plume height to stop *upward* diffusion). Again assessing the growth of σ_y and σ_z, in a near-neutral atmosphere of relatively high turbulence level, one could estimate the history of ground level concentrations during the fumigation episode. The inherent inaccuracies of similar estimates are compounded by such a two-stage procedure and a healthy skepticism is in order in viewing the results. Nevertheless, such two-stage diffusion calculations should show lower ground level concentrations than the single-stage ones discussed further in the next section.

(3) *'Regular' dilution.* In moderate to strong winds with low to moderate turbulence level and under near-neutral conditions the $\frac{2}{3}$ power law may be used to represent plume center line and empirical data on σ_y, σ_z exploited to provide diffusion estimates. In lighter winds buoyancy dominated plumes rise high, while in very strong winds a given quantity of pollutants is stretched out along a long piece of the x-axis resulting in both cases in relatively low concentrations. The worst conditions occur at some intermediate wind speed, which we shall call the 'critical' wind speed, given 'regular' dilution in the above sense. Such 'regular' conditions occur fairly frequently and averaged concentration distributions are usually dominated by them. In the following section we shall discuss a quantitative model to predict maximum ground-level concentrations under these conditions, because they appear to provide reasonable yardsticks of ground level pollution.

(4) *'Convective' dilution.* Given light to moderate winds and high turbulence intensity (a combination associated mainly with convective conditions, i.e., strong thermal turbulence), we have seen above that buoyant plumes break up relatively close to the source and may terminate their upward motion for all practical purposes. Given a sufficiently high turbulence level to cause plume breakup, highest ground level concentrations are again associated with a 'critical' wind speed, which is rather lower than

that appropriate to regular conditions. While less frequent than 'regular' conditions, 'convective' dilution occurs sufficiently frequently and persists for sufficiently long periods to pose an important (and in some cases the dominant) pollution problem. We shall develop a quantitative model also for this situation in the next section.

6.19. Effects of Plume Rise on Ground-Level Concentration

A satisfactory approximate model of the mean concentration field downwind of a chimney is a Gaussian distribution along both y and z, centered at plume height $h(x)$ above ground, which includes the physical chimney height h_s, plus the thermal rise $Z(x)$:

$$h = h_s + Z\,. \tag{6.137}$$

Given the Gaussian model, the concentration varies at ground level, $z = -h$, along the axis of the plume, $y = 0$, as:

$$\chi_a = \frac{q}{\pi U \sigma_y \sigma_z} \exp\left\{ -\frac{h^2}{2\sigma_z^2} \right\}. \tag{6.138}$$

The standard deviations σ_y, σ_z are fairly complex functions of x, but at relatively short distances may be approximated by

$$\begin{aligned} \sigma_y &= i_y x \\ \sigma_z &= i_z x\,, \end{aligned} \tag{6.139}$$

where i_y, i_z are relative turbulence intensities, cross wind and vertical. We shall regard σ_z a rescaled distance variable (replacing x in $Z(x)$) and write

$$a = \sigma_y/\sigma_z = i_y/i_z = \text{const.} \tag{6.140}$$

The approximations made in Equations (6.139) and (6.140) are not acceptable under pronouncedly stable conditions, and are also unsatisfactory in a strongly unstable atmosphere at larger distances x, but they should be adequate for the purposes of the two models discussed below.

In a given wind the maximum ground level concentration χ_m occurs where

$$\partial\chi_a/\partial\sigma_z = 0\,. \tag{6.141}$$

Carrying out the differentiation we find from Equation (6.138) for this point

$$\left(\frac{h}{\sigma_z}\right)^2 = 2 + \frac{h}{\sigma_z}\frac{dh}{d\sigma_z} \quad (\chi_a = \chi_m)\,. \tag{6.142}$$

Should the slope $dh/d\sigma_z$ be negligible at the point where the maximum concentration occurs we find

$$\sigma_z = 0.707\,h \quad (\chi_a = \chi_m)\,. \tag{6.143}$$

In this case the maximum concentration is

$$\chi_m = 2q/\pi U e a h^2\,. \tag{6.144}$$

When $dh/d\sigma_z$ is not negligible, different constants enter Equation (6.144), but these differences are relatively unimportant, χ_m remaining essentially proportional to $qU^{-1}h^{-2}$. The critical quantity from an engineering point of view is the 'effective ventilation,' defined as q/χ_m, which is proportional to Uh^2. The aim of our quantitative models is to predict this product, its variation with wind speed, and especially its value in the 'critical' wind speed.

From Equation (6.144) it would appear that χ_m depends only in a minor way on atmospheric turbulence, except insofar as the latter may influence the plume rise $Z(x)$. The variations of $a=\sigma_y/\sigma_z$ are not very significant in practice, a good rule of thumb being $a=2$. However, it should be noted that the *area* included by given concentration isopleths is proportional to $(i_z U)^{-1}$. Consider, e.g., the contour at ground level where $\chi=0.5\ \chi_m$ and let the area included within this contour be $A_{0.5}$. It is then easily shown from the Gaussian model that

$$A_{0.5} = \text{const.}\ \frac{h^2}{i_z} = \text{const.}\ \frac{q}{\chi_m i_z U}. \tag{6.145}$$

The same average concentration χ_m would be produced if we were to transport the pollutants away from ground level through a duct of cross-sectional area $A_{0.5}$, with an air velocity proportional to $i_z U$. By increasing plume height h we may increase the size of the duct, while by choosing a suitable site we can ensure that the transport velocity $i_z U$ is sufficiently high most of the time.

In order to construct a ground level concentration model for 'regular dilution' we assume that the $\frac{2}{3}$ power law describes plume rise to a satisfactory approximation:

$$h = h_s + C\,(F/U^3)^{1/3}\,x^{2/3}. \tag{6.146}$$

Equation (6.138), with Equation (6.146) substituted for plume height, is conveniently written down in a nondimensional form:

$$K = 1/\pi V D^2 \exp[-(1 + C D^{2/3} V^{-1})^2/2D^2], \tag{6.147}$$

where

$$\begin{aligned} K &= a\chi_a U_s h_s^2/q && \text{(nondimensional concentration)} \\ D &= \sigma_z/h_s = i_z x/h_s && \text{(nondimensional distance)} \\ V &= U/U_s && \text{(nondimensional velocity)} \\ U_s &= (F/i_z^2 h_s)^{1/3} && \text{(velocity scale)}. \end{aligned}$$

At a given nondimensional wind speed $V=$const. the maximum concentration K_m occurs at $D=D_m$, which may be found by solving the implicit Equation (6.142) applied to this case:

$$V = C D_m^{2/3}/(\sqrt{6D_m^2 + 1} - 2). \tag{6.148}$$

For $V=0.68$, for example, the distribution $K(D, 0.68)$ is illustrated in Figure 6.16, using $C=0.68$. The maximum value K_m may be found by substituting Equation (6.148) into (6.147) and is illustrated in Figure 6.17 for arbitrary values of C, because Equa-

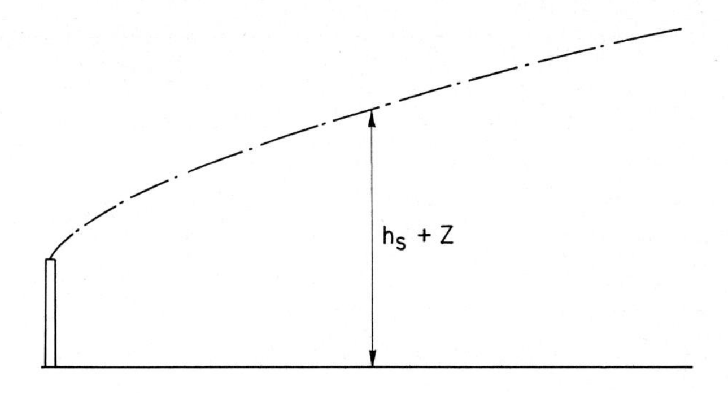

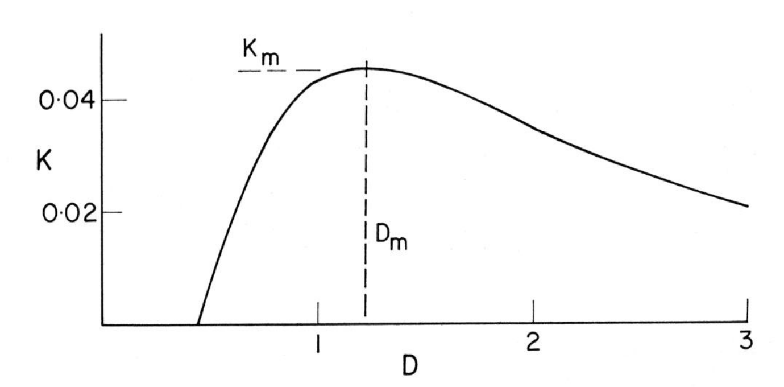

Fig. 6.16. Distribution of concentration along ground level, downwind of an elevated source in a given wind ('regular' dilution).

tions (6.147) and (6.148) can be written as a relationship between CK_m and V/C.

In Figure 6.17 we observe a flat maximum of the $K_m(V)$ curve at the 'critical' wind speed

$$V_c = 1.644\ C\,.$$

At this critical value the maximum concentration (maximized now *both* with respect to distance and wind speed) is

$$K_{mm} = 0.0532\ C^{-1}\,.$$

The distance at which this double maximum occurs is

$$D_{mm} = 0.980\,.$$

On substituting back into Equation (6.146) one also sees that the total plume height

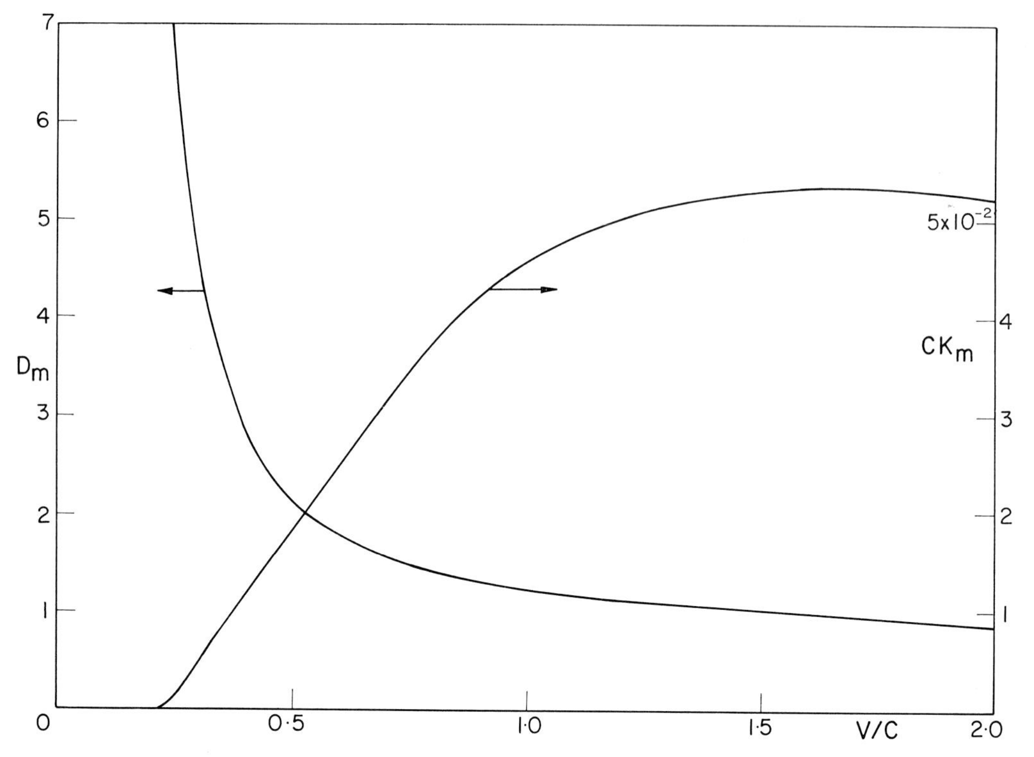

Fig. 6.17. Maximum ground-level concentration vs. wind speed (nondimensional variables) under 'regular' dilution.

at D_{mm} in the critical wind speed V_c is

$$h = h_c = 1.600\ h_s$$

or just 60% more than the physical chimney height. The dimensional value of the maximum concentration in the 'critical' wind speed is, by substituting back into the definition of K:

$$\chi_{mm} = \frac{0.0532}{aC} \frac{q}{U_s h_s^2} \quad (U = U_c = 1.644\ CU_s). \tag{6.149}$$

If we replace U_s by U_c in the last equation we may write the maximum concentration as

$$\chi_{mm} = \frac{0.0875}{a} \frac{q}{U_c h_s^2} \tag{6.149a}$$

which is exactly the same as Equation (6.144) with $U = U_c$ and $h = 1.6\ h_s$. Thus the 'effective ventilation' corresponding to the critical conditions is $U_c (1.6\ h_s)^2$, depending strongly on *physical* chimney height and rather weakly on flux of buoyancy. To illustrate this point better, we substitute the definition of the critical wind speed U_s into

Equation (6.149) and arrive at:

$$\chi_{mm} = \frac{0.0532}{aC} \frac{i_z^{2/3} q}{F^{1/3} h_s^{5/3}} \tag{6.150}$$

which also shows that high turbulence levels tend to increase maximum concentrations in the critical wind. It is further worth noting that, owing to the flatness of the $K_m(V)$ curve (see again Figure 6.17) much the same maximum concentrations are reached already in 70% of the critical wind speed.

Under *convective* conditions we assume that the plume levels off at a height given by Equation (6.114), i.e.,

$$h = h_s + l \frac{2}{3(\alpha b)^2 i_z^2} = \text{const.} \quad (l = F/U^3). \tag{6.151}$$

Defining again a velocity scale U_s, nondimensional concentration K, distance D and velocity V, as above following Equation (6.147), the distribution of K along ground level becomes, if we assume our previously found tentative value for the product αb, viz. $\alpha b = 1$:

$$K = \frac{1}{\pi V D^2} \exp\left\{ -\frac{(1 + \frac{2}{3}V^{-3})^2}{2D^2} \right\}. \tag{6.152}$$

The maximum value K_m of this for given V and its point of occurrence D_m are

$$K_m = 0.234 \frac{V^5}{(V^3 + 2/3)^2}$$
$$D_m = 0.707 \left(1 + \frac{2}{3V^3}\right). \tag{6.153}$$

These relationships are illustrated in Figure 6.18. The critical nondimensional wind speed is

$$V_c = 1.50\,.$$

Given this wind speed the maximum concentration and its point of occurrence are

$$K_{mm} = 0.110$$
$$D_{mm} = 0.85\,.$$

The effective chimney height corresponding to these conditions is $h = 1.2\, h_s$. The dimensional maximum concentration in the critical wind speed is

$$\chi_{mm} = \frac{0.165}{a} \frac{q}{U_c h_s^2} = \frac{0.110}{a} \frac{i_z^{2/3} q}{F^{1/3} h_s^{5/3}} \quad (U = U_c = 1.5\, U_s). \tag{6.154}$$

On comparing this result with the $\frac{2}{3}$ power law model (Equations (6.149a) and (6.150)) it is clear that maximum concentrations are predicted to be higher under convective conditions than under 'regular' ones for several reasons. Firstly the con-

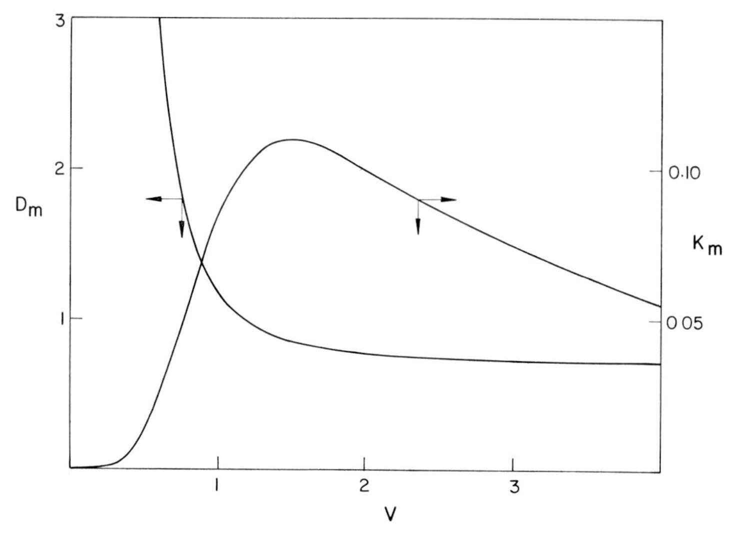

Fig. 6.18. Maximum ground-level concentration vs. wind speed in convective conditions.

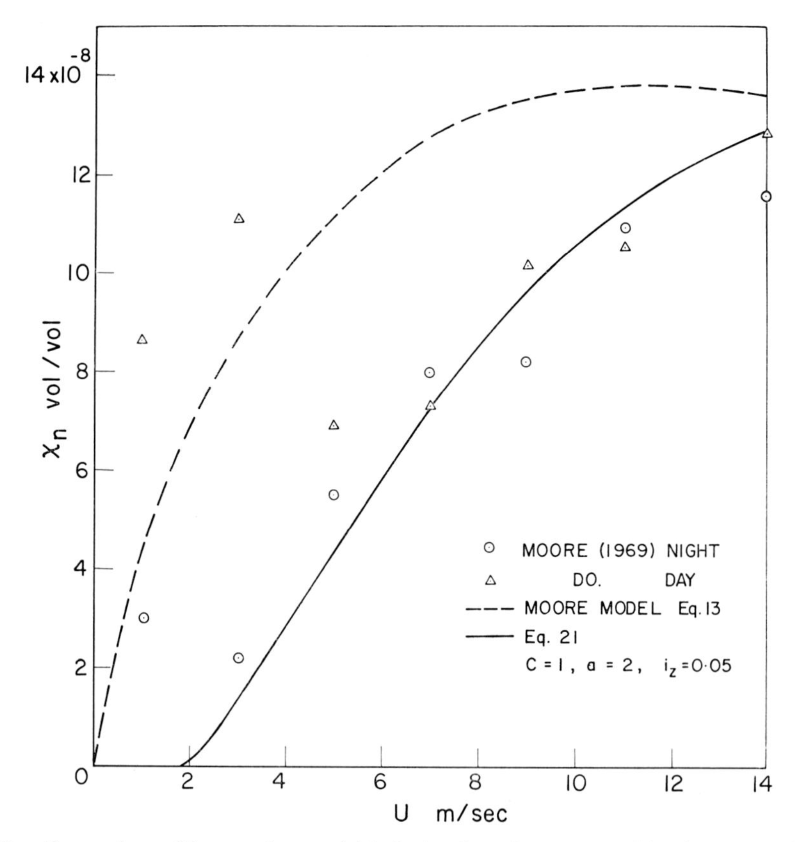

Fig. 6.19. Comparison of $\frac{2}{3}$ power law model (calculated maximum ground-level concentrations vs. wind speed) with averaged observations of Moore (1969).

stant 0.165 is higher than 0.0875; secondly the value of a is closer to 1.0, when turbulence is of thermal origin, than to $a=2.0$ which is characteristic of near-ground mechanical turbulence; thirdly the critical wind speed is lower if i_z is higher, varying as $i_z^{-2/3}$. Indeed from the two formulas one may estimate crudely that the typical ratio χ_{mm} (convective) :χ_{mm} (regular) may be as high as an order of magnitude.

Experimental evidence confirms the above predictions to a quite satisfactory degree, although this finding is quite recent and must at present be regarded 'sub judice.' An illustration of a comparison between the above theory ($\frac{2}{3}$ power law model) and some observations of Moore (1969) near Northfleet power station are shown in Figure 6.19 (Csanady, 1972). The observational points are in fact averages of many maxima observed in given wind speed categories. The influence of convective conditions on such averaged concentrations is evident by the scatter of points at light winds. For similar large heat sources with tall chimneys the 'critical' wind speed turns out to be of order 20 m s^{-1} under 'regular' conditions and about half that under 'convective' ones, although one would not expect convective conditions to prevail at wind speeds much in excess of 5 m s^{-1}.

Perhaps the practically most significant point brought out by the above analysis is that the 'effective' chimney height is proportional to the physical height h_s, being about 20 to 60% higher. Given the rule of thumb that $h_c = 1.5\ h_s$ and an acceptable estimate of the critical wind speed it is possible to make quick and fairly accurate estimates of ground level concentrations.

APPENDIX TO CHAPTER VI

A6.1. Momentum Plumes

While it does not strictly speaking belong in a chapter dealing with effects of buoyancy, for completeness' sake we shall briefly discuss the behavior of a chimney plume in a cross wind which is dominated by its initial upward momentum. Consider that portion of such a plume which has already bent over into the wind sufficiently to be nearly horizontal and assume that successive cross sections in this portion behave very much as would a flow generated by an instantaneous line source of vertical momentum. This is a two-dimensional problem again and the equation of vertical momentum may be written:

$$\frac{\partial w}{\partial t} + \frac{\partial (vw)}{\partial y} + \frac{\partial (\frac{1}{2}w^2)}{\partial z} = -\frac{1}{\varrho}\frac{\partial p}{\partial z} + \nu_T \nabla_1^2 w. \tag{A6.1}$$

Integrating over the entire y, z plane we find:

$$\frac{\mathrm{d}}{\mathrm{d}t}\iint_{-\infty}^{\infty} w\,\mathrm{d}y\,\mathrm{d}z = 0 \tag{A6.2}$$

or that

$$M = \iint_{-\infty}^{\infty} w \, \mathrm{d}y \, \mathrm{d}z = M_0 = \text{const.} \tag{A6.3}$$

We suppose, subject to later check of compatibility with the equations of motion, that a self-similar distribution of velocity develops, i.e., that

$$w = Vw^* \qquad y = Ly^* \quad \text{etc.} \tag{A6.4}$$

Then the integrated momentum equation becomes

$$M_0 = \gamma V L^2 = \text{const.}, \tag{A6.3a}$$

where $\gamma = \iint w^* \mathrm{d}y^* \mathrm{d}z^*$ is again a momentum constant. By an appropriate choice of the length scale L we may ensure that $\gamma = 1$. We assume that the growth of the plume is caused by its own self-generated turbulence, i.e., that a 'law of growth' holds:

$$\mathrm{d}L/\mathrm{d}t = \alpha V. \tag{A6.5}$$

On multiplying this equation by L^2 and noting Equation (A6.3a) we obtain (using $\gamma = 1$)

$$\mathrm{d}L^3/\mathrm{d}t = 3\alpha M_0. \tag{A6.6}$$

We take $t=0$ to be an 'effective origin' of the momentum plume such that $L=0$ at $t=0$. Then

$$L^3 = 3\alpha M_0 t. \tag{A6.7}$$

We also choose the velocity scale to be the bulk vertical velocity

$$V = \mathrm{d}Z/\mathrm{d}t, \tag{A6.8}$$

where Z is the displacement of the center of gravity of the plume above its effective origin. Equations (A6.5) and (A6.8) imply, again as for buoyancy dominated plumes, that

$$L = \alpha Z. \tag{A6.9}$$

Substituting now into Equation (A6.7) we find the equation for plume displacement:

$$Z = (3/\alpha^2)^{1/3} (M_0 t)^{1/3}. \tag{A6.10}$$

To check for the possibility of self-similarity, we note that all terms in Equation (A6.1) are proportional to V^2/L, except the time dependent and the viscous terms. When the equation is multiplied by L/V^2 these two terms come to contain the coefficients

$$(L/V^2)(\mathrm{d}V/\mathrm{d}t) \quad \text{and} \quad \nu_T/VL$$

respectively (in addition to α). Self-similarity is possible if both these are constant in time. From Equations (A6.8) and (A6.10) we may show that

$$\frac{\mathrm{d}V}{\mathrm{d}t} = -\frac{2}{9\alpha}\frac{L}{t^2} \tag{A6.11}$$

so that

$$\frac{L}{V^2}\frac{\mathrm{d}V}{\mathrm{d}t} = -\,2\alpha = \text{const.}$$

The eddy Reynolds number v_T/VL is constant if $v_T=\text{const.}\ VL$, which is the case provided that mixing is due to the self-generated turbulence of the plume. In that event self-similarity is therefore compatible with the equations of motion.

The above time dependencies may be translated into relationships to distance x by the approximate relationship

$$x = Ut. \tag{A6.12}$$

This relationship is approximate because plume slices still possess some deficiency of horizontal momentum compared to the ambient fluid. Thus the approximate plume path is

$$Z = \left(\frac{3}{\alpha^2}\right)^{1/3}\left(\frac{M_0}{U}x\right)^{1/3} \tag{A6.13}$$

or a $\frac{1}{3}$ power law, instead of the $\frac{2}{3}$ relationship characterizing the buoyancy dominated plume's path. This may also be stated by introducing a 'momentum length scale':

$$l_m = (M_0/U)^{1/2} \tag{A6.14}$$

in which case we may write

$$Z = C_m l_m^{2/3} x^{1/3}, \tag{A6.15}$$

where $C_m = (3/\alpha^2)^{1/3}$ is a constant, characteristic of 'momentum plumes.' The 'momentum' M_0 was here expressed in kinematic units (divided by density). When the exit gases have density ϱ_g, issuing with a vertical velocity w_0 from a chimney of radius R_0 into ambient air of density ϱ_a the initial momentum per unit length of the nearly horizontal plume is to be calculated from

$$M_0 = \frac{\varrho_g}{\varrho_a}\frac{w_0^2 \pi R_0^2}{U} \tag{A6.16}$$

so that the plume rise formula is also

$$Z = C_m\left(\frac{w_0}{U}R_0\right)^{2/3}\left(\frac{\varrho_g}{\varrho_a}\pi x\right)^{1/3}. \tag{A6.17}$$

Briggs (1969) has reviewed empirical data on jets in a smooth cross wind. He finds that Equation (A6.17) agrees with observations rather well, much better in fact than the above theoretical argument would lead one to expect. Close to the chimney, where plume inclination is considerable and the flow rather complicated, the formula cannot be valid, but apparently this is an unimportant complication. Briggs recommends the constant $C_m \pi^{1/3} 2^{-2/3} = 1.44$.

From the point of view of ground level concentrations of pollutants discharged from a chimney the 'momentum rise' given by Equation (A6.17) is likely to be unimportant in practice. We observe that the momentum length l_m is proportional to chimney radius. Where the slope dZ/dx reduces to 0.1 or less the displacement Z is about 3–4 chimney diameters above the chimney top, the exact distance being proportional to w_0/U. Not too far from here the atmospheric turbulence may be expected to break up the momentum plume and a reasonable rule of thumb would be to take a final momentum rise to be

$$Z_m = 6R_0 (w_0/U). \tag{A6.18}$$

In most practical cases an addition to the chimney height of this magnitude is not very significant, being usually of the order of 10% of the physical stack height. An exception is provided by oil refinery flare stacks, the velocity ratio w_0/U of which can be as high as 10 or more.

In the case of gases discharged with considerable buoyancy a momentum dominated phase may still occur if for an appreciable portion of the plume the buoyancy generated momentum (which increases linearly in time) remains well below the initial momentum. This occurs for distances such that

$$\frac{F}{U}\frac{x}{U} \ll M_0$$

i.e., for

$$x \ll U w_0/b_0, \tag{A6.19}$$

where, if θ_0 is excess temperature at the chimney top, $b_0 = \beta g \theta_0$ is buoyant acceleration there. One may express this criterion also in terms of the initial Froude number Fr:

$$\mathrm{Fr} = w_0/\sqrt{b_0 R_0}. \tag{A6.20}$$

The initial momentum dominates where

$$\frac{w_0}{U}\frac{x}{R_0} \ll \mathrm{Fr}^2. \tag{A6.19a}$$

An appreciable momentum-phase thus occurs for initial Froude numbers of, say, 3 and greater. Given $w_0 = 10$ m s^{-1} and $b_0 = 10$ m s^{-2}, for example, this is the case for chimneys smaller than $R_0 = 1$ m in radius.

EXERCISE

Analyze ground level concentrations caused by a 150 m tall chimney emitting 1 m^3 s^{-1} of SO_2, accompanied by a flux of buoyancy F of 200 m^4 s^{-1}. Consider 'regular' as well as 'convective' conditions and devote some attention to inversion lids at the 300 m level.

References

Batchelor, G. K.: 1953, *Homogeneous Turbulence*, Cambridge Univ. Press, London, 197 pp.
Batchelor, G. K.: 1967, *An Introduction to Fluid Dynamics*, Cambridge Univ. Press, London, Press, 615 pp.
Briggs, G. A.: 1969, *Plume Rise*, U.S. Atomic Energy Commission, 81 pp.
Bringfelt, B.: 1969, *Atmos. Environ.* **3**, 609.
Chandrasekhar, S.: 1961, *Hydrodynamic and Hydromagnetic Stability*, Oxford Univ. Press, London, 652 pp.
Csanady, G. T.: 1957, *Ind. Eng. Chem.* **49**, 1453.
Csanady, G. T.: 1961, *Int. J. Air Water Pollut.* **4**, 47.
Csanady, G. T.: 1964, *J. Atmospheric Sci.* **21**, 439.
Csanady, G. T.: 1965, *J. Fluid Mech.* **22**, 225.
Csanady, G. T.: 1970, *Water Res.* **4**, 79.
Csanady, G. T.: 1972, Effect of Plume Rise on Ground-Level Pollution, to be published.
Hewson, E. W.: 1945, *Quart. J. Roy. Meteorol. Soc.* **71**, 266.
Hilst, G. R. and Simpson, C. L.: 1958, *J. Meteorol.* **15**, 125.
Lilly, D. K.: 1964, *J. Atmospheric Sci.* **21**, 83.
Lumley, J. L. and Panofsky, H. A.: 1964, *The Structure of Atmospheric Turbulence*, Interscience Publishers, New York, 239 pp.
Meteorology and Atomic Energy: 1955, U.S. Atomic Energy Commission Report, 169 pp.
Moore, D. J.: 1969, *Phil. Trans. Roy. Soc.* **A265**, 245.
Morton, B. R., Taylor, G. I., and Turner, J. S.: 1956, *Proc. Roy. Soc.* **A234**, 1.
Priestley, C. H. B.: 1959, *Turbulent Transfer in the Lower Atmosphere*, Univ. Chicago Press, Chicago, 130 pp.
Richards, J. M.: 1963, *Int. J. Air Water Poll.* **7**, 17.
Scorer, R. S.: 1958, *Natural Aerodynamics*, Pergamon Press, 312 pp.
Sherlock, R. H. and Stalker, E. A.: 1940, *Mech. Engr.* **62**, 455.
Slawson, P. R. and Csanady, G. T.: 1967, *J. Fluid Mech.* **28**, 311.
Slawson, P. R. and Csanady, G. T.: 1971, *J. Fluid. Mech.* **47**, 33.
Starr, V. P.: 1968, *Physics of Negative Viscosity Phenomena*, McGraw-Hill Book Co. New York, 256 pp.
Steward, N. G., Gale, H. J., and Crooks, R. N.: 1958, *Int. J. Air Water Pollut.* **1**, 87.
Townsend, A. A.: 1956, *The Structure of Turbulent Shear Flow*, Cambridge Univ. Press, London, 315 pp.
Tsang, G.: 1971, *Atmos. Environ.* **5**, 445.

CHAPTER VII

THE FLUCTUATION PROBLEM IN TURBULENT DIFFUSION

7.1. Introduction

It may be seen from the past few chapters that a fairly well developed and reasonably satisfactory theory now exists for the prediction of stochastic mean concentrations in turbulent diffusion. The academic respectability of this theory is apt to obscure the fact that it only treats a rather restricted aspect of the turbulent diffusion process, the first moment of the probability distribution of the random variable, concentration. In experimental work the inadequacy of the mean-field theory is quite obvious: short-term ('instantaneous') concentration profiles differ markedly from theoretical ones, as we have illustrated in Chapter IV, for example. From an engineering point of view this means that the prediction or assessment of pollution nuisance or hazard can only be very incomplete if based on mean-field theory alone.

Rational hazard, etc., assessments are based on known effects of a given pollutant on vegetation, animal, and human health, etc. Such effects are usually dependent on both concentration and exposure time. For example, SO_2 damages vegetation in a concentration of 1 ppm, if the exposure is for several hours, but a ten times higher concentration is needed to produce observable damage in an exposure lasting merely minutes. Although this whole field of pollution effects is not very widely explored yet, much similar information is available for example in handbooks devoted to air pollution problems. In order to take advantage of such biological and medical knowledge we have to be able to make predictions on short-term as well as long-term concentrations of pollutants.

Pollution models based on the elementary statistical theory, which we discussed in Chapter III, provide predictions of time-average concentrations at fixed points for periods of the order of $\frac{1}{4}$ to 1 hr. We have already seen that even these time averages are in fact random variables, but at least their variance is small. As we shorten the averaging period, the fluctuations become larger and the important question becomes, what maximum concentration is reached with a frequency which we choose to regard as 'significant.' In practical terminology this is described as the problem of the 'peak to mean ratio.' Clearly this ratio is a function of the averaging period and for so-called instantaneous measurements it depends on the time-constant of the instrument. The shortest interval of practical interest in pollution problems is a few seconds and 'instantaneous' readings are often taken to be averages over a period of order 3 to 30 s.

The theoretical basis of fluctuation predictions is so far very incomplete. The first important contribution to the problem seems to have been a paper of Gifford (1959).

In the following few sections we delineate the problem and develop it systematically as far as possible at present. The practical use of the theory to be developed is made difficult mainly by a lack of suitable empirical input.

7.2. Probability Distribution of Concentration

Consider the instantaneous release of marked fluid into some prescribed region of a turbulent field. Let the resulting 'instantaneously' measured concentration field be $N(\mathbf{x}, t)$ (note also that a value 'at' position vector $\mathbf{x}$ is in fact an average over a sample volume of the order of 1 l, see our remarks in Chapter III). On repeating the experiment under identical environmental conditions, and identical conditions of release, one finds that the measured concentration $N(\mathbf{x}, t)$ (at *fixed* position $\mathbf{x}$ and a fixed time t after release) is a random variable. The results of a large number of trials may be conveniently summarized in a probability distribution $W(n, \mathbf{x}, t)$ which specifies the probability that the observed concentration in a given trial at $(\mathbf{x}, t)$ is less than n, $N(\mathbf{x}, t) \leqslant n$. The value of $W(\ )$ is zero for negative n (however small in absolute value) and $W \to 1$ as $n \to \infty$. Because $N(\mathbf{x}, t) = 0$ may occur with finite probability, at least for certain $(\mathbf{x}, t)$, a saltus in $W(n, \mathbf{x}, t)$ at $n = 0$ is quite likely. At higher values of the concentration the probability distribution is normally continuous, and a density $\mathrm{d}W/\mathrm{d}n$ may be defined. Another saltus to $W = 1$ may occur at the maximum initial value of the concentration, N_0.

The above definition of $W(n, \mathbf{x}, t)$ implies statistical processing of data at a fixed point in space and characterizes therefore the process we described as 'absolute diffusion' in earlier chapters. The only measure of this process we have so far dealt with is the expected or stochastic mean concentration which is related to $W(\ )$ by:

$$\chi(\mathbf{x}, t) \equiv \bar{N}(\mathbf{x}, t) = \int_{W=0}^{1} n \, \mathrm{d}W(n, \mathbf{x}, t) . \tag{7.1}$$

The range of locally observable concentrations may be characterized by their variance:

$$\overline{N'^2}(\mathbf{x}, t) = \int_{W=0}^{1} (n - \bar{N})^2 \, \mathrm{d}W(n, \mathbf{x}, t) \tag{7.2}$$

and by other higher moments of the distribution. The integrals occurring in Equations (7.1) and (7.2) are of the Stieltjes type (Cramer, 1946), meaning that at a saltus ΔW_i they contribute $\Delta W_\mathrm{i} n_\mathrm{i}$, where n_i is the location of the saltus on the n-axis.

As we have discussed in Chapter IV in greater detail, there are certain advantages in processing measured concentration data in such a way that readings at a fixed distance $\mathbf{y}$ from the center of gravity of a diffusing cloud are regarded as members of a given ensemble. We have dealt with the properties of the ensemble-average concentration field in relative diffusion at length in earlier chapters. Here we recognize that a fuller statistical description of the same ensemble is provided by a distribution $\Omega(n, \mathbf{y}, t)$

such that $\Omega(\ )$ is the probability of $N(\mathbf{y}, t) \leqslant n$. The relations analogous to Equations (7.1) and (7.2) are:

$$\bar{N}(\mathbf{y}, t) = \int_{\Omega=0}^{1} n \, \mathrm{d}\Omega(n, \mathbf{y}, t)$$
$$\overline{N'^2}(\mathbf{y}, t) = \int_{\Omega=0}^{1} (n - \bar{N})^2 \, \mathrm{d}\Omega(n, \mathbf{y}, t). \tag{7.3}$$

The above definitions hold generally, although in an inhomogeneous field $\Omega(\ )$ is also a function of the position of the center of gravity. In most of our following discussion we shall tacitly suppose a homogeneous field for simplicity.

'Absolute' and 'relative' diffusion differ to the extent that the center of gravity of a cloud meanders irregularly. The position of the center of gravity is a random vector $\mathbf{c}(t)$ with which we may associate a spatial probability density $\mathbf{P}_c(\mathbf{x}, t)$ such that $\mathbf{P}_c(\mathbf{x}, t)\mathrm{d}\mathbf{x}$ is the (small) probability of the vector $\mathbf{c}(t)$ having its endpoint at time t in the volume element $\mathrm{d}\mathbf{x}$ surrounding point $\mathbf{x}$. This spatial probability distribution is more akin to the particle-displacement probability introduced in Chapter II, than to the distributions $W(\ )$ and $\Omega(\ )$ we have defined so far in this chapter. If the diffusing substance is dynamically neutral, $\mathbf{P}_c(\mathbf{x}, t)$ is not affected by whether there is or is not marked fluid in the neighborhood of $\mathbf{x}$. In other words, $\mathbf{P}_c(\ )$ and $\Omega(\ )$ are statistically independent, so that the following relationship holds between the three distributions:

$$W(n, \mathbf{x}, t) = \int \Omega(n, \mathbf{x} - \mathbf{c}, t)\, \mathbf{P}_c(\mathbf{c}, t)\, \mathrm{d}\mathbf{c}. \tag{7.4}$$

One consequence of this general relationship is a similar connection between ensemble means $\bar{N}(\mathbf{x}, t)$ and $\bar{N}(\mathbf{y}, t)$. If we distinguish the latter function by the subscript r we find

$$\bar{N}(\mathbf{x}, t) = \int \bar{N}_r(\mathbf{x} - \mathbf{c}, t)\, \mathbf{P}_c(\mathbf{c}, t)\, \mathrm{d}\mathbf{c}. \tag{7.5}$$

We have for simplicity discussed the case of instantaneous release of marked fluid. Little reflection shows, however, that the entire argument applies to any specific release method, repeated exactly in successive trials, over a finite period. When the release takes place continuously and at a steady rate (whether from point, line, area or volume sources), the probability distribution $W(n, \mathbf{x})$ becomes independent of time and it may be extracted from long records of the concentration at fixed $\mathbf{x}$. 'Relative' diffusion in such situations is defined as relative to the center of gravity of a continuous plume in a cross-wind plane, so that $\mathbf{y}$ becomes a two-dimensional vector. The corresponding probability distribution $\Omega(n, \mathbf{y})$ is also independent of time, but of course a time series $N(\mathbf{y}, t)$ at fixed $\mathbf{y}$ cannot directly be observed and the distribution $\Omega(\ )$ can only be obtained from a succession of instantaneous cross sections.

In the wake of continuous sources it is not only important to know with what frequency high concentrations occur, but also how long they persist. A full statistical description could be given in terms of a joint probability distribution of concentrations realized at a fixed $\mathbf{x}$ but at two different times, say $N(\mathbf{x}, t_1)$ and $N(\mathbf{x}, t_2)$, for arbitrary (t_2-t_1). A mixed moment of such a distribution provides the covariance $\overline{N'(\mathbf{x}, t_1) \times N'(\mathbf{x}, t_2)}$ (where $N'=N-\bar{N}$). Further below we will make some remarks on this covariance, but will not otherwise concern ourselves with joint probabilities, because they have not so far made their way into the literature of turbulent diffusion.

We should also remark that a probability distribution of *total dosage* may be defined entirely analogously to $W(n, \mathbf{x})$ above. Because dosages are already time-integrals, such probabilities are not functions of time and are in fact very similar to the fields of $W(n, \mathbf{x})$ in continuous plumes.

7.3. The Functional Form of the Probability Distribution

It would clearly be of considerable practical importance to be able to specify the functional form of the distributions $W(n, \mathbf{x}, t)$ and $\Omega(n, \mathbf{y}, t)$. No serious theoretical discussion of the problem has been published so far, but a heuristic argument (akin to Kolmogoroff's (1941) treatment of the size distribution of ground materials) provides a fairly useful model.

Consider marked fluid released in an initial concentration N_0 into a turbulent field. The initial cloud is quickly deformed into a contorted irregular shape of large surface area. Dilution takes place by molecular diffusion across the boundary between marked and unmarked fluid where sharp gradients are constantly reestablished by the larger eddies. The cloud of marked fluid appears to 'entrain' ambient fluid, by which it is diluted in random spurts and starts. If we follow the history of each small parcel of the original marked cloud, we may legitimately describe it as consisting of a number of 'diluting impulses,' in each of which the parcel is mixed with a certain proportion of ambient fluid. Each such impulse produces a drop in concentration by a random factor $\xi_i<1$, leading to a succession of concentration values in between 'impulses' such as

$$N_0, \quad \xi_1 N_0, \quad \xi_2\xi_1 N_0, \quad \xi_3\xi_2\xi_1 N_0, \text{ etc.}$$

Whatever intermediate concentration has been reached at any stage, the next diluting impulse ξ is likely to be of the same order of magnitude as the preceding ones, owing to the self-similarity of the turbulent diffusion process at successive stages of cloud growth (which we have established in other contexts before).

Therefore the *logarithm* of the concentration observed at a sampling instrument which catches some of our original parcel will be

$$\ln N = \ln N_0 + \ln \xi_1 + \ln \xi_2 + \cdots + \ln \xi_n . \tag{7.6}$$

It may be seen that $\ln(N/N_0)$ is a sum of a number of independent random variables (of the same order of magnitude) and should by the Central Limit Theorem tend to a normal distribution for a large number of diluting impulses.

While the above argument is certainly not rigorous, it is a plausible one, probably no worse than the vague justification for the Gaussian spatial distribution of mean concentration. There is, however, one important qualification: there has to be an initial element with concentration N_0. In experimental studies of atmospheric diffusion one often encounters periods of zero concentration at a given instrument. This means that the sampled element contains no portion of the initial cloud and of course there can be no question of dilution. To describe this state of affairs an 'intermittency factor' $\gamma(\mathbf{x}, t)$ for concentration may be introduced, similar to that used by Townsend (1956) for turbulent energy. This factor specifies the fraction of the ensemble in which the concentration is non-zero:

$$1 - \gamma(\mathbf{x}, t) = W(0, \mathbf{x}, t). \tag{7.7}$$

An entirely analogous definition may be introduced in relative diffusion:

$$1 - \gamma_r(\mathbf{y}, t) = \Omega(0, \mathbf{y}, t). \tag{7.7a}$$

For continuous sources of course γ and γ_r are again not functions of time. The intermittency factor approaches zero at the fringes of a diffusing cloud. At the center it often approaches 1.0, but this is not necessarily the case. The non-zero fraction of the ensemble of concentration readings (either at fixed $\mathbf{x}$ or at fixed $\mathbf{y}$) should have according to the above argument a logarithmico-normal probability distribution.

An early pioneering investigation of concentration probability distributions has been carried out by Gosline (1952) who measured 'instantaneous' (10 s) ground level NO and NO_2 concentrations downwind of a 24 m tall chimney at distances of 5 to 10 chimney heights. Zero concentration readings were eliminated from the processing of the data, but Gosline noted that only 14 to 34% of the time was there a measurable concentration, i.e., $\gamma(\mathbf{x})$ at the sites chosen ranged from 0.14 to 0.34. Also the duration of each NO bearing eddy at a given site was between 30 and 90 s. A replot of some of Gosline's data is shown on log-probability paper in Figure 7.1. A straight line in this figure indicates a log-normal distribution, which is indeed seen to be a very good fit.

Similar conclusions have been derived by an analysis of the Fort Wayne experiments (Csanady *et al.*, 1968) in which dosages were measured in the wake of an instantaneous elevated line source. In these experiments some 50 ground-level samples were arranged along lines parallel to the release line: these provided a population of dosage readings from which frequency distributions of total dosage could be estimated. Because a dosage measurement is a rather long concentration sample, some blurring of the distributions may be expected in cases where zero readings are frequent, the sharp saltus at $W(0, \mathbf{x})$ being smoothed out by the method of sampling. Figure 7.2 shows an observed histogram and fitted log-normal curve for a case without zero readings. Figure 7.3 shows two cumulative frequency distributions on log-probability paper for the case of Figure 7.2 and another, similar case. To illustrate what happens to a few zero readings, Figure 7.4 shows a histogram with 44 'normal' readings and 6 'near-zero' ones, with a lognormal curve fitted to the former and a Poisson distribution to the latter.

Much the same conclusions may be derived on reanalyzing some recent data of

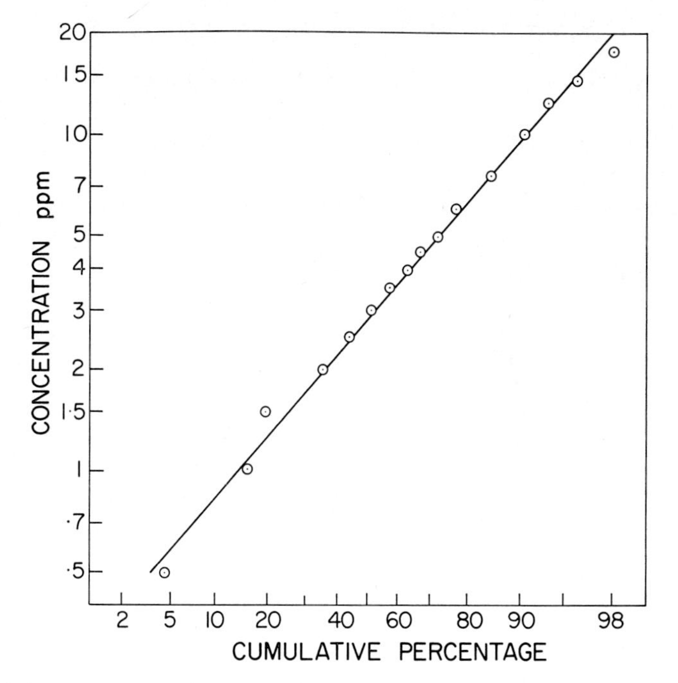

Fig. 7.1. Statistics of ground-level concentrations observed by Gosline (1952), plotted on log-probability paper. Zero readings have been removed from the population.

Ramsdell and Hinds (1971) to remove zero readings. Ramsdell and Hinds present instantaneous (38.4 s) concentrations in the wake of a continuous ground level point source of Krypton gas. Near-zero readings occupied from 35 to 80% of the time at the locations investigated (200 and 800 m from the source). A replot of the non-zero readings again approximates a log-normal distribution.

The most serious gap in our knowledge at present would seem to be the almost complete lack of information on the spatial distribution of the intermittency factor γ or γ_r. In a frame of reference moving with the centerline of a continuous plume $\gamma_r(\mathbf{y})$ is known to be near unity in the center portion of the plume and to be zero outside the plume, the distribution of γ_r across the plume being probably much like the distribution of intermittency of turbulent velocities across a jet (Townsend, 1956). In a *fixed* frame of reference, however, we have already seen that $\gamma(\mathbf{x})$ can be as low as 0.65 even at the axis of the plume and much lower at the fringes. The effect of meandering is thus seen to be particularly serious in this context. A study of the intermittency factors would seem to offer a rewarding topic for future research.

7.4. Hazard Assessment on the Basis of Concentration Probabilities

As we have remarked in the introduction to this chapter, taking into account concentration fluctuations enables one to make a more satisfactory assessment of nuisance

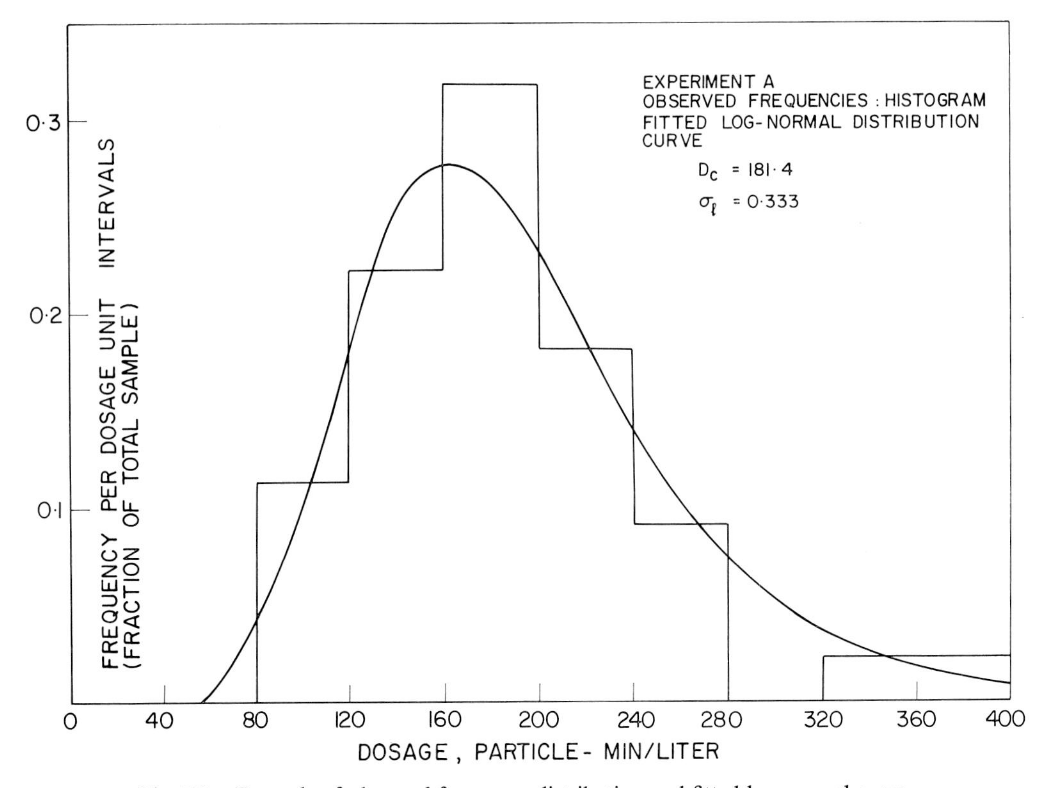

Fig. 7.2. Example of observed frequency distribution and fitted log-normal curve.

or hazard in pollution problems than is possible on the basis of mean-field theory alone. To support this proposition we shall illustrate in the present section how the probability distributions $W(n, \mathbf{x})$ or $\Omega(n, \mathbf{y})$ may be exploited to such practical ends (following mainly the paper by Csanady (1969)). Unfortunately a serious flaw in the discussion below is that we are forced to neglect spatial variations of the intermittency factor γ (in lack of data) so that our results cannot be relied upon quantitatively. The discussion should nevertheless illustrate the kind of hazard assessment which will hopefully be used in future, once more extensive empirical information on fluctuations becomes available.

Consider the concentration field of a continuous plume at ground level $\chi(x, y)$, or the very similar dosage-field of an instantaneous cloud, referring both to a coordinate system attached to the center line of the plume. We choose a description in terms of relative diffusion theory, because in a moving frame the intermittency factor is likely to be near unity for much of the field. Assuming $\gamma=1$ everywhere (as a crude approximation) we take the concentration probability distribution to be log-normal, and write

$$\Omega(n, x, y) = \tfrac{1}{2}\left\{1 + \operatorname{erf}\left[\frac{\ln n/n_c}{\sqrt{2}\,\sigma_l}\right]\right\}, \tag{7.8}$$

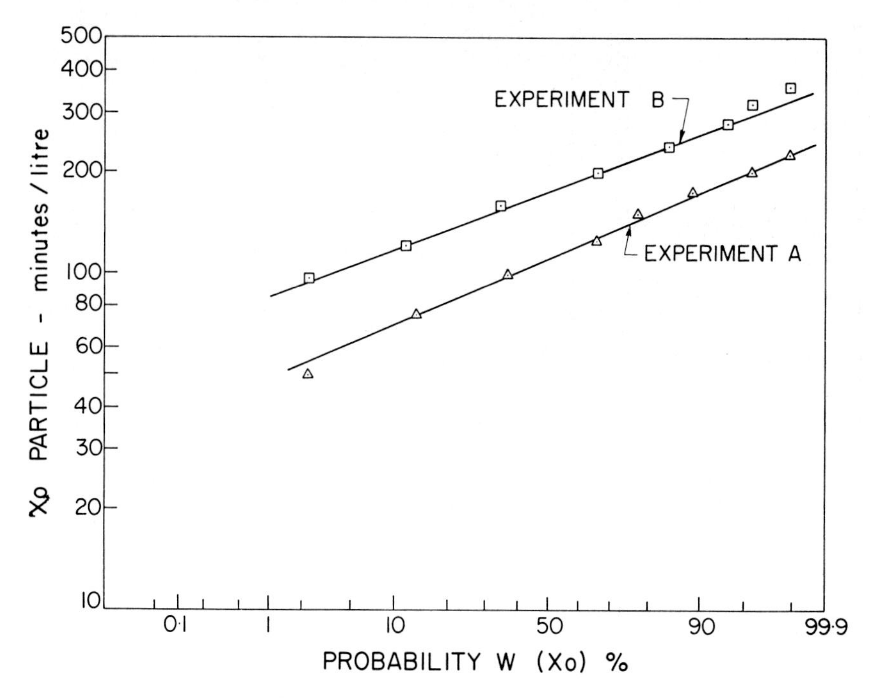

Fig. 7.3. Cumulative frequency distribution of observed dosages at the first sampling line ($x=1600$ m): stable trials.

where

$$\operatorname{erf}(\lambda)=\frac{2}{\sqrt{\pi}}\int_0^{\lambda}\mathrm{e}^{-t^2}\,\mathrm{d}t$$

is again the error function. The probability *density* $\mathrm{d}\Omega/\mathrm{d}n$ is correspondingly

$$\frac{\mathrm{d}\Omega}{\mathrm{d}n}=\frac{1}{\sqrt{2\pi}\,\sigma_1 n}\exp\left\{-\frac{(\ln n/n_c)^2}{2\sigma_1^2}\right\}. \tag{7.9}$$

The log-normal distribution is seen to have two parameters, the median concentration n_c and the logarithmic standard deviation σ_1. Both these may be functions of location (x, y), but may be related to the local mean and variance of observable concentrations by Equation (7.3):

$$\bar{N}(x, y)=\chi(x, y)=\int_{\Omega=0}^{1} n\frac{\mathrm{d}\Omega}{\mathrm{d}n}\,\mathrm{d}n=n_c\exp\left(\frac{\sigma_1^2}{2}\right) \tag{7.10}$$

$$\overline{N'^2}(x, y)=n_c^2\exp(\sigma_1^2)\left[\exp(\sigma_1^2)-1\right].$$

The rms to mean concentration ratio depends only on the logarithmic standard deviation:

$$i_c(x, y)=\frac{\sqrt{\overline{N'^2}}}{\bar{N}}=\sqrt{\exp(\sigma_1^2)-1}\,. \tag{7.11}$$

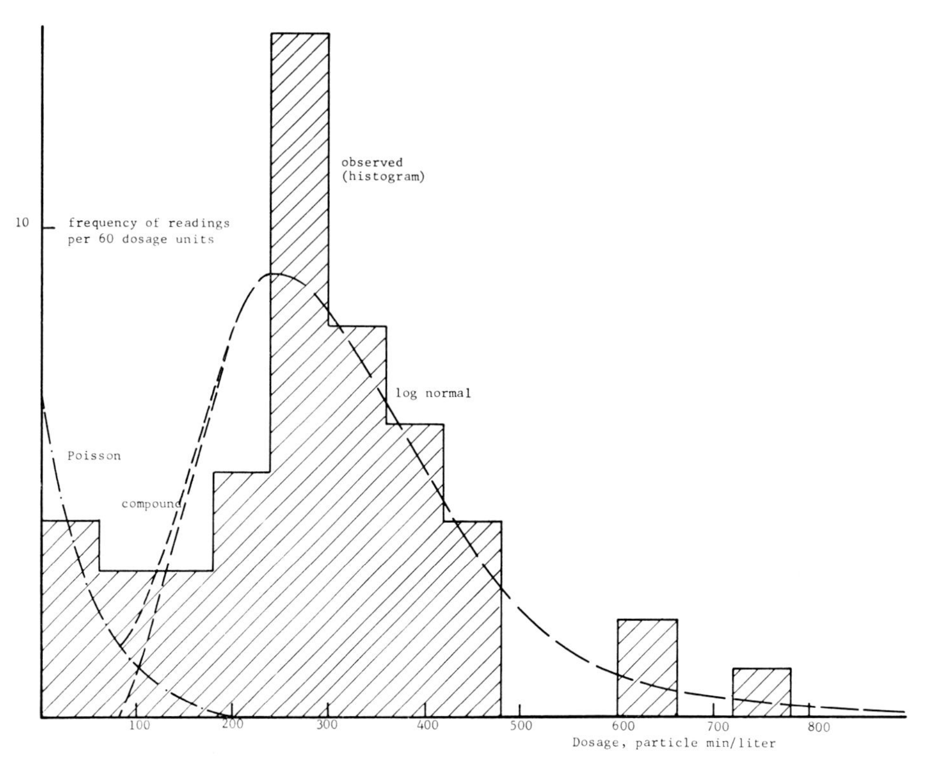

Fig. 7.4. Fitted log-normal distribution to 44 readings and fitted Poisson distribution to 6 readings. (log. normal $D_c = 300$ $\sigma_1 = 0.42$). (Ft. Wayne trial 02G2)

Once this 'relative intensity' of concentration fluctuation, $i_c(x, y)$, is determined or estimated we can find the median concentration as a fraction of the mean:

$$n_c(x, y) = \frac{\chi(x, y)}{\sqrt{1 + i_c^2(x, y)}}. \tag{7.12}$$

The mean concentration distribution $\chi(x, y)$ may be regarded as known, on the basis of information in previous chapters. We shall treat the distribution of the fluctuation intensity $i_c(x, y)$ in the next few sections. For the present we shall regard $i_c(x, y) = $ constant for a given plume, depending on atmospheric conditions and the roughness of the terrain. We shall see that such an approximation is quite reasonable in central portions of a plume. At the fringes $i_c(x, y)$ changes rapidly, but there also the intermittency factor drops and our simple approach of the present section breaks down. We note, however, that a more accurate analysis on exactly the lines adopted below is possible if we incorporate actual $\gamma(x, y)$ and $i_c(x, y)$ distributions in our calculations. Given the validity of the log-normal distribution for non-zero concentration readings, the three distributions $\chi(x, y)$, $\gamma(x, y)$ and $i_c(x, y)$ are sufficient to give a probabilistic description of the whole concentration field.

An analysis of experimental data obtained by Cramer *et al.* (1965) has yielded the following 'typical' values of σ_l (Csanady, 1969):

Conditions	Smooth terrain, stable atmosphere	Smooth terrain, neutral atmosphere	Rough terrain
σ_l	0.2	0.35	0.70

Consider now the 'peak to mean concentration ratio,' which is sometines presented as a result of observational studies of atmospheric diffusion, along with the mean concentration. The 'peak' value of a random variable is of course an uncertain concept but we may lend it precision by specifying that the 'peak' is that value of the concentration which exceeds 99% of the ensemble, i.e., if N_p is such a peak

$$\Omega(N_p, x, y) = 0.99. \tag{7.13}$$

From Equations (7.8) and (7.10) we may now deduce that

$$N_p/\bar{N} = \exp\{\sigma_l(2.326 - 0.5\,\sigma_l)\}. \tag{7.14}$$

Thus the 'peak to mean ratio' is a function of σ_l alone (provided that the intermittency factor is unity), the relationship being illustrated in Figure 7.5. For the above three values of σ_l (0.2 – 0.7) we find $N_p/\bar{N}$ between about 1.5 and 3.5. Such values are indeed characteristic of the *core* regions of diffusing plumes.

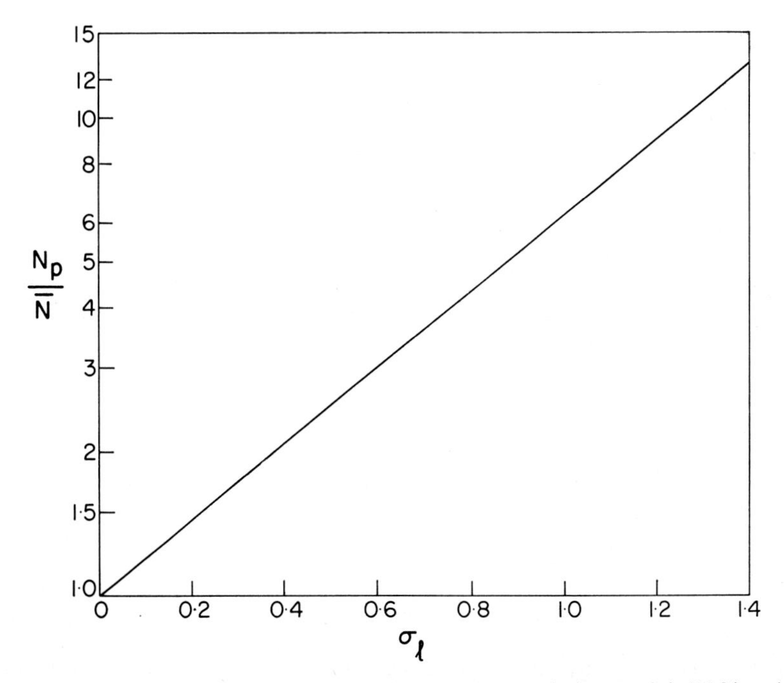

Fig. 7.5. Peak to mean ratios vs. logarithmic standard deviation. Peak is '99% peak'.

In some practical applications the question is whether we have produced some 'critical' or 'threshold' dosage or concentration at a given point. Let such a threshold be N_t. We avoid concentrations higher than N_t 90% of the time if

$$\Omega(N_t, x, y) = 0.9\,.$$

Depending on the seriousness of the effects of a pollutant, we may wish to restrict the probability $P = 1 - \Omega$ of reaching N_t even more. For any given choice of P, we find from Equation (7.8), also using (7.10)

$$\bar{N}/N_t = \exp\{-\sqrt{2}\,\sigma_l\,\mathrm{erf}^{-1}(1-2P) + \sigma_l^2/2\}\,, \tag{7.15}$$

where $\mathrm{erf}^{-1}(\lambda)$ is the inverse function of the error function such that $\mathrm{erf}^{-1}(\lambda)$ gives the *argument* of the error function when the latter's value is λ. From Equation (7.15) we can find the mean value of the local concentration, consistent with exceeding N_t only with probability P. The ratio $\bar{N}/N_t$ can be considerably below unity, reflecting the 'price' one has to pay for avoiding N_t with a given degree of certainly. Take, for example, $P = 0.05$ and $\sigma_l = 0.70$: this yields $\bar{N}/N_t = 0.4$, so that our mean concentration has to be $2\frac{1}{2}$ times less than the threshold, if we want to avoid the threshold 19 times out of 20 observations. The ratio $\bar{N}/N_t$ is illustrated in Figure 7.6 for the three 'typical' values of the logarithmic standard deviations given before. For relatively

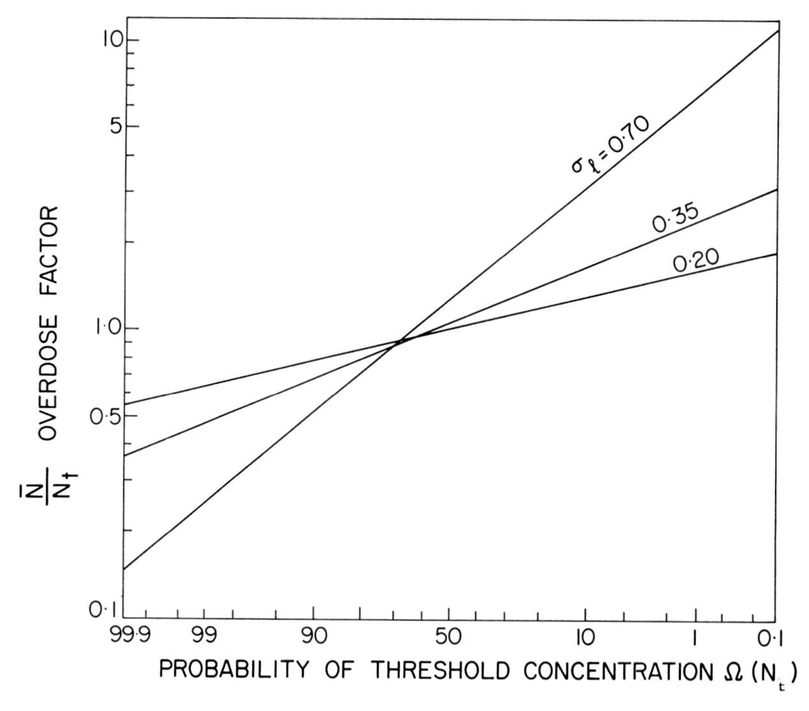

Fig. 7.6. Overdose factors required for reaching a threshold concentration N_t with prescribed probability, for three 'typical' values of the logarithmic standard deviation.

high P (say $P=0.9$) the practical meaning of the ratio is an 'overdose factor' required in order to achieve N_t with given probability (e.g., in the spraying of insecticide, when we want to 'cover' a certain area).

It is also of interest to point out what one accomplishes by the application of the mean-field theory alone to such a problem. If we prescribe that $\bar{N}=N_t$ somewhere, this means by Equation (7.15) or Figure 7.6 that $\Omega=0.53-0.64$, depending on σ_l, so that the prescribed 'threshold' is exceeded 36 to 47% of the time.

In judging the nuisance or 'coverage' attained by releasing a cloud of marked fluid we are not only interested in what happens at isolated points (x, y), but also in the entire concentration pattern. This is described quite well by the *locus* of points where $\Omega(N_t, x, y)$ has a given constant value. A 'well covered area' by this level of concentration is where $\Omega(N_t, x, y) \geqslant 0.1$ (say) so that N_t is exceeded at least 90% of the time. A relatively 'unaffected' area is where $\Omega(N_t, x, y) \geqslant 0.9$ so that N_t is avoided at least 90% of the time. The boundaries of these areas are of course the contours $\Omega=0.1=\text{const.}$ and $\Omega=0.9=\text{const.}$, respectively.

Under the simple assumptions we have made above ($\gamma=1=\text{const.}$ and $\sigma_l=\text{const.}$) we see at once from Equation (7.15) that a $\Omega(N_t)=\text{const.}$ line is at once a $\bar{N}(x, y)= =\text{constant}$ contour. We have illustrated $\chi=\bar{N}=\text{const.}$ contours before: they are long, cigar-shaped contours. We may now relabel such plots as lines showing constant probability of attaining some chosen threshold concentration. Such an illustration is reproduced here in Figure 7.7, from Csanady (1969), showing ground level *dosage*-probabilities in ground-level puff-release trials. The mean dosage to threshold dosage ratios (analogous to our earlier $\bar{N}/N_t$) is also shown on the contours.

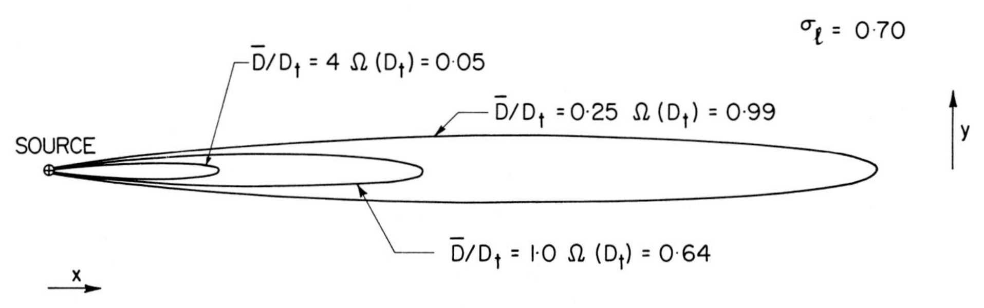

Fig. 7.7. Constant mean dosage contours relabeled with dosage probabilities.

7.5. The Variance of Concentration Fluctuations

As we have just seen, the field of concentration *variance* $\overline{N'^2}$ in either a fixed or a moving frame is important evidence in pollution problems, probably just as much so as the field of mean concentration which we have discussed in most of the rest of this monograph. Work on $\overline{N'^2}$ has however so far been quite meager. Some important fundamental ideas were established by Obukhov (1949), Corrsin (1951), and Batchelor (1959). These analytical studies have shown that the mean-square fluctuation of a scalar quantity such as temperature or concentration in a turbulent fluid behaves somewhat analogously to turbulent kinetic energy, in that it can be transferred in physical space as well as in wave-number space (across the spectrum, between Fourier components). Batchelor indeed uses the phrase '$\overline{\theta^2}$-stuff' when referring to temperature fluctuation intensity. The analogy with turbulent energy extends to the dissipation of mean square fluctuations, which occurs at high wave numbers due to molecular conductivity or diffusivity much as energy dissipation does due to viscosity.

Most of the early theoretical work related to the transfer of mean square fluctuations in wave number space and provided much insight for turbulence theory. In diffusion problems we are concerned with the *spatial* transfer and distribution of $\overline{N'^2}$. An approximate theory of this process is developed below, following Csanady (1967a, b). The theory is mainly applicable to relative diffusion (or absolute diffusion without significant meandering) on account of its reliance on K-theory to represent spatial transfer processes. Presumably, transport by meandering is not very well described by such a theory.

Consider the diffusion equation written for the details of the realized concentration field $N(y, t)$:

$$\frac{\partial N}{\partial t} + v_{\mathrm{i}} \frac{\partial N}{\partial y_{\mathrm{i}}} = D\nabla^2 N . \tag{7.16}$$

Resolve velocities and concentration into mean and fluctuating components:

$$v_{\mathrm{i}} = \bar{v}_{\mathrm{i}} + v'_{\mathrm{i}}$$
$$N = \bar{N} + N' .$$

Take ensemble means on both sides of Equation (7.16) and subtract the averaged from the original equation. The result is a relationship for the fluctuating part of the N-field:

$$\frac{\partial N'}{\partial t} + v'_{\mathrm{i}} \frac{\partial \bar{N}}{\partial y_{\mathrm{i}}} + \bar{v}_{\mathrm{i}} \frac{\partial N'}{\partial y_{\mathrm{i}}} + \left(v'_{\mathrm{i}} \frac{\partial N'}{\partial y_{\mathrm{i}}} - \overline{v'_{\mathrm{i}} \frac{\partial N'}{\partial y_{\mathrm{i}}}} \right) = D\nabla^2 N' . \tag{7.17}$$

We multiply this equation by $2N'$, take means and simplify some of the terms with the aid of the continuity equation:

$$\partial v'_{\mathrm{i}} / \partial y_{\mathrm{i}} = 0 . \tag{7.18}$$

The result is a balance equation for $\overline{N'^2}$:

$$\frac{\partial \overline{N'^2}}{\partial t} + \bar{v}_{\mathrm{i}} \frac{\partial \overline{N'^2}}{\partial y_{\mathrm{i}}} = - \overline{2 v'_{\mathrm{i}} N'} \frac{\partial \bar{N}}{\partial y_{\mathrm{i}}} + \frac{\partial}{\partial y_{\mathrm{i}}} \left(D \frac{\partial \overline{N'^2}}{\partial y_{\mathrm{i}}} - \overline{v'_{\mathrm{i}} N'^2} \right) - \Phi , \tag{7.19}$$

where Φ is a dissipation function:

$$\Phi = 2D \overline{\frac{\partial N'}{\partial y_{\mathrm{i}}} \cdot \frac{\partial N'}{\partial y_{\mathrm{i}}}} \tag{7.20}$$

clearly a positive definite quantity. The terms of Equation (7.19) may be interpreted physically much as those of the turbulent energy equation: on the left we have local and advective change of $\overline{N'^2}$, on the right a production term consisting of flux of N times gradient of $\bar{N}$, then a spatial transfer or flux for $\overline{N'^2}$, having the form of a divergence, and then the rate of dissipation of $\overline{N'^2}$. To simplify the notation somewhat we shall write $S=\overline{N'^2}$ for the quantity that is being produced, transferred and dissipated; in the chemical engineering literature this is sometimes called the intensity of 'segregation'. Also we shall use the symbol χ for $\bar{N}$, as before.

We shall specialize our discussion to continuous point or line source plumes in a uniform fluid of constant mean velocity, i. e., we assume

$$v_{\mathrm{i}} = U\delta_{\mathrm{i}1} \qquad \partial/\partial t = 0 . \tag{7.21}$$

As we have seen earlier, the flux of N in such plumes may be described in terms of an eddy diffusivity K, which is a function of distance from the source:

$$\overline{v_{\mathrm{i}}'N'} = -\, K \frac{\partial \chi}{\partial y_{\mathrm{i}}} . \tag{7.22}$$

We ignore any differences in K along the y_2 and y_3 axes. The variation of K with y_1 may be calculated from

$$K = \frac{U}{2}\frac{\mathrm{d}\sigma^2}{\mathrm{d}y_1}, \tag{7.23}$$

where $\sigma=\sigma_{\mathrm{y}}=\sigma_{\mathrm{z}}$ is the standard deviation of the relative dispersion, a quantity we have frequently dealt with before.

A reasonable hypothesis in regard to the flux of S (second term on the right of Equation (7.19)) is a gradient transport relationship akin to Equation (7.22):

$$\overline{v_{\mathrm{i}}'N'^2} - D\frac{\partial S}{\partial y_{\mathrm{i}}} = -\, K \frac{\partial S}{\partial y_{\mathrm{i}}} . \tag{7.24}$$

One could refine this hypothesis by using a different K (one, however, that is proportional to K determined from Equation (7.38)) but we shall use the simplest possible approach here.

To relate the dissipation Φ to the intensity of segregation S we must make use of some evidence on the cross-spectral transfer of contributions to $\overline{N'^2}$ in isotropic turbulence. Such turbulence is usually generated by placing a coarse grid at the entrance to a wind tunnel. When the grid is heated, temperature as well as velocity fluctuations are produced. It turns out (Gibson and Schwarz, 1963) that the kinetic energy and the mean square temperature fluctuations decay at the same rate, i.e.

$$\begin{aligned} \frac{\mathrm{d}\overline{u_1^2}}{\mathrm{d}t} &= -\frac{3}{2}\frac{\overline{u_1^2}}{t} \\ \frac{\mathrm{d}\overline{\theta^2}}{\mathrm{d}t} &= -\frac{3}{2}\frac{\overline{\theta^2}}{t}, \end{aligned} \tag{7.25}$$

where $t = x_1/U$ travel time behind the grid, $u_1 =$ a turbulent velocity component and $\theta =$ temperature fluctuation. One may express this by saying that the rate of decay of either type of fluctuation is proportional to the intensity $\overline{u_1^2}$ or $\overline{\theta^2}$, the constant of proportionality being the reciprocal of a 'decay time scale' $t_d = \frac{2}{3}t$. The decay time scale is seen to be $\frac{2}{3}$ of the 'age' of both velocity and temperature fluctuations.

In a continuous plume, generated by a concentrated line or point source in a field of homogeneous turbulence we may also expect the rate of dissipation of concentration fluctuations to be proportional to fluctuation intensity, because essentially the same physical factors must govern across-the-spectrum transfer of contributions to $\overline{N'^2}$ regardless of the manner in which the fluctuations were generated. However, the 'ages' of the velocity and the concentration fluctuations differ in this instance and the decay time-scale may vary in an unknown manner. We write therefore

$$\Phi = S/t_d, \tag{7.26}$$

where $t_d = t_d(y_1)$ is some undetermined function of y_1.

With the above substitution, and returning to an x, y, z notation we derive from Equation (7.19) the following equation, governing the S-field in a continuous point or line-source plume:

$$U(\partial S/\partial x) = 2K(\nabla\chi)^2 + K\nabla_1^2 S - S/t_d, \tag{7.27}$$

where ∇_1^2 is a Laplacian in the y, z plane and diffusion along x has been ignored as usual ('slender' plume approximation).

7.6. Self-Similar Fluctuation Intensity Distribution

We know that the mean concentration distribution in a continuous plume in a uniform field is self-similar at different sections and it is plausible to postulate self-similarity also for the S-field, with the same concentration and length scales as apply to the χ-field. If the mean concentration along the axis $y=0$, $z=0$ is $\chi_0(x)$, the two scales are conveniently taken to be χ_0 and σ. For a *point* source plume axial symmetry applies in view of our previous assumption that K (and therefore σ) is the same along y as along z and the two distributions may be written:

$$\begin{aligned} \chi &= \chi_0 f(\xi) \\ S &= \chi_0^2 g(\xi), \end{aligned} \tag{7.28}$$

where $\xi = r/\sigma = \sqrt{y^2 + z^2}/\sigma$ is nondimensional radius.

Continuity requires that the same quantity passes through each section:

$$\frac{q}{U} = \iint \chi \, dy \, dz = \text{const.}\ \chi_0\sigma^2 \tag{7.29}$$

so that

$$\frac{d\chi_0}{dx} = -\frac{2\chi_0}{\sigma}\frac{d\sigma}{dx}. \tag{7.30}$$

Substitution of the self-similar forms of Equation (7.28) into (7.27) now results in the following equation governing the S-field for an axisymmetric point-source plume:

$$\frac{d^2 g}{d\xi^2} + \frac{1}{\xi}\frac{dg}{d\xi} + \frac{U\sigma}{K}\frac{d\sigma}{dx}\left(4g + \xi\frac{dg}{d\xi}\right) - \frac{\sigma^2}{Kt_d}g = -2\left(\frac{df}{d\xi}\right)^2. \tag{7.31}$$

This equation has self-similar solutions if the two nondimensional parameters it contains are independent of x. The first one of these is, using Equation (7.23):

$$\frac{U\sigma}{K}\frac{d\sigma}{dx} = 1. \tag{7.32}$$

For the second parameter we write

$$\sigma^2/Kt_d = \alpha = \text{const.} \tag{7.33}$$

subject to experimental confirmation of self-similarity. For the mean concentration distribution we use the Gaussian distribution

$$f(\xi) = e^{-\xi^2/2}. \tag{7.34}$$

Equation (7.31) then becomes:

$$\frac{d^2 g}{d\xi^2} + \left(\frac{1}{\xi} + \xi\right)\frac{dg}{d\xi} + (4 - \alpha)\, g = -2\xi^2 e^{-\xi^2} \tag{7.35}$$

which is a nonhomogeneous ordinary linear differential equation of the second order. As boundary conditions we impose axial symmetry and a decay to zero at large distances:

$$\begin{aligned} \frac{dg}{d\xi} &= 0 \quad \text{at} \quad \xi = 0 \\ g &\to 0 \quad \text{as} \quad \xi \to \infty. \end{aligned} \tag{7.36}$$

Solutions of Equation (7.35) subject to (7.36) are not difficult to find, e. g. numerically, if the value of the constant α is known*. One may assume an arbitrary value $g(0)$ and determine $g(\xi)$ step by step in small ξ-increments. At large ξ, g should tend to zero: if it does not, one repeats the calculation with a different $g(0)$ until the boundary condition at infinity is satisfied. A relatively simple analytical solution is obtained for the special case of $\alpha = 4$; this is:

$$g(\xi) = 2E_1(\xi^2/2) - 2E_1(\xi^2) - e^{-\xi^2}, \tag{7.37}$$

where $E_1(\lambda)$ is the exponential integral, tabulated in Abramowitz and Stegun (1964):

$$E_1(\lambda) = \int_\lambda^\infty \frac{e^{-t}}{t}\, dt. \tag{7.38}$$

* An analytical solution may also be found in terms of integrals involving tabulated functions, if we observe that the transformation $\eta = -\xi^2/8$ reduces Equation (7.35) to Kummer's equation.

This solution is illustrated in Figure 7.8. The relative fluctuation intensity $i_c(\xi)$ which we introduced before in Equation (7.11) is, in our present notation

$$i_c = \sqrt{g}/f\,. \tag{7.38}$$

This is also shown in Figure 7.8. One may see that the '*absolute*' fluctuation intensity $g(\xi)$ decreases from the center of plume to the fringes much as the mean concentration does, while the *relative* intensity increases markedly at the fringes. To regard i_c constant in the core region of a plume is not exactly accurate, but is a tolerable approximation.

For different values of α different $g(\xi)$ profiles are obtained. The center value $g(0)$ is a rapidly varying function of α, shown in Figure 7.9. For $\alpha>4$ a 'saddle' appears at the origin (because $d^2g/d\xi^2$ is positive) and a full section across the plume shows a double-peaked distribution. Physically the reason is that the maximum rate of production occurs in the region of steepest gradients ($\xi=1$ to 2), from where S diffuses both inward and outward. At relatively low K or t_d (or both), i. e., relatively high α, the peaks occur where production is a maximum. High diffusion and low dissipation (i. e., a low α, $\alpha<4$) smoothes these out and results in a single-peaked distribution, the peak being at $\xi=0$.

In comparing the theory with experimental data on concentration fluctuations, we have of course no direct information on the parameter α, so that the center value of the

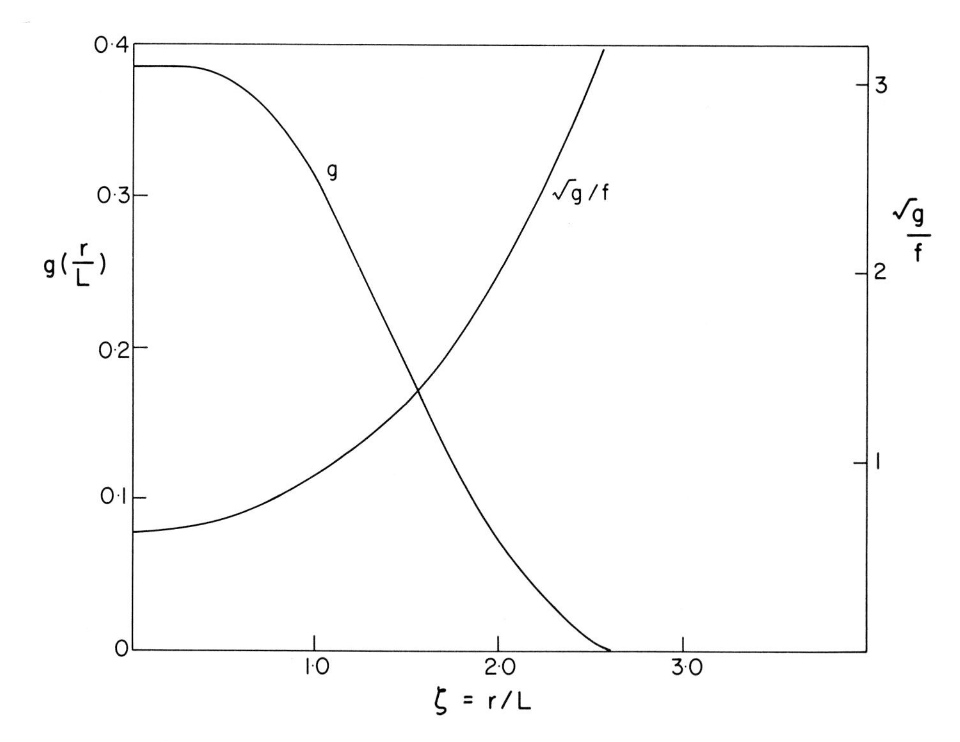

Fig. 7.8. Fluctuation intensity in axisymmetric point source plume, $\alpha=4$.

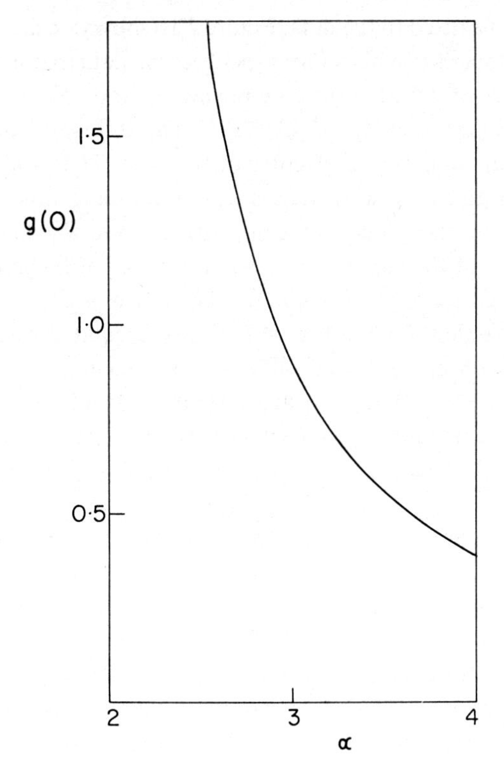

Fig. 7.9. Variation of center fluctuation intensity with parameter α.

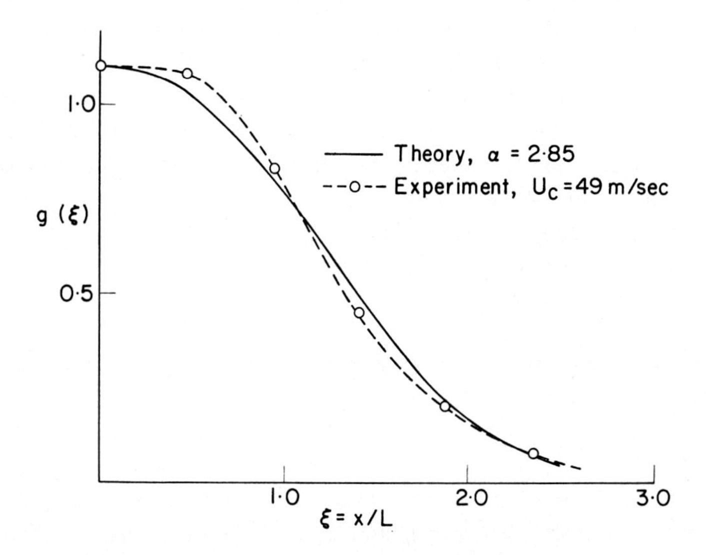

Fig. 7.10. Nondimensional concentration variance in continuous point source plume.

intensity $g(0)$ has to be fitted to the data. Figure 7.10 shows a calculated nondimensional S-profile, $g(\xi)$, together with some experimental data reported by Becker *et al.* (1966). The data were obtained in the core-region of a pipe of 20 cm diam. in which air was flowing at a center velocity of 49 m s^{-1}. The diffusing substance was oil fog released from a point source at the center of the pipe. The experimental profile $g(\xi)$ was self-similar (for given U_c) at distances from the source varying from 2 to 12 pipe diameters. At other center velocities the $g(\xi)$ distribution was markedly different, not so much in the shape of the curve as in absolute values: $g(0)$ changed from a value of 1.3 at $U_c=41$ m s^{-1} to 1.1 at 49 s^{-1} and to 0.88 at 61 m s^{-1}.

The theoretical profile shown in Figure 7.10 was calculated using $\alpha=2.85$, which by Figure 7.9 gave the observed center value. At other center velocities similarly good fits were obtained by $\alpha=2.70$ ($U_c=41$ m s^{-1}) and $\alpha=3.0$ ($U_c=61$ m s^{-1}). We see that with one adjustable constant the agreement between theory and experiment is quite respectable: specifically the spread of the $g(\xi)$ distribution is faithfully reflected by the theory.

Somewhat similar data on diffusing dye plumes in the Great Lakes have been reported by Csanady (1966) and Murthy and Csanady (1971). An illustration from the latter paper is shown in Figure 7.11. The dots represent data obtained from concentration cross-sections at four different distances from a continuous point source. The abscissa is distance from the plume's center of gravity, divided by σ, i.e., the presentation is in the framework of 'relative' diffusion theory. The ordinate is local rms fluctuation to mean ratio, and the observed distribution of this quantity is seen to have the character of the theoretical $\sqrt{g\,/\,f}$ curve shown in Figure 7.8. The center value $g(0)$ is in this case rather lower than in the case shown in Figure 7.10. Very similar atmospheric data have been reported by Ramsdell and Hinds (1971) with $g(0)$ values above 1.0, as in some of the laboratory experiments. Laboratory observations of Lee and Brodkey (1964) on the diffusion of dye in water have also shown self-similar $g(\xi)$ distributions with a rather low value of $g(0)=0.025$.

The experimental data therefore offer strong evidence for self-similarity of the S-distribution in continuous point-source plumes. We may therefore accept that α defined by Equation (7.33) is indeed constant in such cases. From Equations (7.33) and (7.23) we may then make certain inferences on the variation of the decay time scale t_d with distance from the source. The observations all refer to distances where the asymptotic stage of diffusion should have been reached, i.e.:

$$\sigma^2 = 2i_y^2 x_0 (x - x_0) \tag{7.39}$$

with $x_0 = Ut_L$, t_L = Lagrangian time-scale. From Equations (7.33) and (7.23) it follows with Equation (7.39) that

$$t_d = 2(x - x_0)/\alpha U \tag{7.40}$$

or again a linear growth with distance, measured from the effective origin $x=x_0$. For $\alpha=3$ the same relationship between t_d and 'age' of the fluctuations is obtained as in isotropic turbulence. We have already seen that this value indeed characterizes a few

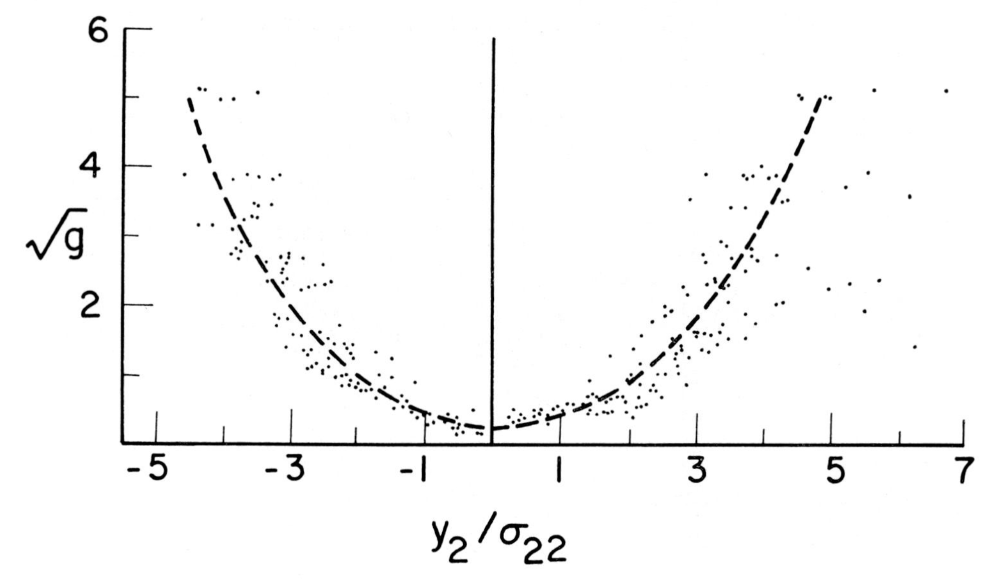

Fig. 7.11. Rms to mean concentration ratio at fixed *nondimensional* distances from center of gravity (distance scale: standard deviation of concentration distribution). Dots represent data at four different sections (Murthy and Csanady, 1971).

observed data, but that considerable variations in α exist between different experiments. There the matter rests at present: we do not know what causes variation in α (or $g(0)$).

Very similar considerations apply to the line-source case. The distributions of χ and S are then functions of $\zeta = z/\sigma$ only, the nondimensional S-distribution being subject to the equation

$$\frac{\mathrm{d}^2 g}{\mathrm{d}\zeta^2} + \zeta \frac{\mathrm{d}g}{\mathrm{d}\zeta} + (2 - \alpha)\, g = -\, 2f'^2 . \tag{7.41}$$

For the special case $\alpha = 2$ this has the solution:

$$g(\zeta) = \frac{\pi}{2}\left[1 - \operatorname{erf}^2\left(\frac{\zeta}{\sqrt{2}}\right)\right] - \mathrm{e}^{-\xi^2} . \tag{7.42}$$

This distribution is shown in Figure 7.12 together with some experimental data of Crum and Hanratty (1965) on temperature fluctuations behind a heated wire stretched across a 3 in. pipe in which air was flowing. This experimental arrangement is not favorable to the establishment of self-similar profiles, because it is not clear whether a two-dimensional distribution can be established in a turbulent field which is axially symmetric. Also the measurements were taken rather close to the source. Self-similarity was *not* observed, although the $g(\xi)$ profiles are much as one calculates from Equation (7.41) for different α-values, $\alpha > 2$. Some atmospheric line-source data of Dunskii and Evseeva (1965) and of Csanady *et al.* (1968) have shown a $g(0)$ value

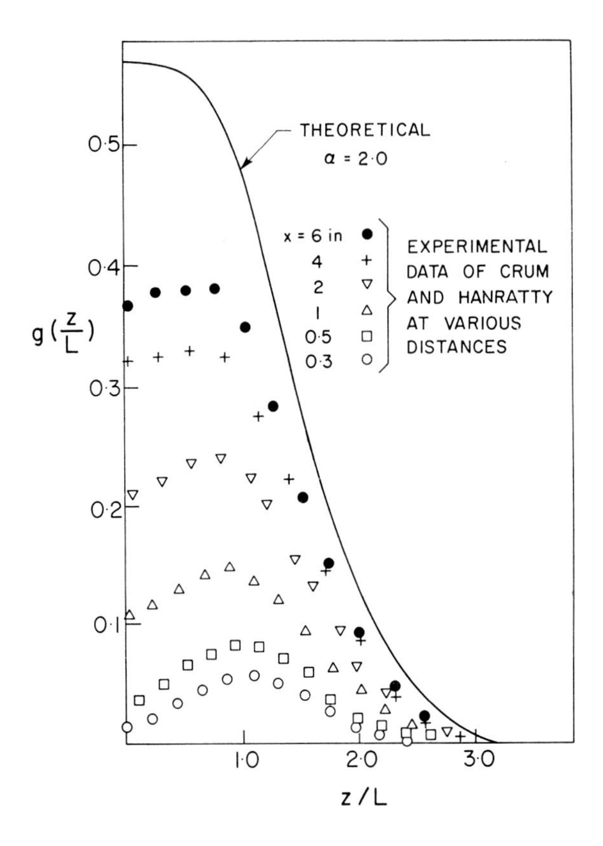

Fig. 7.12. Nondimensional mean square fluctuation profile in the wake of a line source.

constant with distance from a line-source (in both cases $g(0)=0.6$ or so). Further detailed evidence on this subject is again lacking.

7.7. Fluctuating Plume Model

In the past few sections we have deliberately ignored the effects on fluctuations of the large-scale bodily plume movements we have called before 'meandering.' This was done in order to simplify the problem because it is clear that meandering, if at all pronounced, must contribute considerably to locally observable fluctuations and may indeed be mostly responsible for the intermittency of pollution we have noted in connection with Gosline's (1952) observations downwind of short stacks.

Recognizing the importance of meandering in this context, Gifford (1959) has developed a theory which may be regarded as predicting fluctuation probabilities due to meandering alone, without any contribution from the randomness also involved in relative diffusion. This is the 'fluctuating plume model' which we discuss in this present and final section. Presumably it would be possible to discuss the effects of relative diffusion *plus* meandering on a roughly similar basis starting with our basic Equation (7.4) and using some model for $\Omega(n, y, t)$, preferably one that includes an intermittency factor $\gamma_r(\mathbf{y}, t)$, but no such theoretical treatment has come to light yet.

Gifford's model is based on the approximation that the concentration distribution in the moving frame is nonrandom, i.e., that

$$N(\mathbf{y}, t) \equiv \chi_r(\mathbf{y}, t) \tag{7.43}$$

in all realizations. The approximate nature of this assumption should be quite clear from our earlier results and has been also emphasized by Gifford. Nevertheless the theory based on it should provide important insight.

In terms of the probability distribution $\Omega(n, \mathbf{y}, t)$ assumption (7.43) means that there is a saltus in Ω from zero to one where $n = \chi_r$. Stating this in terms of the probability *density* we may write

$$\mathrm{d}\Omega/\mathrm{d}n = \delta(n - \chi_r), \tag{7.44}$$

where $\delta(n)$ is a delta function defined over the concentration axis. We shall only discuss the case of a continuous, point-source plume, for which χ_r is a function of the two-dimensional vector $\mathbf{y}(y_2, y_3)$ in the cross-wind plane, and of the distance x_1 of this plane from the origin. Similarly, the spatial probability density of meandering $\mathbf{P}_c(\mathbf{c})$ is a function of (c_2, c_3) in the cross-wind plane, and of $c_1 = x_1$. The dependence on the latter variable enters in both cases through scale factors of length and concentration.

From our fundamental Equation (7.4) we find now for the fixed-frame probability density (after differentiation with respect to n and substitution of Equation (7.44)):

$$\mathrm{d}W/\mathrm{d}n = \iint \delta(n - \chi_r)\, \mathbf{P}_c(\mathbf{c})\, \mathrm{d}c_2\, \mathrm{d}c_3, \tag{7.45}$$

wherein χ_r is a function of $\mathbf{y} = \mathbf{x} - \mathbf{c}$ and $\mathrm{d}W/\mathrm{d}n$ depends on n and $\mathbf{x}$. Clearly, contributions to the integral in Equation (7.45) only arise from that locus of points which yield $\chi_r = n = \text{const}$. For an arbitrarily chosen fixed-frame point, (x_2, x_3) this locus is illustrated in Figure 7.13 (the locus defines the plume-center positions which produce a given concentration $n = \chi_r$ at one chosen fixed point).

The integral is conveniently evaluated by transforming the coordinates from (c_2, c_3) to (l, p), where l is length of arc along the $\chi_r = n = \text{constant}$ locus and p length of arc along lines everywhere perpendicular to that locus. Let the next locus, on which $\chi_r = n + \mathrm{d}n = \text{constant}$, be $\mathrm{d}p$ apart and write

$$\mathrm{d}p = \lambda(l)\, \mathrm{d}n \tag{7.46}$$

so that $\lambda(l)$ is a scale factor connecting length of perpendicular arcs p to changes in concentration n. The integration with respect to n can be carried out and leaves only an integration along l:

$$\mathrm{d}W/\mathrm{d}n = \oint_{\chi_r = n} \mathbf{P}_c(l)\, \lambda(l)\, \mathrm{d}l. \tag{7.47}$$

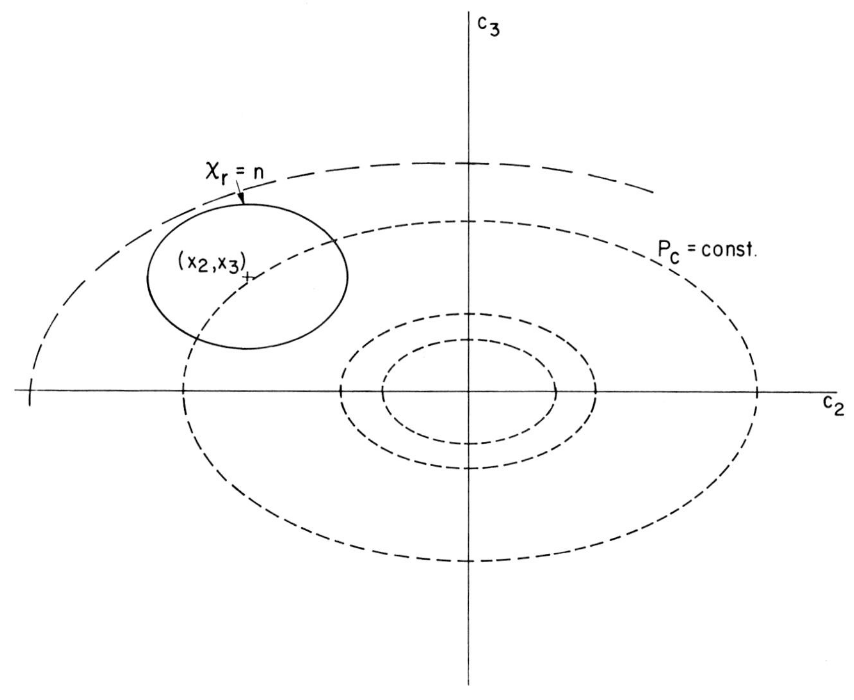

Fig. 7.13. Locus of center positions of a meandering plume which produce a given concentration n at a fixed location (x_2, x_3).

The dependence of P_c on l may of course be determined if the equation of the locus $\chi_r = n$ is known.

As a particularly simple illustration we shall consider the case when both χ_r and P_c are axisymmetric Gaussian distributions:

$$\chi_r = \chi_0(x_1)\exp\left\{-\frac{y_2^2}{2s^2} - \frac{y_3^2}{2s^2}\right\}$$

$$= \chi_0(x_1)\exp\left\{-\frac{(x_2 - c_2)^2}{2s^2} - \frac{(x_3 - c_3)^2}{2s^2}\right\} \tag{7.48}$$

$$P_c = \frac{1}{2\pi m^2}\exp\left\{-\frac{c_2^2}{2m^2} - \frac{c_3^2}{2m^2}\right\}.$$

Here clearly $\chi_0 = q/(2\pi U s^2)$ is the axial mean concentration in the moving frame and s, m are standard deviations of relative diffusion and meandering, respectively (both along y and z, $s = s_y = s_z$, $m = m_y = m_z$). Parenthetically we remark here that, as may be shown from Equation (7.5), also the fixed-frame mean concentration distribution is now Gaussian, and axisymmetric with standard deviation σ, where $\sigma^2 = s^2 + m^2$. The connection of the mean concentration fields was discussed in Chapter IV and we will not dwell upon it further here.

Given the axial symmetry, it is convenient to use radii

$$r^2 = y_2^2 + y_3^2 \tag{7.49a}$$

$$R^2 = x_2^2 + x_3^2 \tag{7.49b}$$

$$\varrho^2 = c_2^2 + c_3^2 .$$

The $\chi_r = n =$const. locus is the circle of radius r where

$$r^2 = 2s^2 \ln(\chi_0/n) . \tag{7.50}$$

The relationship of the three radii along this locus is (see also Figure 7.14):

$$\varrho^2 = r^2 + R^2 - 2rR\cos\phi . \tag{7.51}$$

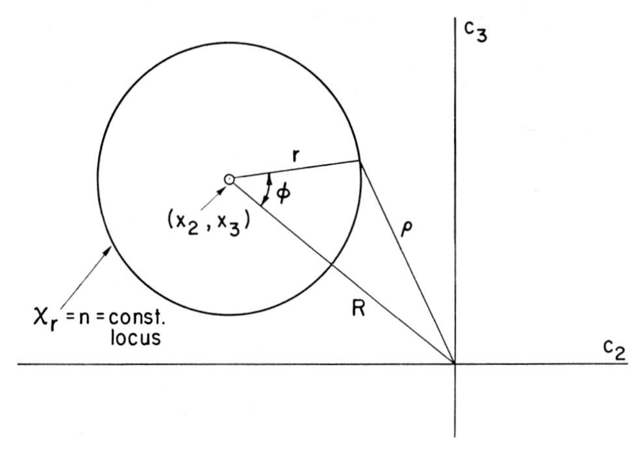

Fig. 7.14. Locus of plume center positions giving constant concentration at (x_2, x_3) for axisymmetric plume.

The line element dl may be taken to be $r\,\mathrm{d}\phi$, dp as dr, giving an area element for the integration indicated by Equation (7.45) of

$$r\,\mathrm{d}r\,\mathrm{d}\phi = -\frac{s^2}{n}\,\mathrm{d}n\,\mathrm{d}\phi . \tag{7.52}$$

In consequence the formulation indicated in Equation (7.47) becomes:

$$\frac{\mathrm{d}W}{\mathrm{d}n} = \int_0^{2\pi} \mathbf{P}_c(\phi)\frac{s^2}{n}\,\mathrm{d}\phi , \tag{7.53}$$

where $\mathbf{P}_c(\phi)$ is evaluated on the $\chi_r = n$ locus. Substituting Equations (7.48) and (7.51) we find without difficulty:

$$\int_0^{2\pi} \mathbf{P}_c\,\mathrm{d}\phi = \frac{1}{m^2}\,\mathrm{e}^{-(r^2+R^2)/2m^2}\, I_0\left(\frac{rR}{m^2}\right), \tag{7.54}$$

where $I_0(\gamma)$ is the modified Bessel function of order zero. Substituting for r from Equation (7.50) we arrive at the result:

$$\frac{\mathrm{d}W}{\mathrm{d}n}\chi_0 = \frac{s^2}{m^2}\left(\frac{n}{\chi_0}\right)^{s^2/m^2-1} \mathrm{e}^{-R^2/2m^2} \times I_0\left(\frac{\sqrt{2}\,sR}{m^2}\left[\ln\frac{\chi_0}{n}\right]^{1/2}\right). \tag{7.55}$$

It is to be noted that $n \leqslant \chi_0$ according to the hypotheses of the Gifford model. The result is seen to be stated in terms of the nondimensional concentration n/χ_0, and is a function also of the ratio of the standard deviations s/m, as well as of the nondimensional fixed-frame radius, R/m. At the center of the plume $R/m=0$ the simple result holds:

$$\frac{\mathrm{d}W}{\mathrm{d}n}\chi_0 = \frac{s^2}{m^2}\left(\frac{n}{\chi_0}\right)^{s^2/m^2-1} \quad (R=0). \tag{7.56}$$

On integrating this with respect to n/χ_0 we find the cumulative frequency distribution

$$W = \left(\frac{n}{\chi_0}\right)^{s^2/m^2} \quad (R=0). \tag{7.57}$$

These distributions are illustrated in Figure 7.15 for the ratio $s/m=0.5$. At $n/\chi_0=1$

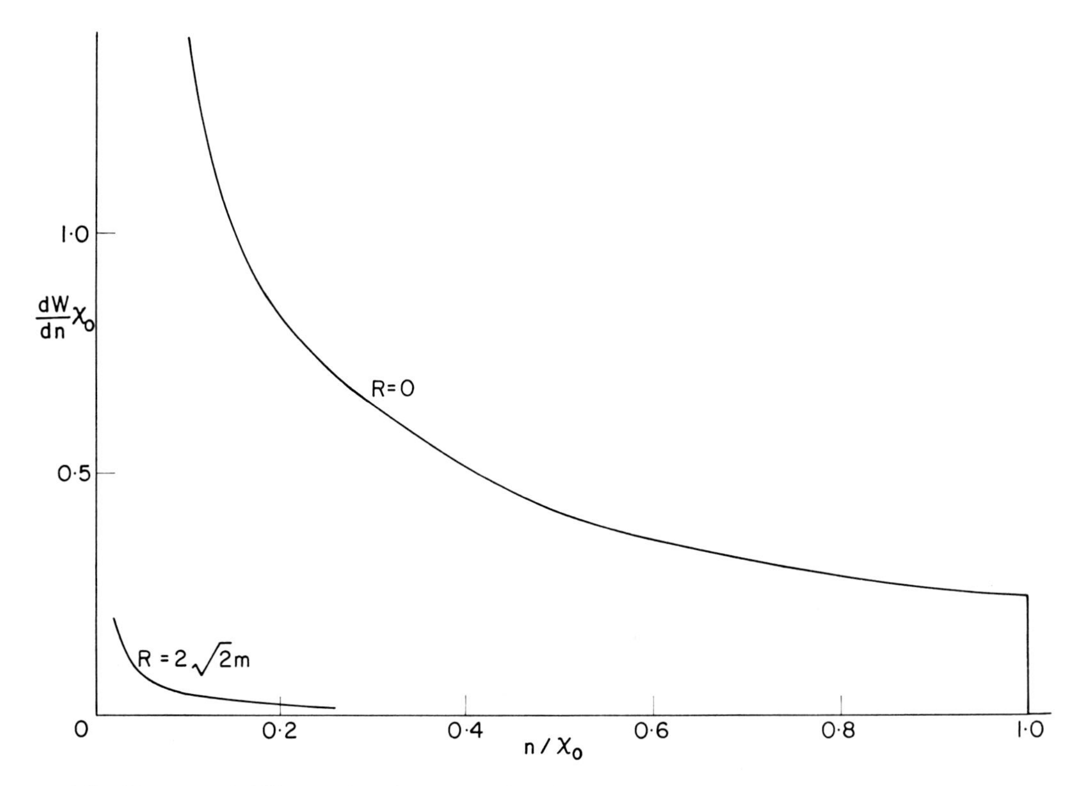

Fig. 7.15a. Probability density of concentration fluctuations at fixed points in continuous plume calculated from Gifford model for $s/m=0.5$.

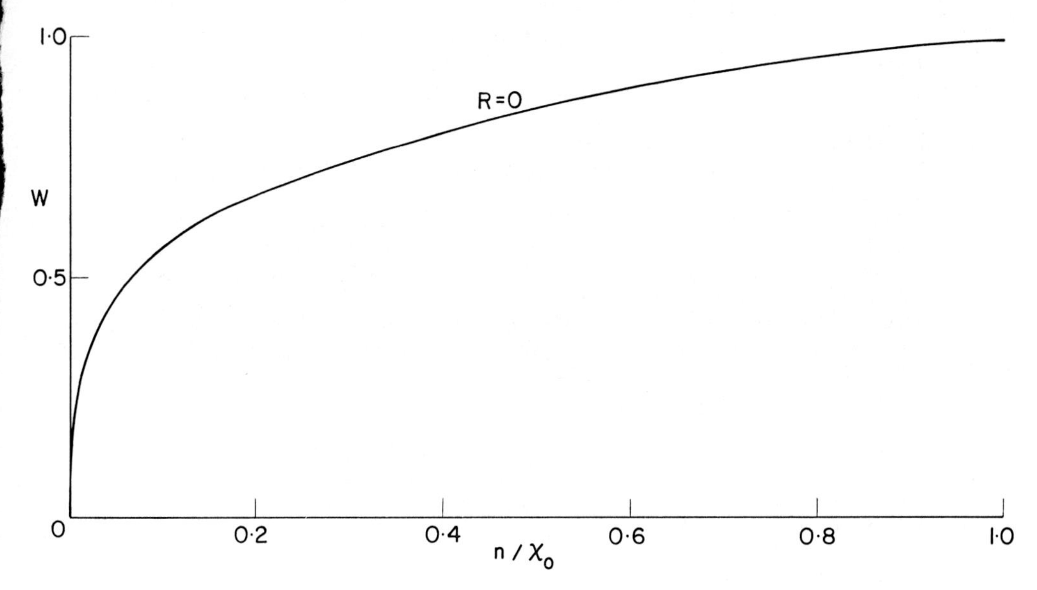

Fig. 7.15b. Fixed point cumulative frequency distribution at plume center, $s/m=0.5$.

there is an abrupt drop of the density $\chi_0 \mathrm{d}W/\mathrm{d}n$ to zero. Both the density and W are shown for $R=0$, with a small portion of the density distribution at $R=2\sqrt{2}m$ also sketched in, the latter location being very much on the fringes of the plume. It is worth noting that even at the center of the plume the concentration is less than 7% of the maximum (or about 16% of the mean) half the time. At the fringe location concentration greater than 10% of the peak χ_0 are present less than 1% of the time. The Gifford model thus provides a realistic qualitative description of the intermittency phenomenon. Its use in predicting 'peak to mean' ratios is a more dubious step, because the 'peak' value is fixed at χ_0 by hypothesis, which may easily be, say 3 times less than the true peak.

It remains to add a few remarks concerning the persistence of concentration fluctuations at fixed points. In cases where meandering is the major cause of fluctuations ($m>s$, i.e., close enough to sources of initially small clouds) it is clear that the duration of a fumigation episode will equal the duration of the gust which brings the cloud from the source to the receptor point. This interval is equal in order of magnitude to the Eulerian time scale t_{E} of turbulence, which is typically 30 s. We have already seen that in Gosline's (1952) observations individual puffs of high concentration lasted 30 to 90 s, in agreement with our reasoning. When meandering is negligible ($m \ll \sigma$, far from the source), the relevant time scale may be either t_{E} or the dissipation scale t_{d} or something in between. Because usually $t_{\mathrm{d}} \gg t_{\mathrm{E}}$, the dissipation time scale may in fact be a less important parameter and the typical persistence time may remain close to t_{E} for all phases of diffusion. Clearly we have little experimental or theoretical guidance on this whole question, as on most others concerning the fluctuation problem.

References

Abramowitz, M. and Stegun, I. A.: 1964, *Handbook of Mathematical Functions*, National Bureau of Standards, 1046 pp.

Batchelor, G. K.: 1959, *J. Fluid Mech.* **5**, 113.

Becker, H. A., Rosensweig, R. E., and Gwozdz, J. R.: 1966, *A.I. Ch. E. J.* **12**, 964.

Corrsin, S.: 1951, *J. Appl. Phys.* **22**, 469.

Cramer, H.: 1946, *Mathematical Methods of Statistics*, Princeton University Press, 575 pp.

Cramer, H. E., DeSanto, B. M., Dumbauld, R. K., Greene, B. R., Morgenstern, P., and Swanson, R. N.: 1965, GCA Technical Report No. 65-9-G, GCA Corporation, Bedford, Mass., 184 pp.

Crum, G. F. and Hanratty, T. J.: 1965, *Appl. Sci. Res.* **A15**, 177.

Csanady, G. T.: 1966, Dispersal of Foreign Matter by the Currents and Eddies of the Great Lakes, Pub. No. 15, Great Lakes Div. Univ. Michigan (Proc. 9th Conf. on Great Lakes Res.), 283-294.

Csanady, G. T.: 1967a, *J. Atmospheric Sci.* **24**, 21.

Csanady, G. T.: 1967b, *Phys. Fluids Suppl.* **1967**, S76.

Csanady, G. T.: 1969, *Atmos. Environ.* **3**, 25.

Csanady, G. T., Hilst, G. R., and Bowne, N. E.: 1968, *Atmos. Environ* **2**, 273.

Dunskii, V. F. and Evseeva, S. A.: 1965, *Izv. Atmos. Oceanic Phys.* **1**, 501.

Gibson, C. H. and Schwarz, W. H.: 1963, *J. Fluid Mech.* **16**, 365.

Gifford, F. A.: 1959, *Adv. Geophys.* **6**, 117.

Gosline, C. A.: 1952, *Chem. Eng. Progr.* **48**, 165.

Kolmogoroff, A. N.: 1941, *Compt. Rend. Acad. Sci. U.S.S.R.* **31**, 99.

Lee, J. L. and Brodkey, R. S.: 1964, *A.E.Ch.E. J.* **10**, 187.

Murthy, C. R. and Csanady, G. T.: 1971, *J. Phys. Oceanography* **1**, 17.

Obukhov, A. M.: 1949, *Bull. Acad. Sci. U.S.S.R. Geogr. Geofiz* **13**, 58.

Ramsdell, J. W., Jr., and Hinds, W. T.: 1971, *Atmos. Environ* **5**, 483.

Townsend, A. A.: 1956, *The Structure of Turbulent Shear Flow*, Cambridge University Press, London, 315 pp.

GEOPHYSICS AND ASTROPHYSICS MONOGRAPHS

AN INTERNATIONAL SERIES OF FUNDAMENTAL TEXTBOOKS

1. R. Grant Athay, *Radiation Transport in Spectral Lines*. 1972, XIII + 263 pp.
2. J. Coulomb, *Sea Floor Spreading and Continental Drift*. 1972. X + 184 pp.

Forthcoming:

4. F. E. Roach and Janet L. Gordon, *The Light of the Night Sky*
5. R. Grant Athay, *The Solar Chromosphere and Corona*
6. J. Iribarne and W. Godson, *Atmospheric Thermodynamics*
7. Z. Kopal, *The Moon in the Post-Apollo Era*
8. Z. Švestka and L. De Feiter, *Solar High Energy Photon and Particle Emission*
9. A. Vallance Jones, *The Aurora*
10. R. Newell, *Global Distribution of Atmospheric Constituents*
11. G. Haerendel, *Magnetospheric Processes*